Two Manuals Combined into One Book

Religious Support

And

Public Affairs Tactics, Techniques and Procedures

Department of Defense

Pentagon
Publishing

Field Manual
No. 1-05

Headquarters
Department of the Army
Washington, DC, 18 April 2003

Religious Support

Contents

Distribution Restriction. Approved for public release; distribution is unlimited.

*This publication supersedes FM 16-1, 26 May 1995.

Preface

The Chaplain Corps is the Army's primary agency for practically ensuring the free exercise of religion for America's soldiers. This responsibility also applies to the Chaplain Corps' religious support to Army family members, Department of Army (DA) civilians, and other authorized personnel. The Chaplain Corps ensures free exercise of religion by executing the command religious programs of Army commanders.

PURPOSE

FM 1-05 establishes the Army's definitive doctrine for the execution of comprehensive religious support across the full spectrum of operations. It is founded upon the Army's operational doctrine articulated in FM 3-0 and FM 101-5 and other supporting documents. FM 1-05 is compatible with the joint religious support doctrine contained in JP 1-05. This doctrine pronounces that the personal delivery of religious support is the primary focus of the Chaplain Corps and places greater emphasis on the chaplain's role as an advisor to the commander. It takes into account emerging technology, possesses a broader focus on religious support at echelons above brigade, and speaks clearly to the interface between religious support operations and civil military operations. It is the authoritative, but not prescriptive, guide to the missions, roles, responsibilities, and duties the Chaplain Corps must execute to deliver comprehensive religious support.

APPLICABILITY

FM 1-05 provides guidance to commanders at all echelons responsible for providing command religious support programs to their units and organizations, as well as, to Chaplains, Chaplain Assistants and other personnel responsible to plan, prepare, and provide religious support to The Army. This guidance is the principle tool for professional education in the art of religious support operations. Trainers at all echelons should employ it as the authoritative guide for the planning and execution of religious support training. For overseas theaters, this doctrine applies to U.S. unilateral operations only, subject to applicable host nation laws and agreements.

USER INFORMATION

The proponent for this publication is the U.S. Army Training and Doctrine Command (TRADOC). Send comments and recommendations on DA Form 2028 (Recommended Change to Publications and Blank Forms) to Commandant, U.S. Army Chaplain Center and School, ATTN: ATSC-DCD, Fort Jackson, SC 29207.

Unless this publication states otherwise, masculine nouns and pronouns do not refer exclusively to men.

Chapter 1

RELIGIOUS SUPPORT FUNDAMENTALS

CONTENTS

The First Amendment of the United States Constitution reinforced by Title 10 U.S. Code (USC) and Army Regulation (AR) 165-1, guarantees every American the right to the free exercise of religion. Commanders are responsible for those religious freedoms within their command. On behalf of the commander, Chaplains provide and perform Religious Support (RS) in the Army to ensure the free exercise of religion. Chaplains are obligated to provide for those religious services or practices that they cannot personally perform. Chaplains perform religious support when their actions are in accordance with the tenets or beliefs of their faith group. Chaplain assistants assist the chaplain in providing or performing this religious support. Religious support includes—

- Religious services.
- Rites.
- Sacraments.
- Ordinances.
- Pastoral care.
- Religious education.
- Family life ministry.
- Institutional ministry.
- Professional support to the command and staff.
- Management and administration.
- Religious/humanitarian support.
- Religious support planning/operations and religious support training.

1-1. The Chaplain's statutory (10 USC) and regulatory (AR 165-1) understanding of religious support includes—

- Provide and/or perform worship services, rites, sacraments, ordinances, pastoral and spiritual care and religious education to nurture the living, care for the dying, and honor the dead.

- Advice to the commander on morals and morale as affected by religion, the impact of local religions on the military mission, and the ethical impact of command decisions.

1-2. Examples of religious support include—

- Worship, in accordance with one's faith, mindful of Army regulations and mission.
- Religious guidance, care, and counsel.
- Observation of Holy days and observances.
- Participation in rites, sacraments, and/or ordinances of one's faith.
- Practice of dietary laws.
- Medical treatment according to one's religious belief.
- Resolution of medical treatment challenges, religious apparel issues and dietary restrictions arising from one's religious belief. (See AR 600-20 for further guidance on religious accommodation.)

1-3. The Unit Ministry Team (UMT) is a task-organized, mission-based team designed to support the religious, spiritual, and ethical needs of soldiers and their families, members of other services, and authorized civilians.

1-4. The UMT consists of at least one chaplain and one chaplain assistant. UMTs are identified according to either a Modified Table of Organization and Equipment (MTOE) or a Table of Distribution and Allowances (TDA). Certain roles, functions and expectations of these UMTs depend on whether they are authorized by the MTOE or by the TDA. Continued mission and task analysis determines the distribution and size of UMT assets. The UMT is embedded into the Army Force Structure as organic and central to the organization.

1-5. Religious Support in the Army has no civilian equivalent. Chaplains execute their distinct religious support mission for soldiers, members of other military services, family members and/or authorized civilians in a variety of geographical locations, operational situations, and circumstances.

1-6. The religious support model in Figure 1-1 depicts supporting functions and tasks the UMT performs or provides for the free exercise of religion, as stated in 10 USC.

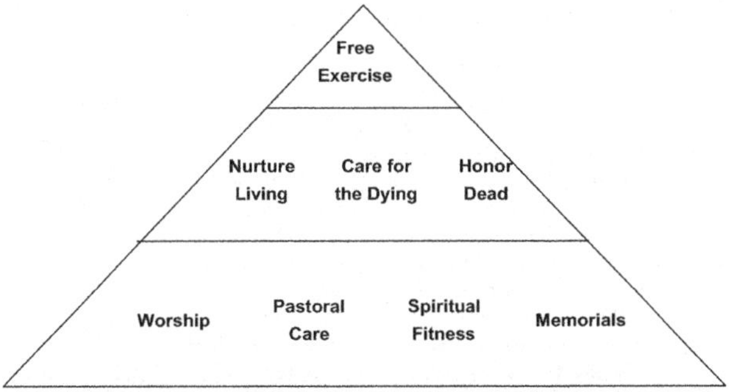

RS estimate, RS Plan, Critical Event Debriefing, Staff Activities, Ethical Decision Making, Security, Navigation, Communication/Digital Competency.

Figure 1-1. Religious Support Model

> Religious Support in the Army has no civilian equivalent.

HISTORICAL FOUNDATIONS

1-7. Chaplains have served in the U.S. Army since the Revolutionary War. The Continental Congress enacted regulations and salaries governing chaplains. George Washington, as Commander-in-chief, was resolute in his view that chaplains were necessary for the good ordering of the military and steadfast in his conviction that only well-qualified men should be chosen to serve as chaplains. To Colonel George Taylor he wrote that chaplains should be men of "Character and good conversation…who will influence the manners of the Corps both by precept and example." On 29 July 1775, the Continental Congress provided for the appointment of chaplains for the armed forces. That date is the recognized "birthday" of the Chaplain Corps, making it the second oldest branch in the Army, preceded only by the Infantry. 10 USC , Sections 3073, 3547, and 3581 codify the position of Chaplain in the Army. The 10 USC and AR 165-1 prescribe the duties of chaplains and require commanders to provide for the religious needs of soldiers. Religious support to families and authorized civilians is a derived task of this duty.

1-8. The War Department established the position of chaplain assistant by General Order Number 253, on 28 December 1909. The commanding officer detailed one enlisted soldier to special duty to any organization in which a chaplain was assigned. This detailed soldier assisted the chaplain in the performance of his official duties. The Army established the military occupational specialty (MOS) chaplain assistant as an official MOS in 1968.

THE CHAPLAIN'S AUTHORITY FOR MINISTRY

1-9. The government recognizes that the primary role of religious support belongs to the religious community. The federal government also recognizes the special religious support needs of armed forces personnel and the difficulty of providing organized, effective religious support. The isolation, unique environment, and training for combat zones throughout the world generate special requirements for religious support. The federal government provides the chaplaincy to meet these needs. Chaplains serve in the Army as clergy representing the respective faiths or denominations that endorse them. A chaplain's call, ministry, message, ecclesiastical authority, and responsibility come from the religious organization that the chaplain represents. Chaplains preach, teach, and conduct religious services, in accordance with the tenets and rules of their tradition, the principles of their faith, and the dictates of conscience. They also perform ecclesiastical rites and ceremonies and administer the sacraments and ordinances of their faith community. When a conflict arises between the standards of a chaplain's faith tradition and the requirements of a religious support mission, chaplains are required to provide for the religious needs of the soldier by obtaining other chaplains or qualified people to perform the needed religious support.

1-10. Chaplains, ordained and endorsed by distinctive faith groups, must be people of integrity who consistently live according to the highest principles of the Army and the distinctive faith groups they represent. Chaplain assistants are also expected to be people of integrity whose lives reflect high moral standards and the values of the Army. The UMT must mirror the Army's values – Loyalty, Duty, Respect, Selfless Service, Honor, Integrity, and Personal Courage.

1-11. As religious leaders, chaplains continue to meet the vision of General George Washington by setting the standard for moral and ethical behavior by upholding the Army Chaplaincy Values: Spirituality, Accountability, Compassion, Religious Leadership, Excellence, and Diversity (SACRED).

Army Chaplaincy Values
Spirituality
The bedrock of all faiths, providing meaning and direction in life.
Accountability
The means to measure how we practice what we preach.
Compassion
The love that sees beyond the surface and discerns our common humanity.
Religious leadership
Interprets the role of leadership prophetically.
Excellence
The standard for staff work and religious responsibilities.
Diversity
Being respectful of different views and ideas that are not like our own.

RELIGIOUS SUPPORT MISSION

1-12. The mission of the UMT is to provide and perform religious support to soldiers, families, and authorized civilians as directed by the commander. Chaplains serve as personal staff officers to commanders at all levels of the command providing essential information on troop and unit morale, quality of life matters, free exercise of religion issues, ethical decision-making, and the impact of religion on the operation.

RELIGIOUS SUPPORT FUNCTIONS

1-13. Comprehensive religious support includes the following three major functions:

- ***Nurture the Living.*** In preparation for missions that span the full spectrum of operations UMTs develop and provide religious support activities to strengthen and sustain the spiritual resilience of soldiers and family members. During the battle, UMTs bring hope and strength to those who have been wounded and traumatized in body, mind, and spirit, assisting the healing process.
- ***Care for the Dying.*** UMTs provide religious support, spiritual care, comfort, and hope to the dying. This focus of religious support affirms the sanctity of life, which is at the heart of the chaplaincy. Through prayer and presence, the UMT provides the soldier with courage and comfort in the face of death.
- ***Honor the Dead.*** Our Nation reveres those who have died in military service. Religious support honors the dead. Funerals, memorial services, and ceremonies, reflect the emphasis the American people place on the worth and value of the individual. Chaplains conduct these services and ceremonies, fulfilling a vital role in rendering tribute to America's sons and daughters who paid the ultimate price serving the Nation in the defense of freedom.

RELIGIOUS SUPPORT ACTIVITIES

1-14. Commanders provide opportunities for the free exercise of religion. UMTs provide and/or perform the following major religious activities, within the unique military context, in support of the command religious plan:

- ***Religious Services:*** Worship; Funerals; Memorial Services and Ceremonies; Prayer Breakfasts.
- ***Rites, Sacraments, Ordinances:*** Marriages; Burials; Baptisms; Confirmations; Blessings; Daily Prayers; other ministrations and religious requirements.
- ***Pastoral Care/Counseling:*** Taking part in command activities; visiting soldiers; calling on families; activities and unit "ministry of presence"; individual and group pastoral counseling; and similar pastoral activities. This may also include elements of the spiritual fitness training program. .
- ***Religious Education:*** Study of religious teachings including activities of faith sustainment (enrichment, formation, study or instruction), marriage enrichment, problem solving, communication skills, parenting skills, youth programs, religious literature.

- ***Family Life Ministry:*** Training UMTs, directing the Chaplain Family Life Center (CFLC). Family Life Ministry is implemented through the CFLC programs and formal training.
- ***Institutional Ministry:*** Preserving soldiers' fighting strength by maintaining their health and supporting family members receiving medical services. Institutional ministry occurs in hospitals and confinement or correction facilities.
- ***Professional Support to the Commander and Staff:*** Support to the commander on matters of religion, morals, and morale, as affected by religion and the impact of indigenous religions on the military mission. Assesses the unit climate by visiting soldiers and providing advice on the religious needs of soldiers and families.
- ***Management and Administration:*** Beyond the normal administration requirements and actions required by the Command Religious Plan, the UMT manages funds, facilities, manpower, supplies, and equipment.
- ***Religious/Humanitarian Support:*** Support programs to Joint Military Operations. Such support includes humanitarian support programs on issues of religion, morale, morals, and ethics.
- ***Religious Support Planning/Operations:*** Providing commanders with Religious Support plans and programs for all mobilization requirements spanning the full spectrum of military operations, as well as, planned ministries for families of deployed soldiers.
- ***Religious Support Training:*** Training that includes spiritual fitness (Spiritual fitness training is the training and development of personal qualities needed to sustain a soldier in times of danger, hardship, and tragedy. These qualities are based in religious, philosophical and human values and they form the basis of character, decision-making and integrity.), moral leadership (see AR 165-1, Chapter 11), soldier and family member suicide prevention, violence in the work place, professional leadership, and training required by UMTs, commanders, staff, soldiers and family members.

ROLES AND RESPONSIBILITIES

1-15. Commanders, chaplains, and chaplain assistants have distinct roles and responsibilities in providing and performing religious support to soldiers, their family members and authorized civilians.

COMMANDER

1-16. The commander provides for the free exercise of religion of soldiers and families thereby enhancing the spiritual health of the command. To assure readiness across the full spectrum of operations, commanders provide for a wide range of religious support activities. Commanders use the Commanders' Religious Program (CRP) to resource the religious support plan.

1-17. The commander accommodates religious practices, in accordance with Army regulations (See AR 600-20). The Army policy is to accommodate religious practices when those accommodations will not have an adverse impact on military readiness, unit cohesion, standards, health, safety or discipline, or otherwise interfere with the performance of the soldier's military duties. Accommodation of a soldier's religious practices must be weighed against military necessity and cannot be guaranteed at all times.

CHAPLAIN

1-18. Chaplains personally deliver religious support. They have two roles: religious leader and staff officer.

- The chaplain as a religious leader executes the religious support mission, which ensures the free exercise of religion for soldiers and authorized personnel.
- The chaplain is a non-combatant and will not bear arms (AR 165-1).

> The Chaplain is a non-combatant and will not bear arms in accordance with AR 165-1.

1-19. The chaplain is a personal staff officer responsible for coordinating the religious assets and operations within the command. The chaplain is a confidential advisor to the commander for religious matters. A chaplain is located at every echelon of command above battalion. Besides his common staff responsibilities, the chaplain's specific responsibilities include —

- Advise the commander on issues of religion, ethics, and morale (as affected by religion), including the religious needs of all personnel for whom the commander is responsible.
- Provide commanders with pastoral care, personal counseling, advice, confidentiality, and sacred confidence.
- Develop and implement the commander's religious support program.
- Exercise staff supervision and technical control over religious support throughout the command.
- Provide moral and spiritual leadership to the command and community.
- Coordinate religious support with unit ministry teams of higher and adjacent headquarters, other services, and multinational forces or coalition partners.
- Translate operational plans into battlespace ministry priorities for religious support.
- Help the commander ensure that all soldiers have the opportunity to exercise their religion.
- Advise the commander and staff, with the G5 (S5), on the impact of the faith and practices of indigenous religious groups in an area of operation.
- Perform or provide religious rites, sacraments, ordinances, services, and pastoral care and counseling to nurture the living, care for the dying and honor the dead.
- Provide religious support to the command and community to include confined or hospitalized personnel, enemy prisoners of war (EPWs), civilian detainees, and refugees.
- Support the commander in the execution of Civil Military Operations (See appendix A).

1-20. The Chaplain trains, equips and supports the subordinate chaplain and chaplain assistant (FM 101-5).

1-21. As a minimum, the chaplain requires a SECRET clearance. This will allow the chaplain to have access to the command post and enable him/her to be part of the planning process.

1-22. Although chaplains do not have command or Uniform Code of Military Justice (UCMJ) authority, they do exercise staff supervision and functional direction of religious support personnel and activities.

1-23. The proper title for a chaplain is "chaplain" regardless of military rank and professional title.

CHAPLAIN ASSISTANT

1-24. The chaplain assistant (CA) is a combatant, trained in religious support and staff skills; who possesses a SECRET clearance; manages force protection, provides security for the UMT on the battlefield, and accompanies the chaplain in the area of operation. The CA is a soldier trained to assist the chaplain in providing religious support. Under the direction of the chaplain, the CA coordinates and synchronizes religious support operations. In the absence of the chaplain, the CA continues the religious support mission for the commander as a staff representative, but does not assume the religious leadership role of the Chaplain.

1-25. Unit Ministry Team Chaplain Assistant. Under the supervision of the chaplain, the CA—

- Assists in religious support planning and operations.
- Coordinates the UMT movement plan, monitors the tactical situation, and attends briefings.
- Coordinates and prepares for religious activities.
- Coordinates for and conducts equipment maintenance.
- Assesses soldier morale and provides this information to the chaplain.
- Conducts pre-counseling interviews.
- Assists with combat operations stress response (COSR), identification, intervention, and care.
- Performs tactical, logistical, and administrative support for the UMT.
- Requires access to classified messages, systems, and secure areas to accomplish the religious support mission.
- When an NCO, the CA serves as a liaison for the chaplain to the noncommissioned officers (NCOs) of the command.

> Chaplain Assistants require a SECRET clearance to accomplish the religious support mission.

1-26. The Chaplain Assistant NCO supervises, mentors, and trains chaplain assistants and serves as the religious support section noncommissioned officer in charge (NCOIC) according to the echelon assigned. In addition to executing the roles and functions of a CA of lesser rank, the NCOIC performs many of the following staff functions:

- Participates in battle staff planning, tracking, and execution.
- Procures contracts for goods and services, and coordinates and monitors the CRP.

- Provides training in suicide prevention awareness, intervention for COSR, and performance counseling.
- Plans, briefs, and performs operational training with other staff NCOs.
- Coordinates taskings.
- Serves as advisor to the unit CSM and other key NCO leaders.
- Coordinates religious support during joint, multinational, and inter-agency operations.

ORGANIZATION

UNIT MINISTRY TEAM AND MODIFIED TABLE OF ORGANIZATION AND EQUIPMENT (MTOE) UNITS

1-27. UMTs authorized by the Modified Table of Organization and Equipment (MTOE) are assigned to units whose primary mission is warfighting. The UMT deploys with its unit and provides religious support for all units in the commander's area of responsibility during each stage of force projection.

1-28. At echelons above brigade, the senior chaplain is at times listed in the command section of the MTOE. Additional chaplains, chaplain assistants, and UMT equipment are listed in a separate section of the MTOE.

1-29. Comprehensive religious support requires the assignment and deployment of UMTs in maneuver, maneuver support, and maneuver sustainment units at all echelons, beginning at the battalion. UMTs move forward to provide religious support to all elements of the squadron/battalion: companies, troops, batteries, platoons, squads, and teams.

1-30. In operations where lethality and intensity are high, religious support becomes increasingly important to soldiers. UMTs provide hope, faith, encouragement, and compassion to soldiers experiencing shock, fatigue, isolation, fear, and death. In the chaos and uncertainty of military operations, the UMT is a reminder of the presence of God.

UNIT MINISTRY TEAMS AND TABLE OF DISTRIBUTION AND ALLOWANCES (TDA) UNITS

1-31. UMTs authorized by a TDA are assigned to units whose primary missions vary widely. The missions and responsibilities associated with TDA organizations depend upon the type of organization served. TDA organizations include installations, schools, hospitals, disciplinary barracks, some Major Army Commands (MACOMs), and Headquarters, Department of the Army (HQDA). The Installation UMT provides the training base for all MTOE and TDA UMTs. In addition to providing religious support directly to the units, the TDA UMTs manage, supervise, and support the religious support mission of their organization.

PROVIDING UNIT RELIGIOUS SUPPORT

1-32. Unit religious support is provided in the following three ways:

- ***Unit support:*** support provided to the unit to which the UMT is assigned or attached. The team normally gives first priority to this mission.

- ***Area support:*** support provided to soldiers, members of other services, and authorized civilians who are not a part of the team's unit, but operate within the supporting unit's area of operations (AO).
- ***Denominational support:*** support given to soldiers and other authorized persons of the chaplain's denomination or distinctive faith group. Limited assets affect the availability of denominational support.

1-33. The UMT extends opportunities for soldiers to serve as volunteers in worship, religious education, and service to the unit. The sponsoring UMT is responsible for recruiting, training, and supervising volunteers. Under the supervision of the unit chaplain, soldiers may serve as members of their own faith groups as Distinctive Faith Group Leaders (DFGL). Denominational certification, the approval of the commander, and the final approval of the MACOM/State Area Command (STARC)/United States Army Reserve Command (USARC) or Major Subordinate Command (MSC) chaplain are required before a soldier may serve as a DFGL. (See AR 165-1)

SUMMARY

1-34. The Chaplain Corps and Regiment remains committed to providing for the fundamental freedom of the free exercise of religion as guaranteed by the Constitution. This is accomplished by—

- Performing religious activities according to the chaplain's faith tradition.
- Providing professional and effective advice to the commander, in coordination with commander's staff, regarding the needs of soldiers not of the chaplain's faith tradition.
- Developing effective teamwork within the UMT.
- Through cooperation with other chaplains without compromising the tenets of one's own faith.

1-35. Through it all, the soldier's spiritual health and welfare, as well as the soldier's spiritual freedoms, remain the focal point. Thus the Chaplain Corps' unofficial motto: *"Bringing God to Soldiers, and Soldiers to God."*

RESOURCES

1-36. The following resources were used for writing and updating this doctrine.

- AR 165-1.
- AR 600-20.
- JP 1-05.
- JP 5-00.2.
- JP 3-0.
- FM 1 (100-1).
- FM 3-0 (100-5).
- FM 16-1.
- FM 22-100.
- DA Pam 165-3.
- Military Rules of Evidence 503 UCMJ.
- U.S. Army Chaplaincy Strategic Plan FY 2000-FY 2005.
- Chaplain Support Force XXI White Paper.
- Chaplain Assistant Concept Paper, 2000.

Chapter 2

RELIGIOUS SUPPORT ENVIRONMENT

"We have to remember something simple but important: Making history is messy. Studying Normandy today looks like arrows and unit symbols. Normandy on 6 June 1944 looked like Saving Private Ryan – dangerous and chaotic."

MG James Dubik

The mission and operational environment determine the shape of religious support. The team conducts religious support effectively with a clear understanding of the context and complexities of military operations in which religious support is provided and performed.

CONTENTS

RELIGIOUS SUPPORT AND THE THREAT ENVIRONMENT

2-1. With the collapse of the Soviet Union, the bilateral world of the post WWII era, pitting the political ideologies of communism and capitalism against one another, came to an end. As a consequence, the operational climate and the organizational structure of the U.S. Army began to change. The Army moved away from the concept of a forward deployed Army based overseas to the concept of a force projection Army based in the United States. "Because the Army is a power projection force, its units deploy from and return to home stations." (FM 3-0, pg. 4-9) The changes in operational climate, organization, and deployment posture require adaptations in the methods of providing religious support.

THE THREAT

"The enemies we are likely to face through the rest of the decade and beyond will not be 'soldiers,' with the disciplined modernity the term conveys, but 'warriors' – erratic primitives of shifting allegiance, habituated to violence, with no stake in civil order."

LTC Ralph Peters, Fighting for the Future

2-2. Potential threats to U.S. forces are many and varied. They range from a local populace with small arms to a conventional force possessing weapons of mass destruction (WMD). Global, regional, and local challenges confront the U.S. Army with adversaries who, as a result of their religious beliefs, may interpret such fundamental ideas as right and wrong, the value of human life, or even the concept of victory and defeat very differently.

2-3. Failing states and the associated fragmentation due to religious or ethnic strife will continue to promote regional instability and produce security concerns for the United States and our allies. The security environment is further complicated by challenges that transcend state borders, such as ethnic disputes and religious rivalries.

2-4. Threats can be of short duration or protracted over a long period of time. They can be waged over competing ideology, ethnic identity, economic policies, national aspirations, religious conflict or any number of social and political issues. Struggles can be waged by a variety of means: diplomatic, informational, military, or economic.

2-5. Asymmetric threats (unconventional or inexpensive approaches) to the United States (such as terrorism—both domestic and international, guerilla warfare, and WMD including attacks on domestic targets) are issues of special concern and create challenges to religious support planning. Particularly problematic is the growing threat to our increasing information infrastructure through acts of cyber-terrorism.

2-6. Unlike the Cold War era, when threats were measurable and predictable, Army forces today are likely to encounter conditions of greater ambiguity and uncertainty. The U.S. Army must be prepared to respond to worldwide strategic, operational and tactical challenges either alone, or as part of a joint or multinational team. The U.S. Army's UMTs must also be prepared to provide religious support in every contingency or circumstance.

2-7. Mission, enemy, terrain and weather, troops and support available, time available, civilian considerations (METT-TC) includes "civilians" as an operational planning consideration. The UMT requires skills to deal with the religious support requirements of DA civilians, contractors in the area of operations, and indigenous and host nation civilians. With the expected increased urban terrain missions, the UMT will have more opportunity and responsibility to interface with the local people and religious leaders as it supports the command mission.

FORCE PROJECTION

Force Projection Process	
Mobilization	Sustainment
Deployment	Redeployment
Employment	Demobilization

2-8. The Army performs its mission within the context of national security and military strategies. Power projection is the application of the elements of national power—diplomatic, economic, information, and military—anywhere in the world.

2-9. The military aspect of power projection is force projection. Force projection is the deployment of forces to perform missions spanning the whole range of military operations anywhere in the world. Force projection applies to the entire Army, Active and Reserve Components.

2-10. Force projection encompasses a process in which forces are managed, trained, and projected to the theater of operations for any contingency. The force projection process is organized in phases: Mobilization, Deployment, Employment, Sustainment, Redeployment, and Demobilization.

2-11. Credible force projection depends on the capability of the United States to deploy forces rapidly and to perform missions spanning the whole range of military operations. A unit may deploy with short notice, causing tremendous stress on soldiers and systems.

2-12. When the United States decides to employ its military resources, the nation continues to meet the soldier's rights to free exercise of religion by deploying UMTs with the soldiers. During force projection operations, the religious support mission is to provide religious support to the soldiers, authorized civilians, and families of the deployed unit. The environment and type of operation may change, but the religious support mission remains constant throughout the spectrum of operations. The successful provision of religious support to meet requirements of force projection rests upon four pillars (table 2-1).

Table 2-1. Four Pillars of Religious Support

Religious support plan (RSP) {to include the CMRP}	Coordinated with all echelons of command and included in the operation plan/order.
Communication architecture	UMTs must be able to interact with their commands and RS providers at higher, subordinate and adjacent units. (Plan redundancy when possible).
Split-based operations	Concurrent RS provided from projection platform to theater of operations.
Force tailoring	UMTs assigned to operational units and Task Forces to meet unique religious support requirements, liaison with faith-based Non-Governmental Organizations (NGOs) and International Organizations (IOs), and governmental agencies.

LEVELS OF MILITARY ACTIONS

2-13. Levels of military actions include strategic, operational, tactical, and military operations other than war (MOOTW). Military operations in the 21st century will span all levels of military actions, often simultaneously or in rapid sequence. For example, forces on the ground may be engaged in high-intensity conflict one moment, then with changes in the strategic diplomatic or political circumstances, be leading a MOOTW operation the next moment, or vice versa. The Army Chaplaincy of the 21st Century will be organized and equipped to meet the challenges at all levels of actions at any time.

STRATEGIC ENVIRONMENT

"Although U.S. military forces enjoy superiority in many dimensions of armed conflict, the United States is likely to be challenged by adversaries who possess a wide range of capabilities, including asymmetric approaches to warfare, particularly weapons of mass

destruction. The United States cannot predict with a high degree of confidence the identity of the countries and actors that may threaten its interests and security."

--Quadrennial Defense Review Report, 2001

Strategic Religious Support Mission Essential Task List

2-14. The Army's strategy is to achieve and maintain a capabilities-based, threats-adaptive Army that is postured to support our nation's strategy to respond to the full spectrum of crises. To support the Army's strategy, the Chaplaincy must anticipate changes in the strategic environment, technological advances, and evolving Joint war-fighting concepts. The Army's strategic mission essential task list (METL) describes what "well–trained, superbly led, and well-equipped soldiers do for the nation (FM 3-0)." Likewise, the Army Chaplaincy in support provides strategic METL tasks that shape and sustain the spiritual freedoms of the soldier.

Army and Chaplain Corps Mission Essential Task Lists

Army—

- Shape the security environment.
- Respond promptly to crisis.
- Mobilize the Army.
- Conduct forcible entry operations.
- Dominate land operations.
- Provide support to civil authorities.

Chaplain Corps—

- Provide and perform personal delivery of religious support throughout full spectrum of operations.
- Train and equip the Chaplain Corps.
- Advise the commander on matters pertaining to the religious, moral, ethical, and morale climate of the command.
- Advise the commander on matters pertaining to indigenous religions and their impact on military operations.
- Manage mobilization of the Chaplain Corps throughout all Army components.
- Plan and coordinate for religious support with other services, international chaplaincies, and governmental agencies.
- Provide UMTs to implement Combatant Commanders' religious support requirements. (Combatant Commanders are the commanders of the nine unified commands in DOD i.e. USEUCOM, USPACOM, USCENTCOM, USSOUTHCOM, USSOCOM, USTRANSCOM, USJFCOM, USNORTHCOM, USSTRATCOM)
- Provide and perform religious support during military support to civilian authorities (MSCA).

OPERATIONAL ENVIRONMENT

2-15. The operational level of war links the tactical employment of forces to strategic objectives. The focus at this level is on operational art—the use of military forces to achieve strategic goals. Emerging doctrine from all branches is shaping the religious support environment by making increasing requirements for support in this much wider spectrum of operations.

2-16. On the nonlinear, asymmetrical battlefield, with operations characterized by multiple and simultaneous operational missions, religious support tasks become much more challenging. A sound operational religious support plan, executed skillfully, increases the chances of tactical success. It does this by decreasing the tactical UMTs risks and supports the UMT's time management requirements. Operational religious support planning integrates conditions, sequences of support actions, and wise management of resources throughout the theater, to optimize the delivery of religious support.

TACTICAL ENVIRONMENT

2-17. Tactics is the art and science of employing all available means to win battles and engagements. Tactical operations are rapid and dynamic in nature. Specifically, the subject of tactics comprises the actions taken by a commander to arrange units and activities in relation to each other and the enemy. At this level, UMTs are fully engaged with their units maneuvering to the area of operations delivering religious support.

MILITARY OPERATIONS OTHER THAN WAR ENVIRONMENT

2-18. Military Operations Other Than War (MOOTW) includes a wide range of contingencies. It may involve offensive or defensive operations, but most often the primary focus is on stability operations and support operations. During MOOTW, UMTs focus their religious support planning and execution on spiritual care of the soldier over long durations of deployment and serve, when directed, as the commander's liaison with local religious leaders, and faith-based NGOs and IOs.

DECISIVE FULL SPECTRUM OPERATIONS

2-19. Within the operational and tactical levels of military actions, commanders plan and execute military operations. There are four types of operations: Offense, Defensive, Stability and Support. The Army's role differs with each operation (Figure 2-1).

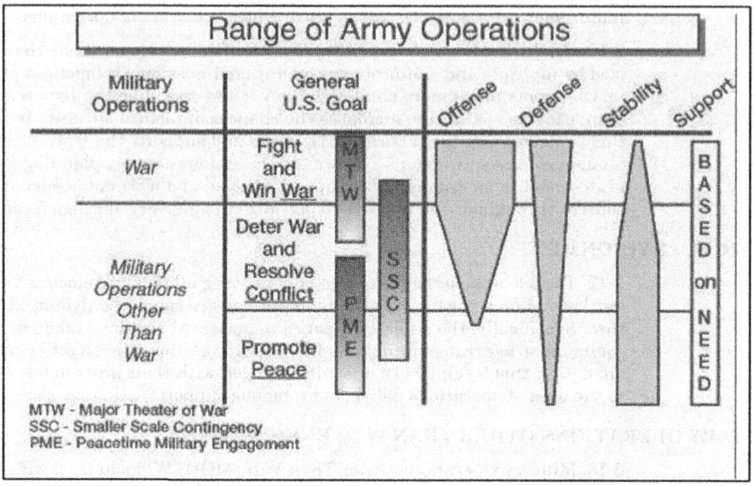

Figure 2-1. Range of Army Operations

OFFENSE

2-20. This is the decisive form of war. Offense is essential to success in all operations—defensive, stability, and support as well as offensive.

DEFENSIVE

2-21. These operations are directed by command to defeat enemy attacks, buy time, economize forces, or develop conditions favorable for the offense.

STABILITY

2-22. Stability operations include a range of actions that Army forces conduct outside the United States and its territories. The main goal is to promote and sustain regional and global stability. Characteristics of stability operations include a diverse and complex environment. These operations are highly demanding on small units and leaders, require intense mental and physical agility, and are often long-term. There are ten types of stability operations:

- Peace operations.
- Foreign Internal Defense.
- Security assistance.
- Humanitarian and civic assistance.
- Support to insurgencies.
- Support to counter drug operations.
- Combating terrorism.

- Noncombatant evacuation operations.
- Arms control.
- Show of force.

SUPPORT

2-23. Support operations are conducted to relieve suffering and help civil authorities prepare for, or respond to, crises. They are divided into two categories: Domestic support operations (DSO) conducted within the United States and its territories, and foreign humanitarian assistance (FHA) conducted outside the United States and its territories. Support Operations usually are conducted to relieve or reduce the results of natural or manmade disasters. They are also conducted to relieve conditions (such as pain, disease, hunger, or privation) that present a serious threat to life or loss of property. There are several forms of support operations.

- ***Relief operations.*** Humanitarian relief focuses on the well being of supported populations. Disaster relief focuses on recovery of critical infrastructure after a natural or manmade disaster.
- ***Support to incidents involving weapons of mass destruction (WMD).*** Assistance is given to civil authorities protecting U.S. territory, population, and infrastructure before an attack by supporting domestic preparedness and protection of critical assets.
- ***Support to Civil Law Enforcement.*** This involves activities related to counter terrorism, counter drug operations, military assistance to civil disturbances, and general support.
- ***Community assistance.*** Provision is given to community assistance for projecting a positive military image, training opportunities, and public relationships.

Religious Support in Peace Operations

2-24. The range of religious support is tailored to support the objectives outlined in the operation plan or order. Unit Ministry Teams must understand each of the four types of military operations in order to plan for, and execute, religious support.

2-25. The Army performs peace operations that range from explicitly peaceful missions to those that have significant applications of combat power. Peace operations are classified as peace making, peace building, and preventative diplomacy. These operations are conducted outside U.S. territories (FM 100-23).

Religious Support in Joint, Combined, and Interagency Operations

2-26. Army units deploy with other military services and agencies of the U.S. government. Increasingly, the U.S. Army operates with other countries as well. Army forces will normally deploy as part of a joint, multinational, and interagency operation. These operations are collectively called unified actions. This is true not only when the military effort is the prime strategic option, but also when other strategic options are preferred.

2-27. Joint operations are operations in which two or more American military departments are employed in the same operation. Joint doctrine is published under the authority of the Chairman of the Joint Chiefs of Staff.

JP 1-05 provides doctrinal guidance for religious support in joint operations. It is important to recognize that Joint publications take precedence over doctrine of the participating service branches.

2-28. Combined operations are operations in which American forces are employed along with forces from other nations. Operational procedures are determined through agreement and coordination by higher headquarters. Provisions for religious support should be included in the operational procedures or policies agreed upon by the participating nations and coalitions; for example, American, British, Canadian, and Australian/New Zealander (ABCA) or the North Atlantic Treaty Organization (NATO).

2-29. Interagency operations are operations in which the military works in tandem with another government agency. The lead agency and organizational relationship are specified in the mission directive. The provision of religious support must be specified for each operation.

2-30. Standardized methods of operating may not be present in many Joint, Combined, or Interagency Operations. Circumstances will occur that cannot be anticipated. UMTs must be flexible and creative in planning religious support for these operations. In each type of operation, the key elements to consider for religious support mission accomplishment are understanding, cooperation, and coordination. These will take place between U.S. chaplains, multinational chaplains, and interagency representatives. This is essential to achieve unity of effort in providing total and comprehensive theater of operation religious support.

Religious Support in Special Operations

2-31. Special Operations Forces (SOF), which consist of specialized Army, Air Force, and Navy units, give the President and the Secretary of Defense additional military options across the full spectrum of operations. These options provide a strategic response capability to prevent future, or stabilize current, international situations with minimum risk to U.S. interests. The theater SOF commander executes special operations (SO) as part of the theater commander's joint special operations effort. Religious support is woven into this specialized Army and joint environment. The religious support planning encompasses all contingencies, whether direct or indirect military actions in support of a Joint Force Command.

2-32. U.S. Army Special Operations Command (USASOC) commands all CONUS-based active and reserve Army Special Operations Forces (ARSOF) units. This command is comprised of five types of units: Special Forces (SF), Rangers, Army Special Operations Aviation (ARSOA), Psychological Operations (PSYOP), and Civil Affairs (CA).

2-33. The command is responsible for readiness and training oversight for all Army Reserve SOF and all outside the continental United States (OCONUS) forward based (deployed) active Army SOF. When directed, USASOC provides mission-ready ARSOF assets, through United States Special Operations Command (USSOCOM), to the regional combatant commanders for employment under their command authority. (See Chapter 5 for religious support tasks.)

Religious Support and Military Support to Civilian Authorities

2-34. Another military environment in which religious support is provided is Military Support to Civilian Authorities (also known as MSCA). These domestic support operations involve the Army in assisting civilian authorities by providing general military assistance. General assistance for domestic support operations includes disaster assistance, environmental assistance, and law enforcement support and community assistance. These operations are conducted within U.S. territories (FM 100-19).

2-35. Religious support in both peace operations and MSCA, as well as Military Assistance to Civil Disturbances (MACDIS), require intentional coordination and cooperation within the Chaplaincy among Army components (active duty, Reserve, and National Guard (NG)), as well as with civil and NGO leaders.

2-36. Civil Support will involve extensive coordination and liaison among interagency, joint, multi-jurisdictional (state and local), and active and Reserve Component (AC/RC) chaplaincies. When state authorities request Federal help, the Federal Emergency Management Agency (FEMA) most often is the lead agency, with DOD supporting. The Secretary of the Army is the executive agent. The Army Chaplaincy works directly for the military command providing RS to soldiers and authorized civilians.

2-37. Reserve Component UMTs provide a unique capability for the Chaplaincy. They have a long-term relationship with state and local religious leaders, institutions, officials, and businesses. They also have special capabilities and are geographically dispersed to conduct religious support operations. UMTs provide and perform the following religious support before, during, and after crises:

- *Planning*. Army National Guard (ARNG) and Reserve component (RC) UMTs are ideally suited to support any locality because of their dispersion throughout the Nation and long-term relationships with local clergy, civil agencies, and institutions. Planning must consist of information management, synchronization with all military and civil assets, equipping and training for all contingency missions, close support of Civil Military Operations tasks, shaping the force religious support structure to meet operational needs and staff integration.
- *Preparing*. Doctrine, organization, training, material, leader development, personnel and facilities (DOTMLPF) issues need to be continuously reviewed by chaplain leadership in all Army components to respond effectively within Civil Support operations. All written guidance and plans need to be continuously updated to meet METT-TC needs and directives. Standard operating procedures (SOPs), UMT battle-books, and other publications need to address all areas to which all UMTs will be directed to respond. Training incorporates many of the collective skills that UMTs already possess and continually update.
- *Responding and executing*. When a crisis occurs, UMTs of selected units, in support of the lead agency, will provide and perform direct religious support to military units, will perform liaison responsibilities with State and Federal agencies pertaining to religious issues, will conduct religious risk assessment for the command, and will liaison with civilian religious leaders.

- *Follow-up*. In the days following a crisis, UMTs provide and perform religious support directly to military forces as they are employed after a hostile event, to include the following tasks: mass casualty religious support, critical event debriefings, liaison with local religious leaders, etc.

2-38. The Medical Command (MEDCOM) plays a key role in response efforts. Existing MEDCOM units may be tailored to meet the METT-TC requirements. MEDCOM UMTs will bring specific skills in mass casualty, trauma, crisis intervention, and critical event debriefing ministry to these events. RS planning must be well integrated within this health care response. Army AC/RC deployable hospitals may move forward to the area of operations. Military medical facilities may be used as a trauma center when that capability does not exist locally, or when called for under National Disaster Medical System (NDMS). Large numbers of casualties resulting from WMD will challenge the UMT and other caregivers, resulting in increased need for critical event debriefing and other types of intervention ministries. Special Medical Augmentation Response Teams–Pastoral Care (SMART-PC) play a unique religious support role in response ministry. SMART-PCs, as well as the chaplain's role in support of Special Medical Augmentation Response Teams–Stress Management (SMART-SM), are discussed in depth in chapter 6 of this FM.

SUMMARY

2-39. Religious support provided during a military operation can occur under widely differing circumstances. The UMT's basic mission to provide religious support to the soldier, however, remains constant. What differs is the operational environment in which the religious support is provided. The differing types of military operations determine the specific nature of religious support.

2-40. The U.S. Army's UMTs must be prepared to provide religious support in every contingency or circumstance. With the expected increase in urban terrain missions, the UMT will have more opportunity and responsibility to interact with the local people and religious leaders in supporting the command mission. Additionally, METT-TC includes "civilians", which include both authorized DA Civilians, civilian contractors on the battlefield, as well as indigenous and host nation civilians. These civilians are included as an operational planning consideration for the UMTs to determine who, what, when, and where to provide and perform religious support.

2-41. In all types of operations across the full spectrum, key elements to consider for the religious support mission accomplishment are: understanding, cooperation, and coordination between U.S. chaplains, multinational chaplains, and interagency representatives. This is essential to achieve unity of effort in providing total and comprehensive theater of operation religious support.

2-42. There are no boundaries between strategic, operational, tactical, and MOOTW environment. Over time in the battlespace of a theater, operational activities link the tactical employment of forces to strategic objectives. The Army strategic METL describes what "well –trained, superbly led, and well-equipped soldiers do for the nation (FM 3-0)." Likewise, the Army

Chaplain Corps, in support, provides a strategic METL that helps shape and sustain the spiritual center of the soldier.

2-43. Civil Support will involve extensive coordination and liaison among interagency, joint, multi-jurisdictional (state and local), and active and reserve component (AC/RC) Chaplaincies. The RC UMT provides a unique capability to the Chaplaincy. They have long-term relationships with state and local religious leaders, institutions, officials, and businesses. Also, RC UMTs have special capabilities and geographic dispersion to conduct religious support operations. Religious Support tasks will be in concert with the Army's role in Civil Support which follow in these categories: force protection, support to crisis management, support to consequence management, protection of critical assets, support to counter-terrorism, deterrence/defense against strategic attack, and MSCA missions.

RESOURCES

2-44. The following is a list of main resources utilized for writing and updating this doctrine.

- AR 165-1.
- JP 1-05.
- JP 5-00.2.
- JP 3-0.
- JP 3-05.
- FM 3-0 (100-5).
- FM 16-1.
- FM 22-100.
- FM 1 (100-1).
- FM 100-19.
- National Military Strategy
- Joint Vision 2010
- Army Vision 2010
- A National Security Strategy For a New Century, May 1997
- U.S. Army Chaplaincy Strategic Plan FY 2000-FY 2005
- Chief of Chaplains MSCA Concept Paper, September 2000
- Chaplain Support Force XXI White Paper
- Quadrennial Defense Review Report, September 2001.

Chapter 3

UNIT MINISTRY TEAM OPERATIONS

"I was determined that whatever else might happen next time [post-W.W.I], at least I myself would be prepared, and trained, and ready when the time came. I had at times a kind of inward feeling that the call would come, to me personally, and in my prayers morning and evening I used to ask that I might be given help and strength so that I might not fail when put to the test."

--Field Marshall Montgomery, The Path of Leadership

"Everything is simple in war, but the simplest thing is difficult."

--Carl Von Clausewitz, On War

This chapter establishes the essential elements of Religious Support (RS) planning, coordination, and execution efforts in all phases of operations. It examines the following key points:

- General Religious Support planning, preparation, coordination and execution overview.
- Religious Support Planning for Force Projection.
- General Responsibilities of the UMT During Religious Support Operations.

CONTENTS

GENERAL RELIGIOUS SUPPORT PLANNING, PREPARATION, COORDINATION AND EXECUTION OVERVIEW

3-1. To meet the religious needs of soldiers and other authorized personnel in all operations, the Unit Ministry Team (UMT) prepares a comprehensive religious support plan that is staffed and integrated into every operational phase. The UMT advises the commander on the plan for accomplishing the religious support mission. The plan encompasses RS required from training and mobilization, to redeployment and demobilization.

3-2. The commander is responsible for the execution of the Religious Support Plan (RSP). The Chaplain, as a member of the commander's personal staff, is responsible to the commander to plan and implement the following:

- Developing the RSP.
- Executing the approved RSP.
- Use of available religious support resources.
- Allocation of assets.
- Location of UMTs.
- Coordination with indigenous religious leaders.
- Coordination with religious NGOs and IOs in coordination with Civil Military Operations leaders.
- Religious support for all authorized civilians in the AO.

PLANNING

3-3. The UMT participates in the collaborative planning process with the other staff sections known as the Military Decision-Making Process (MDMP). Religious support planning is continuous, detailed, and systematic. It examines all factors relating to the religious support of an operation. The planning process builds on the religious support section of the unit's tactical standard operating procedures (TSOP). It becomes specific to an operation through the preparation of a religious support estimate (RSE). The product of this planning process is a religious support plan (RSP). The RSP is included in the operations plan or operations order (OPLAN/OPORD). Inclusion may be in the plan/order as an appendix, annex or TAB. The specific placement of the RSP is dependent upon the particular unit's protocol for constructing operation plans and orders. The RSP informs the command and staff of the concept of religious support, and guides the work of the UMT. When appropriate, the RSP can be contained in a matrix format. A model for such a RSP matrix is contained in Appendix D of this manual. The use of a RSP matrix is appropriate when UMTs are assigned or attached to units that publish their OPLANS/OPORDS in a matrix format. The RSP matrix model is a self-explanatory form that clearly and concisely articulates how the UMT will provide religious support during the operation. The use of a RSP matrix is highly recommended when time constraints impact the planning process and the UMT needs a ready tool to direct and encapsulate its religious support planning.

KEY PLANNING CONSIDERATIONS:

- Commander's Intent
- Concept of Operations
- Concept of Logistics
- Concept of Medical Support
- ROE
- METT-TC of the battlespace/theater of operations

Religious Support Estimate

3-4. The estimate process is a problem solving procedure to assist the commander in determining the best course of action (COA) to accomplish a mission or task. Staff estimates are designed to supplement the commander's estimate within the respective areas of expertise. A staff estimate consists of significant facts, events, conclusions and recommendations on how best to use available resources to accomplish the mission. Upon receipt of the commander's guidance, each staff section develops its own staff estimate relating how its area of interest will affect the mission.

3-5. The chaplain develops the Religious Support Estimate (RSE) that becomes the basis for the Religious Support Plan (RSP). Development of the RSE ensures the careful and intentional planning of religious support. In the development of the RSE, the chaplain considers the —

- Location and movement of the UMT during all phases of the operation.
- Logistical and communication support required for specific missions.
- Replacement and/or augmentation of UMT members in the event UMT members become casualties or are unable to meet all RS needs during especially intense fighting.
- Religious, moral, and morale issues that affect the soldiers and the unit.
- Impact of indigenous religions on the military operation.

BASIC PRINCIPLES OF RS PLANNING AND COORDINATION

- Plan early and continuously.
- Use guidance from higher HQ.
- Exploit all available RS assets.
- Use the lowest echelon capable of furnishing effective RS.
- Use the most effective RS means.
- Furnish the type of RS requested.
- Avoid unnecessary duplication.
- Coordinate RS along command and technical channels.
- Provide adequate support.
- Provide rapid coordination.
- Provide safeguards and survivability.
- Establish RS coordination measures.
- Establish communications support.

3-6. The preparation of the estimate follows a logical, continuous, and methodical process to provide the appropriate religious support for the commander's selected COA. In conjunction with the other staff sections, the UMT prepares the RSE using the following steps:

1. Gather facts and assumptions.
2. Analyze the mission.
3. Develop courses of action.
4. Analyze and compare courses of action.
5. Make a recommendation.

Step 1: Gather Facts and Assumptions

3-7. The staff gathers information concerning the assigned mission and divides the information into two categories: facts and assumptions. Facts are statements of known data concerning the situation, and include the following:

- Enemy and friendly dispositions.
- Available troops.
- Unit strengths.
- Materiel readiness.
- Mission and the commander's intent.

3-8. Special religious needs of soldiers in the unit are key facts for the RSE. The team develops a Religious Preference Profile (RPP) using data from a Standard Installation/Division Personnel System (SIDPERS) ad hoc query. (See AR 680-29 for religious preference codes.) The report is a statistical analysis of the unit's religious demographics and provides the stated religious preference of soldiers assigned to that unit.

3-9. When facts essential to planning for RS are not available, the UMT must make assumptions. An assumption is a supposition about the current or future situation, assumed to be true, though without positive proof. In the absence of facts, assumptions are used to fill gaps. Assumptions are necessary only when they bear directly upon the planning or the execution of the plan. As new information becomes available, some assumptions become facts and some are modified or discarded.

3-10. Mission, Enemy, Terrain and Weather, Troops and support available, Time available, and Civil Considerations, or METT-TC, is a useful outline by which to consider and formulate facts and assumptions essential for developing an RSE.

Mission, Enemy, Terrain and Weather, Troops and support available, Time available, and Civil Considerations (METT-TC)

Mission: The UMT identifies specified or implied religious support tasks from the higher headquarters' OPLAN or OPORD. For example, a requirement to provide area religious support may be implied but not specified by a higher headquarters' task organization. Other considerations may include protection of religious shrines, artifacts or sacred places in the AO.

Enemy: In this part of mission analysis, the team identifies how the enemy may threaten the religious support mission. For example, enemy activity in rear areas may restrict team travel and necessitate additional area support requirements. The team may also consider how religion plays a part in the enemy's capacity or inclination to fight.

Terrain and Weather: The UMT assesses the terrain and weather in the AO. Traffic-ability, weather extremes, and the amount of daylight are factors that will affect the religious support mission. (Consider observation and fields of fire, cover and concealment, obstacles, key terrain, and avenues of approach (OCOKA).

Troops: Religious support is focused upon soldiers. What soldiers are doing often determines their availability for religious support. The UMT adjusts its plan to match the soldiers' schedules to meet their needs. The UMT must also assess the level of anxiety and fatigue among the troops to shape its ministry to the soldiers more effectively. The UMT will also advise the commander on the morale, moral, and spiritual health and welfare of soldiers.

Time Available: The UMT assesses the time required for planning and the time available for executing the religious support mission. The amount of planning time determines the detail of the plan. The team also considers the travel time to each unit, preparation time, and available light.

Civil Considerations: The UMT researches and advises the command on the religious attitudes and activities of the civilian population, religious and cultural dynamics, religious organizations, and religious leaders within the battlespace. Coordination with CMO personnel is essential. The chaplain advises the commander regarding civil concerns and their impact on the mission:

- Understanding the operational religious environment.
- Recognize religious implications for courses of action during planning.
- Anticipate and control negative impacts on the local religious climate.
- Consider long-term effects of operations on the indigenous religious climate.
- Ensure legitimacy and credibility of friendly forces.
- Balance operational priorities with concern for religious freedom.
- Minimize negative effects of operation on unit soldiers belonging to the same faith group as the local population.

Step 2: Analyze the Mission

3-11. Mission analysis begins with a review of the intent of commanders one and two levels higher. From this review, the commander and staff derive the specified tasks that the unit must perform to accomplish the mission. The unit must also identify implied tasks and determine which tasks are essential. A tentative list of essential tasks is presented to the commander for approval. Based upon the approved list, the staff presents the commander with a restated mission for approval. Commanders disseminate the restated mission as a warning order.

3-12. The UMT participates with the rest of the staff in mission analysis by identifying specified and implied religious support tasks. The UMT also considers any religious factors that may directly affect the mission. (FM 101-5, Appendix J)

3-13. During mission analysis, in addition to the RSE, the UMT brings expertise to the staff team in studying the following areas:

- ***Unit Morale.*** The UMT should have a sense of the morale of the unit. Factors such as whether the troops are fatigued, are excited or dreading the mission, and the spiritual health and welfare of the unit all impact on unit morale.
- ***Ethical and moral implications.*** The chaplain can assist the staff in considering the ethical and moral implications of a particular plan as it impacts on the lives of leaders, soldiers, and the local civilian population. For example, this may include compassionate treatment of EPWs, detainees, and displaced persons.
- ***Impact of local religion on the mission.*** The chaplain can assist in the religious assessment of an AO. The UMT should familiarize themselves with the local religions and their significance within the area of operations. What and where are the places that are considered sacred? What are the significant religious holidays and how are they celebrated? How could this impact the operation or the timing of the operation? (See Appendix F).

Step 3: Develop Courses of Action

3-14. Course of action (COA) development is based on mission analysis and on the facts and assumptions developed earlier. After receiving the commander's guidance, the Chief of Staff (CofS) or the executive officer (XO) leads the staff in developing COAs for accomplishing the mission. As the staff begins to develop COAs, the chaplain considers the religious and moral/ethical implications of each proposed COA and how to provide religious support. While the operations officer formulates the tactical possibilities, the chaplain and other staff officers consider how to integrate their functional areas into each COA. In the midst of COA development, the use of Chaplain Detachments (CD) by those units to which they have a doctrinal association must be considered and implemented into the COAs and the eventual RSP. Likewise, units with a doctrinal association with CDs must proactively request CDs through appropriate staff actions to ensure their operational employment.

Step 4: Analyze and Compare Courses of Action

3-15. After the staff develops several COAs, it analyzes each one. COA analysis consists of two parts: war-gaming and COA comparison. Through this process, the staff identifies the best COA to recommend to the commander.

3-16. Note that the chaplain may have a different kind of input depending on the kind of mission that is being planned. For units preparing for offensive or defensive actions, often the chaplain's role is more involved with integrating with the staff and being a team player while developing the RS aspect of the plan. In a stability and support operation, a chaplain may have more to contribute in the plan development process. In both cases, the chaplain will want to anticipate the COA toward which the commander may be leaning.

3-17. Additionally, the chaplain will want to pay particular attention to the casualty flow estimates and concepts of support. During the war gaming phase of the MDMP, chaplains must pay particular attention to where the casualties are expected, remembering that this critical ministry is one of the three major functions of comprehensive religious support. If the chaplain participates in the war gaming phase of the analysis, the chaplain will have a much clearer understanding of the scope of the mission when the COA is finally decided upon and then rehearsed.

Step 5: Make a Recommendation

3-18. After choosing the best COA, the staff briefs the commander. The commander may agree with the staff's recommendation, modify it, or select another. Once the commander decides on a COA, the UMT uses the concepts of support developed during the RSE process to prepare the Religious Support Plan (RSP).

3-19. The commander and staff begin preparations for the mission rehearsals. The chaplain needs to participate in the rehearsals in order to test the effectiveness of the RSP. Afterwards an assessment is made as to whether the plan actually meets the needs of the troops according to the commander's guidance.

Note: The adjutant (S1), supply officer (S4), and the chaplain are often not included in the accelerated MDMP process. However, this does not preclude the UMT from preparing RSEs and RSPs. Staff coordination and proper analysis is always necessary.

Religious Support Plan

3-20. Based on the selected COA, the UMT completes the details of its plan for inclusion in the OPLAN/OPORD. The religious support plan (RSP) assigns religious support responsibilities, defines area support requirements, and authorizes coordination between subordinate units.

3-21. Additional information found in the RSP may include the following:

- Concept of support for special faith groups and special holy days.
- Concepts of support for joint and multinational religious services support.

- Religious support requirements of all authorized civilians.
- Religious support for EPWs and Detainees in accordance with the Geneva Convention.
- Command guidance regarding liaison responsibilities with faith-based NGO/IOs and liaison support, in coordination with CMO personnel, to religious leaders in the AO.

3-22. Priority of religious support is—

- Unit military personnel.
- Other military personnel in AO.
- Authorized civilians such as DOD employees and contractors.
- EPWs and Detainees.

Planning Religious Support in Joint Operations

3-23. Just as in the Army planning process, the UMT is included in the system of joint planning. The Religious Ministry Support Team (RMST) (chaplain and chaplain assistants as designated under JP 1-05, which the Army designates as UMT) at each echelon is intricately involved with the orders process. Planning religious support operations must encompass and consider—

- Denominational assets.
- Command relationships.
- Location of RMSTs.
- Logistical requirements.

(See JP 1-05 for more detail.)

Planning Religious Support in Multinational Operations

3-24. Religious support planning with multinational forces in an operation must include the differences in religious support doctrine (or military custom of providing religious support), capabilities, force structure, logistics, and culture (denominational representation, language, supervision, etc). Communication, cooperation, and coordination are essential in planning multinational religious support operations. Senior force RMSTs need to understand the capabilities and limitations of religious support with each country represented within the coalition. Planning must remain flexible due to these differences while emphasizing the similarities that may include, language common between countries, denominational backgrounds of chaplains, and locations for area support. Senior U.S. RMSTs must be highly sensitive to allocating religious support assets of another country and should only do so with prior consultation and coordination. Planning considerations include—

- Communication requirements and limitations.
- Denominational representation.
- Language difficulties.
- Limitations in formal chaplain structures.

Rules of Engagement

3-25. Rules of engagement (ROE) specify the circumstances and limitations in which forces may engage the "enemy". Many factors influence ROE, including national command policy, mission, operational environment, commander's intent and the law of land warfare. Properly written ROE are clear and tailored to the situation.

3-26. Chaplains at all echelons help commanders apply the ROE by advising on the moral/ethical implications of proposed COAs. Senior-level chaplains consider the restrictions and constraints of the ROE when recommending religious support policy to the commander.

PREPARATION

3-27. Preparation is the link between planning and execution. While the commander and staff are planning, the unit prepares for the operation based on a warning order issued earlier by the commander. Once the plan is complete, preparations unique to the mission may begin. Upon completion of the individual and team preparations, the UMT provides religious support to soldiers while they complete their preparation for the operation.

Rehearsals

3-28. Units conduct rehearsals during preparation for combat. In addition to maneuver rehearsals, units conduct combat service support (CSS) and casualty evacuation rehearsals. Rehearsals provide an opportunity to refine the RSP. The UMT must be present at the rehearsal, and integrate the RSP into it, to fully understand the scheme of maneuver, support, and other critical pieces of information shared. If time permits, UMTs may conduct their own RS rehearsals following those of the unit. Such RS rehearsals offer UMTs an opportunity to resolve communication issues, as well as, build a comprehensive understanding of each UMT's religious support mission.

Individual and Team Preparation

3-29. The chaplain and chaplain assistant prepare themselves for combat like other soldiers of the unit. This preparation must be physical, mental and spiritual. The chaplain and chaplain assistant must also insure that their families are ready for coping with the challenges of deployment.

3-30. Team preparation includes load plans, pre-combat checks, updating information, map reconnaissance, travel planning, rehearsals, and mission coordination. Other preparations include securing equipment for the operation such as worship aids, devotional literature, music, and liturgical items.

Pre-Combat Inspections and Preparation

3-31. The UMT conducts a pre-combat inspection according to the unit TSOP. This inspection ensures the UMT possesses sufficient supplies and equipment to successfully execute the operation. On every mission, the team must carry food, water, and other survival equipment. The chaplain assistant prepares and updates the team's load plan and inventories and arranges the necessary supplies and equipment for movement.

Information Update

3-32. The UMT updates mission information regularly— especially before departing from its operational base. The tactical situation can change rapidly on the battlefield. The team checks the situation map, situation reports (SITREPS), logistics reports, and the latest fragmentary orders (FRAGOs) or warning orders. If there are significant changes in the situation, the team modifies its plan accordingly.

Training

"The key to winning on the battlefield is the understanding of "how we fight" and the demonstrated confidence, competence, and initiative of our soldiers and leaders. Training is the means to achieve the tactical and technical proficiency that soldiers, leaders, and units must have to enable them to accomplish their missions."

FM 7-0, TRAINING THE FORCE

3-33. Training is the cornerstone of preparation for military operations. The Army's primary activity during peacetime is realistic, battle-focused training oriented on the unit's mission essential task list (METL). Installation chaplains, through their training managers, and in coordination with subordinate UMTs and unit operations/training officers (S3/G3), ensure that UMTs have adequate technical and tactical training for deployment and combat. The training objective is operational readiness—units and soldiers that are tactically and technically proficient.

3-34. Tactical proficiency is a demonstrated understanding of the Army's war fighting doctrine and tactics. To provide effective religious support in combat, the UMT must know its unit's standard operational procedures, tactics, and techniques. The UMT must also maintain the requisite skills of field craft essential for survival on the battlefield. Tactical training is primarily the responsibility of the unit to which the UMT is assigned or attached.

3-35. Technical proficiency is the demonstrated ability to perform all tasks related to one's functional area. Technical proficiency is required for both garrison and field missions. Technical proficiency is acquired from schools, self-study and training with other UMTs in performing the unit's and installation's religious support mission. Training and continuing education conferences, functional courses offered by the U.S. Army Chaplain Center and School (USACHCS), and courses offered through the Officer and NCO Education System (OES and NCOES) supplement the installation religious support-training program.

"To understand firepower and its application you have to use it, actually practice with it and see it work. This means maneuvers in the field, maneuvers that reproduce as realistically as possible, within the bounds of prudence, the actual conditions a soldier will encounter on the battlefield. Only so can you condition, not only the human body, but also the human spirit, to face and survive the stresses and strains the soldier will encounter in battle. And in future battle, if it comes, these stresses and strains will be immeasurably greater."

General Matthew Ridgeway

3-36. The UMT must plan for and participate in religious support training for both combat and garrison environments. Furthermore, the UMT must participate fully in both the unit tactical and religious support training to assure their tactical and technical proficiency.

Assessment of Unit Ministry Team Readiness

3-37. As the commander's personal staff officer for religious support, chaplains at brigade and higher levels are responsible for assessing the tactical and technical proficiency of UMTs in subordinate units. They conduct assessments of the UMT during field training exercises (FTXs), external Army Training and Evaluation Program (ARTEP) events, unit training and Command Inspections. Observer controllers provide additional assessment assistance when the UMT is involved in Battle Command Training Program (BCTP) events or Combat Training Center (CTC) rotations. These CTC assessments for UMTs are important for readiness analysis and training.

3-38. Chaplains and chaplain assistants participate in the installation's Command Master Religious Plan (CMRP) to help soldiers develop the personal spiritual readiness needed to sustain them during combat. Spiritual readiness is also essential to sustain soldier families while the unit is deployed. As part of the installation's CMRP, the UMTs conduct worship services, offer religious education, and provide spiritual fitness training and other activities in installation chapels and facilities, and in the theater of operations. Spiritual readiness applies to all levels of the command. Planning for spiritual readiness programs must be part of the preparation for combat for each soldier and authorized personnel.

Religious Support Preparation in Joint Operations

3-39. Coordination between all RMSTs is critical for full implementation of the RSP. Total synchronization between all religious support elements is key to a comprehensive plan. The joint task force (JTF) RMST considers the following additional tasks for operational preparation:

- Identifying RMST logistical requirements.
- Personnel staffing and augmentation requirements.
- RSP integration within the operational plan.
- RSP rehearsals.
- Spiritual Nurture programs for the command and staff.
- Loading of equipment and material.

3-40. The Command JTF RMST works closely with the Joint staff for RMST closure within the Joint Operational Area (JOA), ensuring communication and synchronization of the RSP.

Religious Support Preparation in Multinational Operations

3-41. Each nation brings different capabilities and limitations to the Combined-Joint Task Force (CJTF). Task organizing along functional lines vice national lines can help fill capability gaps, and designs each nation's role so they are full participants. Comprehensive religious support information disseminated to all participating nations describing religious support operational expectations will prepare all combined participants for spiritual and religious coverage. UMTs must logistically prepare themselves to meet the many diverse and varied religious and cultural differences. Communication limitations and differences must be prepared for such as language, terminology, systems, and distance. Liaison and training can offset some of the challenges.

"A chaplain visits our company. In a tired voice, he prays for the strength of our arms and for the souls of the men who are to die. We do not consider his denomination. Helmets come off. Catholics, Jews, and Protestants bow their heads and finger their weapons. It is front-line religion: God and the Garand."

Audie Murphy, To Hell and Back

COORDINATION

3-42. Coordination consists of three categories: mission, internal, and external.

Mission Coordination

3-43. The size of the AO and unit dispersion challenge the UMT's ability to coordinate for religious support. Communications equipment is critical for timely and effective coordination. Also, key communications systems allow the UMT to be available for emergency tasks. Early planning and coordination will help to eliminate potential problems. Coordination is both internal to the participating units and UMTs, and with external units, agencies, and UMTs.

Internal Coordination

3-44. As a personal staff officer, the chaplain works for the commander, advises the commander, and executes the commander's religious support program on his behalf. The UMT coordinates the execution of the commander's religious support program with the Chief of Staff or XO. The team plans cooperatively with all staff sections for the exchange of information.

External Coordination

3-45. Because of the rapidly changing dynamics, conditions, and circumstances on the modern battlefield, coordination for area support is difficult. Although religious support execution is decentralized, UMTs must synchronize religious support within their subordinate units. Establishing synchronization procedures with subordinate commands before deployment reduces the communication and coordination challenge for the UMT.

3-46. Military communication channels streamline information distribution on the battlefield. The team uses command, staff, and technical channels for communication, guidance, and coordination.

- ***Command:*** The command channel is the official communication line between headquarters for passing orders, instructions, and tasks to subordinate units. Command channels link commander to commander. Within the authority granted them, staff officers use command channels when acting for the commander. It is seldom if ever used by the UMT.
- ***Staff:*** The staff channel is the staff-to-staff link between headquarters. It is the primary channel the UMT uses for planning and coordinating religious support. Through this channel, the team transmits and coordinates planning information and provides staff supervision of religious support in subordinate units.
- ***Technical:*** The technical channel is the link between members of the same branch or discipline to convey technical instructions and guidance. Chaplains use the technical channel to communicate with UMTs of other

units about the technical aspects and details of religious support. It is not used to send official reports, to request religious support, or to send routine information between teams. This information is communicated through command or staff channels.

EXECUTION

3-47. The execution of the religious support mission on the battlefield is decentralized and performed by UMTs in accordance with the RSP. Adjustments to the plan are coordinated between affected units.

RELIGIOUS SUPPORT PLANNING FOR FORCE PROJECTION

MOBILIZATION

3-48. United States military strategy rests upon the ability to rapidly respond to crises. Responding to crises requires the Army to deploy its forces and may require mobilization of Reserve Component (RC) resources. Deployment comprises those activities required to prepare and move the force and its sustainment equipment and supplies to the AO.

Religious Support during Mobilization

3-49. Mobilization is the activation of the U.S. Army Reserve (USAR) and federalization of the Army National Guard. Mobilization provides commanders with RC forces, manpower, facilities, and logistics. FM 100-17-3, Reception, Staging, Onward Movement, and Integration (RSOI), provides additional information on RSOI procedures. It expands the Army's capability to respond to crises. Mobilization is conducted in five phases:

- Phase I. Planning.
- Phase II. Alert.
- Phase III. Home Station.
- Phase IV. Mobilization Station.
- Phase V. Port of Embarkation.

Mobilization Phase I: Planning

3-50. Planning begins long before a unit is mobilized. During peacetime, Active component (AC) and Reserve component (RC) UMTs coordinate with AC units for their mobilization planning. This planning takes into account operational and training requirements, equipment status, and readiness. It also includes the impact of mobilization on soldiers, families, and communities.

3-51. The primary planning tool for mobilization is the Unit Ministry Team Mobilization Systems Planbook (MOB Planbook). The MOB Planbook integrates all phases and levels of mobilization, and it provides a detailed guide for integrating mobilization planning.

3-52. Installation, Continental United States Army (CONUSA), STARC, and USARC chaplains are key to Reserve Component UMT mobilization. Each maintains a copy of the UMT MOB Planbook. Organizational MOB Planbooks contain the same basic information, but each is tailored for the particular headquarters. The MOB Planbook specifies what must be accomplished during each phase of mobilization. While many of the tasks in the

MOB Planbooks are unique, they are designed to integrate the mobilization process.

3-53. A unit's success is directly related to the quality of its preparation before it conducts operations. A representative listing of the mobilization planning responsibilities for installation, CONUSA, STARC, and USARC chaplains is detailed in Chapter 7. Planning considerations—

- Does the home station require special planning or preparation?
- Where is the home station in relation to support resources?
- How does the UMT provide family support briefings?
- How will the UMT provide required religious support if some parts of the unit remain at home station for an extended period of time, while others elements are sent forward to the mobilization station (MS)?
- Does the UMT go forward with the advance party or with the main body of the unit?
- How do durations in personnel holding areas, movement, or separation influence soldier morale?
- What religious support activities would address morale problems in the unit?
- What actions are required to close out civilian responsibilities?
- What coordination is required with faith groups for religious coverage during mobilization?
- What supplies are required for religious support?
- What religious support is required as the unit transitions to war?
- What training is required to bring the UMT to validated status for deployment?

3-54. Thorough planning and realistic training before a mobilization reduces confusion between all key UMT participants. Mobilization training gives both AC and RC UMTs an opportunity to practice their mobilization religious support missions, and an opportunity to assess the adequacy of mobilization plans. Installations assist with the following RC UMT training events:

- Annual training (AT).
- Individual Mobilization Augmentee (IMA) training.
- Individual training events, Individual Ready Reserve (IRR), and Chaplain Candidate training.
- AC/RC Sustainment Training.
- Training for UMTs in Garrison Support Units (GSUs).
- AC / RC Annual Chaplaincy Regional Sustainment Training (CREST) as mandated by the Chief of Chaplains.

Planning Considerations:
• Does the home station require special planning or preparation? • Where is the home station in relation to support resources? • How does the UMT provide family support briefings? • How will the UMT provide required religious support if some parts of the unit remain at home station for an extended period of time, while others elements are sent forward to the mobilization station (MS)? • Does the UMT go forward with the advance party or with the main body of the unit? • How do durations in personnel holding areas, movement, or separation influence soldier morale? • What religious support activities would address morale problems in the unit? • What actions are required to close out civilian responsibilities? • What coordination is required with faith groups for religious coverage during mobilization? • What supplies are required for religious support? • What religious support is required as the unit transitions to war? • What training is required to bring the UMT to validated status for deployment?

Mobilization Phase II: Alert

3-55. The Alert Phase begins when the unit receives the alert order of a pending mobilization order. The UMT reports immediately to its home station and the unit conducts the final screening of soldiers and cross leveling of personnel and equipment. The Alert Phase ends on the effective date of the unit's entry on active duty.

Note: RC chaplains cannot be mobilized (i.e. cannot be brought onto Active Duty) until they have completed all phases of the Chaplain Officer Basic Course (CHOBC). AC chaplains are non-deployable until the sustainment phase of CHOBC is completed. (See AR 165-1)

Mobilization Phase III: Home Station

3-56. Phase III begins on the effective date of the unit's mobilization. The effective date of mobilization is when RC chaplains and chaplain assistants come on Active Duty. During this phase the unit begins its transition to active duty. At home station the unit inventories property, dispatches an advance party, and moves to the mobilization station (MS). This phase ends when the unit arrives at the MS.

3-57. UMTs must balance the time spent in their own deployment preparation with providing religious support to soldiers and families.

Mobilization Phase IV: Mobilization Station (MS)

3-58. Phase IV begins when the unit arrives at the mobilization station. Activities at the MS include processing personnel and equipment, accessioning the unit into the active structure, cross-leveling, and soldier readiness processing. The UMT, along with the unit, engages in individual and collective training and validation exercises for survival on the battlefield. The goal of the unit during this phase is to achieve mission capability in the shortest possible time. The MS phase ends when the unit arrives at the port of embarkation (POE).

Note: The unit's line of authority shifts during this phase from the CONUSA to the MS commander.

Mobilization Phase V: Port of Embarkation (POE)

3-59. This phase begins with the arrival of the unit at its POE. Actions at the POE include preparing and loading equipment and manifesting and loading personnel. The POE phase ends with departure of personnel and equipment from the POE.

UNIT RELIGIOUS SUPPORT DURING DEPLOYMENT

3-60. Deployment is the planning, preparation and movement of forces and support base from any location to an area of operations in response to military need or crisis. Deployment may take the form of a forcible or permissive entry. Deployments are conducted in five phases:

- Phase I. Pre-deployment Activities.
- Phase II. Movement to Ports of Embarkation.
- Phase III. Strategic Lift.
- Phase IV. Theater Base Reception.
- Phase V. Theater Onward Movement.

Deployment Phase I: Pre-deployment Activities

3-61. Pre-deployment activities begin when a unit receives a deployment notification. These activities include mission briefings and planning, family pre-deployment briefings, equipment loading, chapel closeout, and the transfer of property accountability.

Mission Planning

3-62. As discussed above, the commander and staff begin the MDMP after they receive the deployment notification and they review contingency plans for completeness and accuracy. The UMT, along with the rest of the staff, begins to prepare an estimate of the situation. The team balances planning and preparation with providing religious support.

3-63. Religious support planners at higher echelons integrate the broader scheme of the operation into the planning process. They assess lines of authority, task organization, and faith group needs to determine the required number and mix of UMTs. Senior planners monitor the tailoring of the deploying force. When a JTF or Army Forces (ARFOR) commander forms an ad hoc headquarters, senior chaplains must monitor its composition to ensure the inclusion of a senior chaplain on the staff.

3-64. If the theater chaplain is not scheduled to deploy immediately, he designates a senior chaplain to be responsible for coordinating religious support in theater during each stage of the deployment preceding his arrival.

Chapel Transitions

3-65. The installation chaplain is responsible for all religious support facilities on the installation. Deploying UMTs coordinate chapel transition with the installation chaplain who decides whether to continue, combine, or suspend specific chapel programs and services. The UMTs transfer property accountability to the installation chaplain before departure. Deploying UMTs, in concert with non-deploying UMTs, provide transition activities for soldiers and their families.

Load Lists and Plans

3-66. The UMT must determine what to ship and what to carry. Some religious support supplies (communion bread, wine or grape juice) are consumable and cannot be pre-positioned with other war stocks. Sacred scriptures, religious literature, rosaries, crosses and medals can be pre-positioned. Equipment and supplies may not be available in theater for an unforeseen length of time. The UMT must include in the TSOP a religious support supply load list and plan.

Family Support Activities

3-67. As information about the deployment becomes available, the unit, as directed by the commander provides pre-deployment briefings to families (typically the S1 coordinates such briefings). The information given will be constrained by operational security requirements. Pre-deployment briefings include information on services that will be available on the installation during the deployment. Information will also be discussed regarding the advantages or disadvantages of family members remaining in the military community or returning to their community of origin. UMTs participate in pre-deployment briefings to provide religious support information to families.

3-68. Religious support provided to soldiers and their families directly contributes to the success of mobilization and deployment. Religious support fosters family wellness, a substantial factor in soldier readiness. Lessons learned from previous deployments and other research establish that soldiers deploying while anxious about personal and family problems are more vulnerable to combat stress. They are more likely to experience panic, poor judgment, battle fatigue, and the loss of the will to fight.

3-69. A lack of sufficient information, concern for coping without a spouse or parent, and the trauma brought on by a soldier's departure can generate stress within the family. Because families need religious support during this period, many will turn to the UMT. The deploying team must balance the needs of soldiers and families with its own preparation for departure. The non-deploying UMTs must assume much of the religious support responsibility for the families of the deploying soldiers.

3-70. It must also be mentioned that while care for the families of fellow soldiers is a responsibility for the UMT, the families of UMTs will also need care and religious support. Non-deploying UMTs can assist these families through the deployment process.

Deployment Phase II: Movement to the Port of Embarkation

3-71. When pre-deployment activities are concluded, the unit begins movement to the port. Units traveling outside of CONUS use both sea and aerial ports of embarkation (SPOE/APOE). Heavy equipment and the soldier operators travel by ship to the theater of operations. The remainder of the unit usually travels by air. Splitting the unit presents a challenge for the UMT.

Since the UMT will normally travel by air, RS for soldiers traveling by ship will require special planning and coordination.

Holding Areas

3-72. While the unit awaits transportation to the SPOE/APOE, it is normally confined to a Personnel Holding Area (PHA). Time spent there can seem long and monotonous. If the holding area is on the installation, this can be a frustrating time for soldiers because families are near but out of reach. This can be an important opportunity to provide religious support.

Deployment Phase III: Strategic Lift

3-73. Strategic lift begins with departure from the POE and ends with the arrival in theater. Transportation Component Commands are responsible for the strategic transportation of forces and their support. Religious support during this phase is oriented toward soldiers traveling with UMTs to the theater of operations.

Deployment Phase IV: Theater Base Reception

3-74. Force reception begins when the unit arrives at the port of debarkation (POD) in theater and ends when the unit departs the POD to begin the onward movement phase. One of the primary requirements during this phase is coordinating the onward movement of the deploying forces to their destination. The theater chaplain is responsible for religious support in the ports of debarkation.

3-75. Except in the case of opposed entry, CSS units may arrive early in the deployment to process combat units through the POD and to establish logistical support systems. Religious support resources in the port-based CSS units are limited. To prepare for operations, units are moved out of the PODs as rapidly as possible, through marshaling or staging areas, and into assembly areas (AA) or logistics bases.

3-76. In assembly areas soldiers prepare for onward movement. The unit may task organize the force, develop and issue tactical plans and orders, coordinate with other units, conduct reconnaissance, conduct training, and rehearsals. The immediacy of the pending mission frequently focuses soldiers' concerns upon matters of faith or religion. When units are in Assembly Areas (AA), UMTs are generally able to provide religious support to soldiers with little risk and with great receptivity.

Deployment Phase V: Theater Onward Movement

3-77. Onward movement begins with the personnel and equipment linkup, the reconfiguration of forces, sustainment, and receipt of pre-positioned war reserve stocks at designated areas. This phase concludes with the unit's arrival at the gaining command's Intermediate Staging Base (ISB) and tactical assembly areas.

Intermediate Staging Base

3-78. Units traveling through the ISB en-route to their combat mission present some religious support challenges. ISBs vary in size depending on the forces involved in the mission, and at times can be large in terms of population and geography. Units remain in the ISB for a minimum time, thus creating a highly transient community.

Tactical Assembly Areas

3-79. Tactical assembly areas (TAAs) are occupied by brigades, battalions, and companies that are forward of the rear area. Enemy contact is likely and the commitment of the unit directly into combat is possible and often anticipated.

3-80. Actions in the TAA focus on preparing the unit for future operations. Common TAA activities include task organization, maintenance, personnel replacement, re-supply, reorganization, rest, and planning for future operations. UMTs monitor changes in the task organization and adjust their plans accordingly. TAAs offer good opportunities for religious services.

Continental United States Replacement Centers

3-81. Replacement centers within the Continental United States (CONUS) receive and certify individuals for deployment to a combat theater. The CONUS Replacement Centers (CRCs) process non-unit related AC soldiers (to include RC soldiers accessioned onto active duty), Department of the Army civilians, contract civilians, Red Cross workers and other civilians.

BATTLEFIELD TRANSITION PLANNING AND EXECUTION

3-82. As with planning for different types of operations, planning for mission transitions must extend throughout the planning process and into operations and redeployment. Transferring control of an operation to another organization requires detail planning and execution. Transition depends upon the situation and will have its own characteristics and requirements. The UMTs should continuously communicate and coordinate among themselves, ensuring continual RS. After a hostile conflict or engagement, regeneration of force capabilities will be a primary consideration for the UMT in the transition plan. Following a stability or support operations mission, the soldier's physical, emotional, family, and spiritual recovery may be the UMT's transition focus. Mission analysis should provide the initial information to commence transition planning. The RSP supports the command's OPLAN with thorough integration. Transition planning and execution considerations include—

- Logistical supplies.
- Synchronized RSP transition with follow-on units.
- Denominational support.
- Transition religious support continuity file.
- Religious and Cultural information of the AO.

3-83. A smooth transition between forces and UMTs is essential for continuous religious support of all the military units. Cooperation among UMTs will ensure that all military and approved personnel receive quality religious support.

RELIGIOUS SUPPORT DURING REDEPLOYMENT AND DEMOBILIZATION

3-84. With the cessation of hostilities or the declaration of a truce, the deployed force in a theater of operations will transition through the last stages of Force Projection. This transition from the post-conflict stage through the redeployment and demobilization stages may begin even if residual combat operations are still underway.

3-85. While some forces are preparing for redeployment, other forces remain in support of Peace Building Operations. While soldiers restore order and reestablish the host nation infrastructure, UMTs provide religious support during the post-conflict stage.

3-86. As operations cease and forces are no longer required for Peace Building, units begin the redeployment stage. Forces and materiel not required for subsequent operations re-deploy to CONUS or home-theater to prepare for future missions. Redeploying forces may deploy to another theater or to an area other than home station on short notice. Redeployment, like deployment, occurs in several phases:

- Reconstitution for strategic movement.
- Movement to redeployment assembly areas.
- Movement to the port of embarkation.
- Strategic lift.
- Reception at a port of debarkation.
- Onward movement from port of debarkation.

3-87. During demobilization the process of transferring RC units, individuals, and materiel from an active posture to a pre-mobilization or other approved posture begins. Demobilization coincides with the first three phases of redeployment, the phases before the units depart the theater of operations. During this stage, RC units return to pre-mobilization status.

3-88. Demobilization requires careful planning which takes into account public relations and what the soldiers are thinking and saying to each other.

3-89. The phases of demobilization are—

- Planning.
- Area of operations demobilization actions.
- Port of embarkation to demobilization station operations.
- Demobilization station operations.
- Home station actions.

Unit Religious Support During Redeployment

3-90. Religious support during redeployment is similar to religious support during deployment. Some soldiers and contractors will be preparing equipment for strategic lift or waiting in holding areas for transportation, while others will be engaged in force projection operations.

> Worship, pastoral care, and religious education help soldiers begin the transition back to normal life.

3-91. UMTs providing religious support in assembly and PHAs during redeployment face challenges similar to those encountered during deployment. Worship, pastoral care, and religious education will help soldiers begin the transition back to normal life. Reunion sessions, an important part of the UMT's work, require early planning to be effective.

Reunion

3-92. Reunion denotes the reuniting of family members after separation as a result of deployment. Army research shows that family problems intensify when spouses return from deployments. Characteristics of the readjustment period, which lasts for about two months, include the emergence of physical symptoms, children's behavior problems, and marital tensions. For many couples, the first weeks of reunion are the most challenging part of the whole deployment experience. Tensions become most acute during the two weeks before and after the soldiers' return.

3-93. Religious support during reunion assists soldiers and family members in dealing with the stress that may occur while attempting to integrate experiences while separated, as well as, personal expectations of family life after deployment. UMTs help to restore and preserve the relationships that couples and families knew before deployment. During the reunion, UMTs join with the Installation Religious Team, Army Community Services, STARCs, United States Army Reserve Command (USARC) Major Subordinate Commands (MSC), family program coordinators, and mental health personnel to provide family assistance.

> By assisting with the stress of reunion, the UMT helps to restore and preserve the relationships that couples and families knew before deployment.

3-94. As married soldiers encounter reunion issues with spouses, single soldiers will experience reunion struggles with family and friends. Soldiers deployed as individuals or in small detachments also experience reunion problems upon return to their units. Reunion briefings and workshops help to minimize the problems for soldiers, families, and units. These events equip soldiers with coping skills to face the challenge of reintegration to families, peer groups, and units.

Reunion Preparation In Theater Before Redeployment

3-95. Before departing a theater of operations, UMTs coordinate with the S1/G1 for reunion briefings and workshops. The workshops teach skills needed to reconnect with younger children or to bond with newborns. During the pre-reunion period soldiers and families wonder, "What will it be like to be together again?" and "How has this experience changed us and our relationship?" While deployed UMTs conduct workshops in theater, non-deployed UMTs conduct similar workshops for waiting families.

> UMTs coordinate with the S1/G1 for briefings and workshops for soldiers in order to prepare them for reunion.

Reunion Preparation at Home Station after Redeployment

3-96. UMTs plan additional workshops at home station. Some workshops will be for soldiers and spouses, others for soldiers or spouses alone. UMTs may coordinate workshops with the Installation Ministry Team (IMT), ACS, Family Readiness Groups (FRG), and Chaplain Family Life Centers. The UMT must be prepared to sustain the effort for several months, depending on the length of the separation and the severity of the mission. From time to time, UMTs conduct assessments of the effectiveness of the workshops and the need to continue them.

3-97. Members of UMTs are cautioned not to ignore their own family reunion process. Being occupied with helping others is not a reason for neglecting one's own family. Installation and Family Life Chaplains conduct workshops specifically for redeploying UMTs.

UMTs are cautioned not to ignore their own family reunion process.

Recovery/Refitting

3-98. The unit's need to reestablish its readiness level sometimes competes with the soldier's family for the soldier's time, energy, and attention. During the recovery and refitting phase, soldiers need to clean, repair, and secure equipment; identify shortages; requisition supplies; and reestablish stock levels. Property accountability is reestablished with the installation Property Book Officer (PBO). UMTs are not exempt from these concerns, even though they are busy helping soldiers and families to reintegrate.

UMTs must balance reunion activities with refitting.

RELIGIOUS SUPPORT DURING DEMOBILIZATION

Planning Actions

3-99. Demobilization planning begins with mobilization and ends with the decision to release units and individuals from active duty. This planning includes—

- Relocating units and individuals.
- Incorporating annual training (AT) for non-mobilized RC units to support demobilization.
- Mobilizing additional RC units.
- Using individual volunteers (by means of temporary tours of active duty) to support critical missions in the demobilization process.
- Basic maintenance of equipment.
- Awards and decorations.
- Evaluation reports.
- After Action Reviews/Lessons Learned.
- Chaplaincy Detachment (CD) and GSU integration for support.

In Theater Activities

3-100. Before a unit departs the theater, it initiates demobilization administrative and logistical actions to include—

- Processing awards, decorations, and performance reports.
- Initiating line-of-duty investigations.
- Updating supply records and equipment accountability.

3-101. Medical and dental actions are started in theater when resources are available. During this phase soldiers may be frustrated as they become anxious to return home. As always, UMTs adapt religious support based on an assessment of soldiers' needs.

Home Station Activities

3-102. The reintegration of soldiers who deployed with those who did not is an important consideration for UMTs. During mobilization, many soldiers may deploy at different times, or with other units as a result of cross leveling. During demobilization, they may also return to home station at different times. Feelings of resentment may build in the unit, and in families, because some soldiers have returned and others have not. Reentry into the civilian work force is of another concern of the soldiers of the RC.

INSTALLATION RELIGIOUS SUPPORT

REINTEGRATION/REUNION

3-103. The Installation Religious Team plans for—

- The reintegration of returning soldiers with their family members and with non-deployed soldiers in their units.
- Support to casualty survivors.
- Non-deploying UMTs to continue their coverage of units and family members normally served by the returning UMTs, so that they will be able to reintegrate without the pressure of immediate duty.

3-104. Chaplain Family Life Center will—

- Assist in preparing families for reunion with soldiers.
- Assist UMTs with training and technical guidance on reunions.

(See Chapter 7 for further discussion of the Installation Chaplain and NCOIC responsibilities.)

GENERAL RESPONSIBILITIES OF THE UNIT MINISTRY TEAM DURING RELIGIOUS SUPPORT OPERATIONS

Staff Supervision

3-105. Chaplains at brigade and higher levels provide staff supervision of religious support by visiting subordinate units, monitoring the tactical situation, and revising estimates and plans based on changes in METT-TC.

3-106. A staff officer does not have the authority to order a subordinate unit to execute directives or orders. Staff officers make recommendations to subordinate commanders. These recommendations may be accepted or rejected. A staff chaplain may contact a subordinate UMT to provide instructions, advice, and recommendations, offer assistance, or exchange information. The commander will establish the nature and degree of the chaplain's actual authority or control in implementing the RSP.

Advising

3-107. Chaplains advise the commander on religion, morals, ethics, and morale as affected by religion. Advice extends to concerns such as the quality of life for soldiers and families, impact of indigenous religions on operations, and the allocation of resources. Commanders rely on the advice of the UMT to help them discern the unit climate. At brigade and higher levels, the UMT advises the command on the assignment of chaplains and chaplain assistants. The chaplain's direct access to the commander allows face-to-face communication and is the most effective means for providing advice.

Administration

3-108. Administrative actions continue in the tactical environment. They comply with unit TSOPs and appropriate regulations and doctrine.

Correspondence

3-109. Battlefield conditions complicate the preparation of correspondence. The necessity for correspondence, however, will not decrease. The UMT must prepare required reports and correspondence despite equipment limitations or the tactical situation. Prior to deployment, UMTs must consider how to adapt their procedures to field conditions.

Records Management

3-110. Army regulations and the unit TSOP determine the procedures for managing and maintaining records. The UMT may need to maintain the following records:

- Baptisms and other sacramental acts.
- Services of worship and attendance.
- Ministration to casualties.
- Memorial services or ceremonies.
- Staff journal (See DA Form 1594).
- After Action Reviews.

Logistics

3-111. Logistic support is essential to religious support in combat. Because support requirements are more difficult to meet in combat, UMTs must become familiar with the military supply system and use it as their first source of re-supply rather than making local purchases. UMTs should know what items are available within the system, which items are authorized for the UMT, and what procedures to use to requisition supplies.

3-112. The Modified Table of Organization and Equipment (MTOE) is the authorization document for ordering equipment. The chaplain is responsible for all team equipment; while the chaplain assistant is accountable for its proper use, security, and operational readiness. Request for changes to the MTOE should be coordinated and staffed through the chain of command using DA Form 2028 (Recommended Changes To Publications and Blank Forms).

Supplies

3-113. The chaplain assistant monitors supply levels, conducts inventories, and requisitions supplies through unit supply channels digitally using Department of Defense on-line requisition sites, or in coordination with the Defense Supply Center Philadelphia (DSCP) Item Manager. This procedure will ensure timely replacement of standard expendable items. The team reports its supply requirements to the supply section by national stock number (NSN), nomenclature, amount, and projected usage rates.

3-114. UMTs should deploy with at least 60 days of ecclesiastical supplies. Normally the unit establishes stockage levels, expressed as "Days of Supply" (DOS). The UMT may need to adjust the amount of supplies carried in accordance with the projected rate of resupply for each mission. A team deploying to an immature theater with limited logistical support may not have resupply available for 90-120 days.

3-115. UMTs at division and higher echelons coordinate directly with the Defense Logistics Agency (DLA) Liaison Officer and the 9925 Item Manager at DSCP to ensure a steady flow of Chaplain Resupply Kits and other religious support supplies to subordinate teams.

3-116. Planning must incorporate religious education/sustainment supplies for long-term continual operations. These supply items are also managed by the 9925 Item Manager. The supplies will support all faith groups' spiritual sustainment needs.

Maintenance

3-117. Maintenance is critical in a combat environment. The team is responsible for the maintenance of its assigned vehicle and other equipment. An effective maintenance program is essential to ensure the UMT's ability to perform religious support. Operator maintenance focuses on preventive maintenance checks and services (PMCS). Both the chaplain and the chaplain assistant must be able to operate and maintain the vehicle. These skills must be acquired and practiced before deployment.

Transportation

3-118. Section 3547 of 10 USC mandates transportation assets be afforded to the Chaplain for the conducting of religious services. The vehicle must be able to transport the team to all unit locations within the AO. The team must take precautions to avoid compromising the unit's location or fighting positions. Both members of the UMT must be trained and licensed to operate the vehicle.

3-119. METT-TC will determine who drives the vehicle. The practice of the chaplain driving in hostile areas ensures that the assistant, who is a combatant, can provide security. When not providing security for the team, the chaplain assistant operates the vehicle.

3-120. The UMT should take full advantage of unit convoys. It must be thoroughly familiar with convoy procedures, unit policy regarding movements, and vehicle densities.

Communication

3-121. As the UMT travels on the battlefield, it must maintain communications with its operational base to keep the base informed of its status, location and travel plans. At the same time, the team receives and transmits situation reports (SITREP). The commander must provide the team with dedicated communications equipment that is compatible with and keeps pace with the commands secure communication equipment. The team may use a variety of other means available within the unit. The team must be flexible and creative to maintain communications.

3-122. Religious support is synchronized and implemented throughout the area of operations, theater of operation, and communications zone (COMMZ) by being fully integrated into the digitized communication and automation systems. The ability to reach from the area of operations to the home station support node (HSSN) is essential for relaying time sensitive information and religious support requests. A mobile communication link between each echelon enables the UMT to allow time-sensitive data to be rapidly communicated to and from each echelon. This is essential for the UMT's mission of providing soldiers with the personal delivery of religious support. It ensures that the UMT has the ability to move to where time-critical religious support is essential. The communication linkage between UMTs provides immediate access to processed critical information needed to ensure the commander's religious support plan is executed successfully.

Assessment

3-123. UMTs continuously identify, assess, and disseminate lessons learned throughout the force and branch. Assessment of lessons learned from operations is the key element for future training objectives. Observation teams deploy with Army forces on every major operation. Religious support is an important element of the operation and lessons learned must be gathered on and integrated into the overall assessment of the operation.

3-124. Lessons learned are also critical during a transition phase between one force and another. UMTs arriving into the theater of operations can better understand the battlefield and METT-TC implications through lessons gathered by departing UMTs.

RELIGION, MORALS, ETHICS, AND MORALE

3-125. Chaplains serve as the "conscience of the command." Chaplains advise the commander on the moral and ethical nature of command policies, programs and actions, as well as, on the impact of command policies on soldiers. They advise the commander on—

- Religious/ethical issues as they bear on mission accomplishment and morale.
- Inter-faith relations among religious groups in the command.
- The role of religion within indigenous population groups.
- Accommodation of special religious needs/requirements.
- Quality of life issues.
- Unit command climate.

INDIGENOUS RELIGIONS

3-126. The chaplain advises the commander and staff on the beliefs, practices, and customs of religious groups in the AO, and the impact that local religions may have on the conduct of the mission. This advice is based on information developed at division and higher echelons and is coordinated with the G5 and Civil Military Operations (CMO) officer. (See Appendices A and F for more information.) Common concerns include—

- Religious organizations and doctrines.
- Religious practices and customs.
- Places of worship, shrines, and other holy places.

PUBLICATIONS

3-127. Prior to deploying, UMTs review the list of publications in the reference section of this manual to determine what to carry. As a minimum, the team carries this manual (FM 1-05), JP 1-05, AR 165-1, and pertinent "go to war" publications of the unit. The chaplain assistant requisitions missing publications and maintains the team's publications.

JOURNALS AND WORKBOOKS

3-128. The UMT should maintain a staff journal covering each 24-hour period (See DA Form 1594). The staff journal is a chronological record of events and serves as a permanent record for operational reviews, training issues, and historical research. The team's journal entries should provide enough detail to describe its activities adequately and to fix the time and place of significant religious support events; however, care must be taken to avoid compromising privileged information by recording it in the staff journal. Journals are useful in reconstructing events, maintaining religious support histories, writing AARs and lessons learned for the operation. Typical entries include—

- Summaries of plans and orders.
- Religious support activities.
- Reports sent and received.
- Reports of contact with other UMTs, missionaries, faith-based NGO/IOs, and humanitarian agencies.

3-129. A UMT may choose to keep a staff section workbook, also known as a "battle book." It is an informal, indexed collection of references and information taken from written and oral orders, messages, journal entries, and meetings. The battle book may include extracts from doctrinal manuals, forms, and other information. It has no specific format and is not a permanent record. (See FM 101-5.) RC UMTs should build battle books in concert with the AC unit they train with in peacetime and/or deploy with for operations.

SUMMARY

3-130. Religious support operations planning, preparation, coordination, and execution must be comprehensive. It must envelop all faith groups and personnel, be synchronized among all commands and UMTs, and be flexible enough to meet the fluid environment. The UMT will be faced with a great deal of information it must manage, implement, and disseminate. Constant and aggressive information gathering and analysis will keep the UMT informed of the religious support requirements for the commander. The chaplain, as a personal staff officer, and the chaplain assistant are a critical assets for ensuring the commander's religious support plan is implemented in a comprehensive and timely manner. Through proper staff procedures, mission rehearsals, and staff coordination, the RSP will meet the mission's requirements.

RESOURCES

3-131. The following is a list of main resources utilized for writing and updating this doctrine.

- AR 165-1.
- JP 1-05.
- JP 5-00.2.
- JP 3-0.

- JP 3-05.
- FM 16-1.
- FM 3-0 (100-5).
- FM 25-100.
- FM 25-101.
- FM 100-17-3.
- FM 101-5.
- FM 55-30.
- National Military Strategy
- Joint Vision 2010
- Army Vision 2010
- Concept for Future Joint Operations, May 1997
- U.S. Army Chaplaincy Strategic Plan FY 2000-FY 2005
- Chaplain Support Force XXI White Paper
- Quadrennial Defense Review Report, September 2001.

Chapter 4

BATTALION, SQUADRON, BRIGADE, REGIMENT AND GROUP RELIGIOUS SUPPORT OPERATIONS

The levels described within this chapter are the smallest echelons to which UMTs are assigned.

CONTENTS

TACTICAL ECHELON DESCRIPTION

BATTALION AND SQUADRONS

4-1. Maneuver battalions can be reinforced with other combat and combat support companies to form task forces for special missions. Task organization increases the capability of maneuver battalions to ensure operational success. A Brigade commander through the staff estimation process may task-organize tank and mechanized infantry battalions by cross-attaching companies between these units. Field artillery (FA) batteries can be reinforced with batteries of any kind to form artillery task forces. Light units and engineer battalions can also be reinforced using this same task organization process.

4-2. Combat support and CSS battalions vary widely in type and organization. They may be separate division or non-divisional battalions but normally perform functional services for a larger unit within that unit's area of operations. Air defense artillery and signal battalions assigned to, or supporting, divisions routinely operate throughout the division area of operations. The Corps Support Battalion (CSB) is part of a Corps Support Command (COSCOM) and a Corps Support Group (CSG) and provides support to units that could belong to a division, or a Corps.

4-3. Battalions, or their equivalent, conduct operations as part of a brigade's operation. Occasionally they are directly under a division or regiment's control. Their capability is increased through task organization. The commander understands the intent of the brigade commander and division commander so that he can properly employ his force.

BRIGADES, REGIMENTS, AND GROUPS

4-4. Brigade size units control two or more battalions or squadrons. Separate brigades of infantry, armor, field artillery, air defense artillery, engineer, and aviation, and armored cavalry regiments can be used to reinforce corps or division and can be shifted from unit to unit to tailor forces for combat.

4-5. Division commanders adjust the organization of their brigades and change their task organization as required by the factors of METT-TC. They normally receive support from FA battalions, engineer battalions (heavy divisions), forward support battalions, and smaller combat, combat support, and CSS units. Their chief tactical responsibility is synchronizing the plans and actions of their subordinate units to accomplish tasks for the division or corps.

4-6. Separate maneuver brigades and armored cavalry regiments have a fixed organization with organic cavalry, engineer, air defense, field artillery, military intelligence, chemical, and CSS units. Separate brigades are capable of operating as independent units.

4-7. Other combat, CS, and CSS brigades and groups are organized to control capabilities for divisions, corps, and larger units. They may also be the building blocks of large unit support structures such as corps and theater army support commands and of combat support commands such as engineer commands. Divisions receive support from an organic brigade-sized, division artillery (DIVARTY), an aviation brigade, and a Division Support Command (DISCOM) comprised of CSS battalions and companies. Heavy divisions have an organic engineer brigade.

Part of a Division

4-8. Brigades or their equivalent usually operate as part of a division. The division is a large, fixed Army organization that trains and fights as a tactical team. Normally, the division commander controls two- to five-ground maneuver brigades. The brigade can be employed in autonomous or semiautonomous operations when properly organized for combat. All brigades must be able to deploy, conduct offensive operations, conduct defensive operations, and conduct retrograde op-

Planning Checklist
Staff: communication, annex, integration, briefings
Specialized:
• Nurture the living:
◆ Spiritual Fitness
◆ Worship
◆ Counseling
◆ Visitation
• Care for the Dying:
◆ Spiritual triage
◆ Leadership
◆ Battle Stress
• Honor the Dead:
◆ Memorial Services
◆ Memorial Ceremonies
◆ Rites
◆ Sacraments
◆ Prayers

erations. Airborne, air assault, and ranger brigades/regiments are capable of conducting forced entry operations. All brigades may deploy to conduct military operations other than war. The brigade UMT is part of the division religious support staff and program.

Part of a Joint Task Force

4-9. Brigades or similar may deploy as part of a JTF with or without its traditional divisional headquarters. In these types of operations, the brigade may work directly for the JTF commander. The demand for experienced liaison will be high for a brigade in a JTF. Liaison may be required with joint, multinational, interagency, and or non-governmental agencies. The brigade's requirement for liaison will exceed its normal personnel and equipment structure.

4-10. Planning Religious Support for this type of operation needs to take into consideration the diversity and complexity of the chain of command, logistics, force structure, and area of operation.

FUNDAMENTALS OF RELIGIOUS SUPPORT OPERATIONS

GOAL OF UNIT MINISTRY TEAMS

4-11. The goal of all Unit Ministry Teams is the personal delivery of comprehensive, timely, and effective religious support to all soldiers.

4-12. Comprehensive, because each soldier has his or her own definition of what constitutes religious support. While not every spiritual need of every soldier can be met in the battlespace, the UMT seeks to meet as many needs as possible. Timely, because time is a limited commodity, and the situation changes in the battlespace. The UMT must be aware of the changing situation and prioritize ministry accordingly. Effective, because religious support must be more than minimal presence and a hurried service by the UMT. Quality effective religious support can provide hope and courage to soldiers who are facing the stress of battle.

ROLE OF THE BATTALION/SQUADRON UNIT MINISTRY TEAM AND SECTION

BATTALION/SQUADRON TACTICAL UMT	
CHAPLAIN (O-3)	**CHAPLAIN ASSISTANT (E-3/4/5)**
Role Commissioned Officer/Staff Religious Professional Title 10 Protected	**Role** Noncommissioned Officer Specialized Assistant Religious Support Specialist Team Security
Mission Advises commander on religious support and quality of life issues to support the unit's mission.	**Mission** Advises Chaplain and NCO Chain. Provides specialized assistance in areas of religious support to support the chaplain.

BATTALION/SQUADRON TACTICAL UMT	
CHAPLAIN (O-3)	**CHAPLAIN ASSISTANT (E-3/4/5)**
Performs/provides/coordinates the religious support mission for the battalion/squadron. Team leader.	Perform battle staff functions. Team member/ NCO.
Parallel Tasks Performs religious support. Provides spiritual, stress, fatigue, suicide preventive counseling to individuals and family members. Plans unit CMRP. Initiates religious support plan. Manages section funds and field offerings. Supervises Chaplain Assistant/Team NCO.	**Parallel Tasks** Prepare and synchronizes religious support and provides emergency religious ministrations in the battlespace. Screens stress, fatigue, suicide individuals for counseling or provide referrals if needed. Implements the unit CMRP. Prepares the Religious Support Annex. Safeguard and accounts for field offerings. Maintains unit equipment.

RESPONSIBILITY AND ACCOUNTABILITY

Staff Responsibilities

4-13. The chaplain serves the commander as a personal staff officer to plan, synchronize, and coordinate religious support, within the commander's area of responsibility. The battalion chaplain is the section leader. The chaplain assistant serves as the section NCOIC.

4-14. Chaplains serve as the "conscience of the command." Chaplains advise the commander on the moral and ethical nature of command policies, programs, actions, and the impact of command policies on soldiers. They advise the commander on—

- Religious and ethical issues as they effect mission accomplishment.
- Indigenous religions and their impact on the mission.
- Accommodation of special religious needs and practices within the command.
- Matters of unit morale as effected by religion.

Relationship to Brigade Unit Ministry Team

4-15. The battalion/squadron UMT works closely with the brigade (BDE) UMT ensuring a comprehensive and coordinated religious support plan for the BDE commander. The BDE UMT technically supervises the subordinate UMT and supports their RSP. During combat operations the battalion (BN) UMTs maintain communications with the BDE UMT sharing information about the religious support mission. In garrison operations, the BN UMT trains with the BDE UMT, coordinates unit and family religious support, and prepares for deployments.

Relationship to the Division Unit Ministry Team

4-16. The battalion/squadron UMT looks to the division UMT as their mentors and advisors during garrison operations. During combat operations, the

BN UMT will be prepared to communicate religious support issues directly if they assume the BDE UMT mission or if they work directly for a division/task force (TF) UMT. The principle of understanding and implementing the guidance of the commander two levels up also applies to the work of the Battalion UMT. The BCT UMT plans its RSP in harmony with the guidance the Division UMT. The Battalion UMT must understand and implement the Division RSP as it plans and executes its RSP.

Supervision Responsibilities

4-17. The chaplain exercises staff supervision and technical control for religious support throughout the battalion AO. The Chaplain is responsible for the development, implementation, and coordination of religious support assets, operations and activities. The chaplain assistant NCOIC is an active member of the NCO support channel in the unit on religious support issues and soldier welfare. The Chaplain assesses the soldiers and other authorized personnel's interpersonal/spiritual well being, which can affect the unit's fighting spirit. The UMT advises the commander on humanitarian aspects of operations, and the impact of command policies on indigenous religions. The Chaplain provides and coordinates privileged and sensitive spiritual counseling and pastoral care to the unit's command, soldiers, authorized civilians, and families. The UMT locates where they can best coordinate, communicate and facilitate religious support across the battlespace.

Staff Supervision During the Operation

4-18. The battalion Chaplain supervises the Chaplain Assistant. The UMT provides direct support and area support to units located within the battalion AO. The UMT coordinates religious support with each unit. The UMT monitors the tactical situation, and plans based on changes in METT-TC. If necessary, they recommend changes to the religious support annex and ensure supported units receive information concerning changes via` FRAGOs. The UMT reports required religious support data to the brigade UMT. The battalion UMT receives operational missions via the Brigade Operational Order/FRAGOs. The brigade UMT is the technical supervisor for the battalion UMT who does prior coordination and establishes contingency plans for the brigade religious support plan.

4-19. Before the battle, the battalion UMT submits a religious support annex that gives guidance to all unit commanders of the religious support provided. The chaplain assistant gathers information that will allow the UMT to battle track the operation, current unit positions, overlays, special operating instruction (SOI) information and successfully navigate the battlespace.

4-20. The UMT before, during, and after a phase of operation will coordinate with the Battalion Task Force for treatment and forward surgery teams to provide comprehensive religious support to soldiers being treated and to the team itself. The UMT coordinates their efforts in conjunction with the movement of the casualties. The UMT maintains contact via radio, casualty feeder cards, etc. with the brigade UMT in the form of SITREPs on their location, health, statistics, and ability to provide religious support such as mass casualty (MASCAL).

4-21. The battalion UMT considers the spiritual condition of the soldiers in follow-on planning. The severity of the previous operation and the number of casualties suffered are key factors in planning, preparing, and executing cycle that is a nonstop process.

Training

4-22. The battalion UMT has the responsibility to be proficiently trained in all areas necessary in order to adequately execute the religious support mission. It is the responsibility of each member of the UMT to be a tactically and professionally proficient team member. They must train in soldier tasks and collective tasks as a team. The battalion UMT trains to the unit's METL utilizing the derived religious support METL. The religious support METL serves as the focal point on which the UMT plans, executes, and assesses training. Training must encompass the following principles to effectively train as a team:

- Train as you provide religious support in the battlespace.
- Use all appropriate doctrine and tactics, techniques, and procedures (TTPs).
- Use performance-oriented training.
- Use mission-oriented training.
- Make training a challenge.
- Train to sustain proficiency.
- Train using multi-echelon techniques.
- Take control and maximize training opportunities.

ROLE OF THE BRIGADE UNIT MINISTRY TEAM

4-23. The religious support section is made up of a minimum of a Unit Ministry Team (UMT) composed of a chaplain (56A) and one enlisted chaplain assistant (56M). The UMT facilitates and coordinates religious support across the battlespace.

4-24. The brigade chaplain is a religious leader, competent staff officer and team leader. The brigade chaplain assistant NCOIC is a leader and advisor, mentor and information manager for the UMT. The UMT must have a keen awareness of the world situation as it bears on potential military operations, must know the role of military religious leaders and must have a broad understanding of the impact of indigenous religions on military operations.

4-25. UMTs must use technology to enhance the personal delivery of religious support. They must be trained in joint and multinational operations and skilled in synchronizing all aspects of religious support during military operations. They will operate in an environment of increasing volatility, complexity and ambiguity, thus the brigade UMT must be innovative and disciplined.

4-26. The brigade chaplain assistant needs to be a graduate of the Battle Staff NCO Course (with the 2S identifier). This will enable the brigade chaplain assistant NCOIC to take part in the MDMP in the absence of the chaplain, to track the battle analytically, and to better utilize the digital enablers that the brigade will share with other staff sections.

4-27. Brigade UMTs require the same level of technology as the command. UMT automation, communication and transportation must keep pace with the command's requirements in order to effectively coordinate religious support, communicate critical information and remain physically flexible to meet requirements across the brigade battlespace. The UMT is an integral part of the brigade.

THE BRIGADE/REGIMENTS/GROUPS TACTICAL UMT	
CHAPLAIN (O-4/5)	**CHAPLAIN ASSISTANT NCO (E-6/7)**
Role Commissioned Officer/Staff Religious Professional Title 10 Protected	**Role** Noncommissioned Officer Specialized Assistant Religious Support Specialist Team Security
Mission Advises commander on religious support and quality of life issues to support the unit's mission. Performs/provides/coordinates the religious support mission for the brigade. Brigade Team Leader.	**Mission** Advises Chaplain NCO Support Channel Provides specialized assistance in areas of religious support to support the chaplain. Performs Battle Staff functions. Team NCOIC.
Parallel Tasks Provides religious support supervision of Bn/Sqd Chaplains. Initiates and develops the religious support plan. Train Chaplains on MOS task. Plan and reviews the BDE CMRP.	**Parallel Tasks** Provide religious support supervision of Bn/Sqd Chaplain Assistant NCOs and Chaplain Assistants. Synchronizes the religious support plan. Trains Chaplain Assistants on MOS task. Prepare and implement the BDE CMRP.

RESPONSIBILITY AND ACCOUNTABILITY

Staff Responsibilities

4-28. The Brigade UMT works directly for the commander. The chaplain serves the commander as a personal staff officer to plan, synchronize, and coordinate religious support, within the commander's area of responsibility. The brigade chaplain is the section leader. The chaplain assistant serves as the section NCOIC. The brigade UMT serves as the technical supervisor and mentor to subordinate UMTs. As the commander's personal staff officer for religious support, chaplains at brigade-level and higher are responsible for assessing the technical proficiency of UMTs in subordinate units. They conduct assessments with the unit during field training exercises (FTXs), Battle Command Training Program (BCTP) events, external Army Training and Evaluation Program (ARTEP) events, Combat Training Center (CTC) rotations, unit training, and command inspections.

4-29. Chaplains serve as the "conscience of the command." Chaplains advise the commander on the moral and ethical nature of command policies, programs and actions, and the impact of command policies on soldiers. They advise the commander on the following, in addition to, the previous battalion/squadron topics:

- UMT personnel issues.
- Division RSP.
- Training issues.

Relationship to the Division Unit Ministry Team

4-30. The Division or TF UMT is the immediate technical supervisor for BDE UMTs assigned or attached operationally to a division or task force. If not assigned or attached, there may be a need operationally for coordination and close communication to ensure comprehensive religious support.

4-31. The BDE UMT or equivalent fully supports the division and/or TF RSP by supervising and providing guidance to BN UMTs RSPs. They work closely supporting the operational information religious support management process for the entire division or task force.

Relationship to CORPS Unit Ministry Team

4-32. The principle of understanding and implementing the guidance of the commander two levels up also applies to the work of the Brigade UMT. It is essential for the BCT UMT to understand and implement the Corps RSP as it plans and executes its RSP.

Staff Supervision During the Operation

4-33. Brigade UMTs provide staff supervision of religious support by visiting subordinate units, monitoring the tactical situation, and planning based on changes in METT-TC. If required, they recommend changes to the religious support annex and ensure subordinate units receive and execute the changes, which are issued as FRAGOs. The brigade UMT should maintain contact with subordinate UMTs by visits, radio spot reports, or any means possible to insure subordinate UMTs status is 100 percent and religious support data is collected and sent to the division UMT.

4-34. The brigade UMT monitors the tactical situation, the status of each battalion UMT, and plans for future operations. If changes in METT-TC warrant adjustments in religious support, the brigade chaplain recommends changes to the brigade commander or executive officer. Reassigning a chaplain or chaplain assistant, or changing the mission of a battalion UMT, requires a FRAGO. The brigade UMT must be prepared to augment TF UMTs in the event of mass casualties. The brigade religious support annex should have a plan to augment the religious support provided at the base support battalion (BSB), forward support medical company (FSMC).

4-35. When the brigade reaches the objective, casualty evacuation increases. The brigade UMT again assesses the requirements for additional religious support in the AO. If the number of casualties is high, a MASCAL is declared. The brigade UMT will issue a FRAGO to notify supporting UMTs. These UMTs will assist the BSB UMT in providing religious support at the FSMC to casualties and by conducting religious support debriefing.

4-36. If a TF UMT has not submitted a current SITREP (See format model in FM 101-5-2), the brigade UMT NCO contacts the team to determine its status. When feasible, the brigade UMT moves forward to the TF UMTs to provide direct spiritual care. Priority goes to the TF executing the main attack. If the brigade continues the mission, time may be very limited. This phase may start with reconnaissance, or it may go directly into deployment for another assault.

4-37. The brigade UMT considers the spiritual condition of the soldiers in follow-on planning. The severity of the previous operation and the number of casualties suffered are key factors in the planning, preparing and executing cycle that is a nonstop process.

4-38. If the brigade is operating as part of a division mission, the brigade UMT of the follow-and-support brigade establishes liaison with the UMT of the lead brigade to facilitate pastoral care casualties and other religious support. The follow-and–support brigade may provide evacuation and treatment for casualties of the lead brigade, so that the lead brigade can continue its tempo. This mission is of particular concern to the brigade UMT because of its complexity, intensity, and momentum.

4-39. In the mobile defense, battalions engaged in offensive, defensive, and retrograde actions present a challenge for the brigade UMT. The UMT plans religious support for each of the units based on METT-TC. The religious support plan is incorporated into the brigade OPLAN as an annex. The area defense differs from the mobile defense in that all units are engaged in defensive operations.

Supervision of Brigade Combat Team (BCT) Religious Support Operations

4-40. All action starts with a vision and that includes religious support. "Vision" is a personal concept of what the organization must be capable of doing by some future point. It is the target. What does the organization need to accomplish today, tomorrow, or years from now to achieve the desired result? How is the organization developed to achieve that capability? Simply put, vision sees the required effort, leadership, and plan to get the job done.

4-41. The transition from battalion to brigade-level religious support requires a transition in focus as well as location. The brigade-level chaplain's focus moves from being primarily an operator to being a staff supervisor of religious support for the commander. A very important part of being a brigade chaplain/supervisor is the development of the vision mentioned above. That vision must broaden to the whole brigade and see supervision as an aspect of religious support proper.

4-42. One component of the vision is certain imperatives for focusing action -- a vision that includes provisions to provide purpose, direction, motivation, and continuity. To provide some insight as to how this applies to the Brigade Unit Ministry Teams, the following few paragraphs will paraphrase FM 22-103, Leadership and Command at Senior Levels.

4-43. Purpose insures there is a clear and worthy object for every action and sacrifice. It provides the UMTs with a reason to withstand the stress of continuous operations. It depends heavily on the ability of the battalion UMTs to understand all requirements. The brigade chaplain does this by giving a sense of priority and importance to necessary tasks. The brigade chaplain explains how the success or failure of the UMTs will impact upon the unit. Teams with a good sense of purpose will succeed in ways that cannot be forecast before the action. Purpose provides the religious support teams of the BCT with goals. It gives the teams a sense of "why" they are doing what they are doing.

4-44. Direction channels the collective efforts of the people on the teams toward the unit's goals, objectives, and priorities. It puts the Unit Ministry Teams on a clear and unambiguous course and thus creates the conditions that allow them to succeed. Setting goals and standards, ensuring discipline, and training the teams are some of the means the brigade chaplain can use to chart a direction for the teams. Goals and standards provide the azimuth. Good discipline provides the confidence that goals and standards are met in the absence of leaders and immediate guidance. Training ensures results. An additional means of providing direction is the SOP. Direction helps the teams know what to do, while purpose helps the teams understand why they do it.

4-45. Motivation engages the will of Unit Ministry Teams to do their utmost to achieve the purposes, goals, objectives, and priorities of the unit, even under the most difficult circumstances of combat. It is not enough to know why and what to do; we need the moral force of a strong will.

4-46. Continuity anticipates future conditions, situations, and needs; it takes steps in the near term to ensure that the BDE Unit Ministry Team can meet future missions. Synchronization of plans and efforts among all Unit Ministry Teams is critical to quality religious support. The Brigade Chaplain is the focal point for this as the staff representative for religious support for the commander.

Supervisory Relationships

4-47. The complex job of the brigade chaplain is to organize the efforts of UMTs in different subunit commands with different battlefield operating system (BOS) responsibilities and support. Some of the teams are in units assigned directly or as attached to the brigade, i.e., the maneuver battalions and field artillery battalion (unit support). To further expand religious support supervision and synchronization, area support to soldiers, members of other services, authorized civilians, and all who have no organic or available religious support is part of the comprehensive plan. The supervisory role expands to denominational coverage to these soldiers. Still, the mandate to UMTs is to provide religious support to all soldiers in the AO.

4-48. The Brigade Chaplain as a staff officer for the commander uses different external coordination channels to ensure the accomplishment of religious support. Three channels of communication command, staff, and technical require development and rehearsal. Too often coordination and guidance are communicated through the wrong channels; and because the Brigade Chaplain has no command authority, he/she must work effectively through the correct coordination channels.

Supervision of Brigade Religious Support Operations

4-49. Task Force UMTs work for their unit commanders. The BDE UMT cannot give guidance to subordinate UMTs that conflicts with their commander's guidance. Proper coordination and solid common sense on the part of the BDE UMT make the entire BDE religious support plan work.

4-50. Proper coordination with the commands and staffs in the brigade opens doors of quality communication and support. A good working relationship builds respect and cooperation with commanders and staffs throughout the brigade combat team. The brigade chaplain and NCOIC need to cultivate this relationship and respect their position on the commander's staff. Proper staffing procedures make religious support a reality.

OBSERVED PRINCIPLES OF EFFECTIVE SUPERVISION
The primary mission of the Brigade UMT is to synchronize Religious Support.
The BDE Unit Ministry Team assigns its highest priority to managing religious support for the whole BDE. The UMT must see supervising as their part of comprehensive religious support. Providing religious support to the tactical operations center (TOC) is secondary to the primary mission of synchronizing religious support throughout the BDE. When the BDE UMT is effectively overseeing religious support, quality religious support reaches all corners of the BDE.
PRINCIPLE 1. The BDE UMT trains as a team before it arrives in a theater of operation. A goal of the BDE UMT training is to work as a synchronized team. UMTs do individual religious support well, but the BDE UMT must work hard at coordinating the religious support that happens in the Brigade.
PRINCIPLE 2. Synchronize Religious Support Appendices at all levels. The BDE UMT ensures that religious support is synchronized on paper and in action. Make certain that the Brigade Religious Support Appendix is accurately reflected in subordinate UMT Religious Support Appendices. A goal for BDE UMT training is to have continuity in religious support appendices throughout the BDE UMT.
PRINCIPLE 3. Plan, rehearse and coordinate casualty replacement plans for members of the BDE UMT. UMT members in the BDE should know who would replace them in the event they become a casualty.
PRINCIPLE 4. Coordinate all Religious Support guidance with the appropriate commanders. Just because something is written in an appendix does not mean that it will happen, it must be coordinated. Coordinate religious support prior to the mission with all subunit UMTs. This effort helps to ensure comprehensive religious support synchronization in the AO.

OBSERVED PRINCIPLES OF EFFECTIVE SUPERVISION
PRINCIPLE 5. Daily communication is vital to Religious Support. UMTs have a crucial piece of the overall religious support plan and need to send that information up and down. UMTs must know what sort of information is necessary to report. BDE UMTs must be aggressive in maintaining disciplined and systematic information flow. Daily communication between all UMTs in the BDE is imperative for up-to-date information. Here are some suggestions for making that work: • Net calls. A net call establishes a specified frequency and time when all UMTs are on the radio at the same time. In this way information can be put out quickly and every UMT becomes knowledgeable of the religious support situation. • Maintain regular hours for monitoring the radio and mobile subscriber equipment (MSE) so subordinate UMTs know when they can directly contact the BDE UMT. • Send daily guidance and updates to subordinate UMTs. • Spot reports to pass on time sensitive information. • Use the BDE RETRANS frequency (coordinate with the signal officer. • Send messages by couriers such as commander's drivers, ambulance drivers, and logistics package (LOGPAC) drivers. • Authorize lateral communication between UMTs. Often one UMT cannot make radio-communication with the BDE UMT, but can pass a message through a neighbor UMT. • Schedule regular meetings with subordinates in the field.
PRINCIPLE 6. Daily updates from the BCT UMT to subordinate UMTs are critical. When the BDE UMT communicates to its subordinate UMTs it builds teamwork. This enables individual UMTs to understand what is happening, and what is expected of them. They feel like part of the team. Always remember that effective communication is a two way street. UMTs must communicate the following items of information: • A daily status report of all UMTs in the BDE. • UMT NCOs update. • Changes in religious support for attached/detached units, and units operating in the brigade AO. • Upcoming religious events in the command, and indigenous religious holidays. • The Brigade Commander's daily guidance and other important issues from the command. • Questions that need answering reference indigenous religions and their potential impact on operations.
PRINCIPLE 7. Daily SITREPs from subunit UMTs to the BDE UMT are critical. The BDE UMT must know what the subunit UMTs know to effectively manage the religious support mission. Emphasis upon effective communication will build teamwork and enhance the religious support mission. Examples of information that must be communicated follows: • The status of the UMT. • The status of ecclesiastical supply. • Daily activity. • Changes in unit RPP. • CA issues. • Information on indigenous religious leaders and sacred sites. • Issues of religion, morals, ethics, and morale.

TRAINING

4-51. To supervise religious support in a combat environment, the BDE UMT creates well-rehearsed coordination, synchronization, and communication techniques through training before deployment. If it does not happen effectively in a garrison/training environment then, quality religious support may not happen in a stressful and fast-paced combat operation. The responsibility for training rests heavily on the BDE UMT. Quality combat survivability training is available at the unit level. Religious support training is available at division and higher level. This does not prevent BDE Unit Ministry Teams from conducting training within the BDE. Training at this level is crucial to support the commander's mission. The BDE UMT needs to be proactive with an approved, staffed and coordinated training plan that takes into account Officer and Enlisted Critical Task lists as part of the total BDE training plan.

4-52. In addition to the BN training topics and individual and collective responsibilities, the BDE UMT must be personally involved in the training of subordinate UMTs. The BDE UMT plays a critical role in the development of the subordinate unit's junior chaplains and chaplain assistants. Success in providing spiritual support to soldiers requires adequate training. The UMTs of a BDE require sufficient training to enable it to mobilize, deploy, and provide quality spiritual support to soldiers anywhere in the world. The battlespace fixes the directions and goals of training. The BDE UMT is the training management team for subordinate UMTs. Their responsibility is to initiate, plan, and coordinate training for all UMTs in the command. The BDE UMT will make training effective by:

- Developing and communicating a clear training vision.
- Requiring subordinate UMTs to understand and perform their roles in unit, technical, and individual training.
- Being proficient on their mission essential tasks.
- Continually assessing training completed.
- Eliminating training distractions.
- Developing near and far term training plans.
- Being systematic and aggressive.

4-53. Suggested training topics include—

- Land navigation—mounted/night/navigation.
- Operations order process—
  - Developing and writing a religious support estimate.
  - Writing a religious support annex.
  - Writing religious support FRAGOS.
- War-gaming a battle as a brigade UMT combat team.
- Communications rehearsal—
  - Daily reports.
  - Communicating vital information.
  - UMT net call.
  - Practicing alternate means of communication.
  - Practicing cross communication between BN UMTs.
- Rehearsing the RSP of succession.

- Re-supply training: ordering and getting ecclesiastical supplies through logistics channels.
- Map reading, military symbology, and battle tracking.
- Conducting Religious Area and Impact Assessments
- UMT TSOP development and rehearsal.
- Chaplain assistant coordination of the RSP in absence of the Chaplain.
- Briefing the RSP.
- Staff Integration.

OPERATIONS PLANNING AND PREPARATION FOR UNIT MINISTRY TEAMS

4-54. Military operations consist of four types: offensive, defensive, stability and support. Each type consists of four phases: pre-deployment, deployment, sustainment, and redeployment. The following is a discussion of the responsibilities of company and field grade UMTs in each type and through each phase of operation.

CREATING THE TASK FORCE RELIGIOUS SUPPORT PLAN

TIMELINE OF CRITICAL EVENTS + PRIORITY OF RELIGIOUS SUPPORT = TF UMTs RSP

Timeline

4-55. The most valuable resource is time. The UMT must manage time so that the RS mission gets accomplished. A TF chaplain will not be able to see every soldier and certainly not solve every problem. Set priority of effort and allocate time to that effort. You must use the tools of time management. First, the UMT extracts all the time critical events from the OPORD and places them on a time line; for example—

- Logistics package (LOGPAC) activation and deactivation.
- Enemy recon expected.
- Friendly recon/counter-recon sent out.
- TF rehearsals/brief-backs.
- CSS rehearsals.
- Engineer movements.
- Company team movements.
- Line of departure (LD) /Be-Prepared-to-Defend times.
- Jump Aid Station (JAS) moves out, and any other time driven events.

4-56. Second, the UMT determines the intent of religious support during this phase of the battle. Knowing the religious support intent will help determine how much time to spend on visitation, where to spend it, and where priorities lay.

4-57. Third, the UMT determines soldier activity level. The Commander's Intent and unit warfighting doctrine must be understood. With these in mind, the UMT will know what tasks the company teams must accomplish. The UMT may then gauge when is the most effective time to visit and provide religious support.

4-58. Fourth, the UMT determines limiting factors like restrictions in the OPORD, ROE, METT-TC, and personal state of readiness.

4-59. The RSP is an informal plan and is flexible to meet mission requirements. It is likely to change during the execution phases of the battle because not everything goes as planned. It is easier to alter an existing plan than create (and communicate) one in a crisis.

Priority

4-60. Priority of religious support is to be identified as early as possible.

4-61. First, the UMT takes into account what has been experienced by each company in the TF, and their role in the coming mission. The UMT must determine what and where the needs are, and which unit has priority for religious support.

4-62. Second, the UMT considers morale factors known before the battle, leadership issues, human sustainment issues (food/rest/water), and interpersonal conflicts.

DEVELOPING PRIORITIES FOR THE TASK FORCE UNIT MINISTRY TEAM

4-63. The driving force behind the planning process of a TF UMT is determining its religious support priorities. Where will you place your main efforts and why? The UMT must create a time line that includes the unit's critical-time-driven-events. Working with this data, the UMT can then select blocks of time for priority religious support.

1. ***Study the Brigade OPORD/OPLAN and extract the following information:***
   - The mission assigned to your battalion.
   - The BDE Commander's intent for this mission.
   - Attachments / Detachments
   - The RS missions assigned to the UMT - Unit and Area Support.
   - Channels of communication with the BDE UMT, and with UMTs in the BDE area of responsibility (AOR).
   - Succession plans in case the BDE UMT becomes a casualty.
2. ***Study the Battalion OPORD / OPLAN and extract the following information:***
   - The missions assigned to the various Battlespace Operating Systems (BOS).

– Intelligence	(S2 / G2/Scouts)
– Maneuver	(combat companies)
– Fire Support	(artillery, mortars, close air support)
– Air Defense	(Air Defense Artillery)
– Mobility & Survivability	(Engineers)
– Logistics	(support, supply, repair, maintenance; medical)
– Battle Command	(Command and Staff)

_3._	***Identify the main effort company(s), the supporting attack company(s), and the Scout's mission.***
_4._	***Identify the first location for the medical units (BAS, JAS, FSMC), and the next several places each will jump by phase. Locate the AXPs and plot them on your map.***
_5._	***Determine the priority of religious support the UMT will perform:***
	• Main effort company(s)
	• Supporting attack and reserve company(s)
	• Ordinances/ Rites / Sacraments / Worship services
	• Visitation / Counseling
	• Senior leader religious support
	• Company / Platoon leader religious support
	• Individual soldiers
	• Morale problem areas
	• Honor the dead - Memorial Ceremony/Service
	• Scouts
	• Area Support requirements
	• Religious support to other UMTs
	• Ethical / Moral counseling
	• Displaced Persons / Detained Person religious support
_6._	***Create a time line for your religious support***
	• Write in the critical events of the day that effect the UMT.
	• Identify blocks of time for priority religious support.
	• Write in your priority religious support objectives in the priority religious support time blocks.
	• Consider flexible time factors in a high operation tempo (OPTEMPO) and fluid combat environment.
	• Choose an alternative plan in case conditions change unexpectedly.

WRITING RELIGIOUS SUPPORT APPENDIX

4-64. ***Consider the audience.*** The religious support appendix to the Service Support Annex is for the entire command. The religious support appendix must contain information that is both specific to subunit UMTs, and important to the whole command. The religious support appendix must support the commander's mission.

4-65. ***Only submit the appendix when it's needed.*** There are many times when there is no critical information to contribute. When appropriate, submit an appendix/matrix or a paragraph for the CSS annex. (See Appendix D)

4-66. ***Content is key, not the format.*** The UMT has no command authority so all tasks must be coordinated with commanders. Working with the S3 / G3, insert tasks for subordinate UMTs into the main body of the OPORD as tasks to subordinate units in paragraph 3. This will ensure command awareness and support for these tasks. Space for the religious support appendix varies from three pages in the BDE OPORD; one page for a matrix; or possibly only have space for a paragraph. Use the correct military format. A matrix can be utilized in keeping with the styles and formats used in the unit (See Appendix D). Or write a concise paragraph for inclusion in the CSS Annex. Ensure the content only states what is critical to support the mission.

4-67. ***A good TSOP reduces OPORD size.*** TSOP material is not repeated in the RS Appendix! The TSOP is an effective, usable, and combat oriented document. The content of the TSOP will help in the preparation of appendices and pre-combat coordination. UMTs must possess the RS portions of higher and lower units TSOPs.

4-68. ***Intent should be clearly written.*** Write the concept of religious support so higher and subordinate UMTs understand what is to be accomplished.

BRIGADE

4-69. The primary mission of the brigade UMT is to coordinate religious support for all soldiers and authorized civilians in the brigade commander's area of operations (AO). They provide religious support to soldiers in the brigade headquarters and others in the AO without religious support assets. They provide critical guidance and technical control to subordinate UMTs.

RELIGIOUS SUPPORT COVERAGE PLAN

4-70. The brigade UMT coordinates the brigade UMTs religious support throughout the brigade AO, thus ensuring that all units and individuals operating within the AO are provided religious support. This includes refugees, displaced persons, civilians in the area of operation, and enemy prisoners of war. The support of civilians and Enemy Prisoners of War is the responsibility of the commander; the commander may or may not require the chaplain to provide or perform religious support. As part of the Army component, civilian employees and contractors will be present providing essential work. Pending higher guidance issued in the division OPORD, the UMT will assess the requirements of civilian employees and contractors deployed in the AO and will include them in the religious support plan, at all levels.

CONTRACTORS

4-71. Chaplains are the staff officers in charge of coordinating religious support for contractors. All contractors who are clergy/lay ministers and readers will submit an application for approval as prescribed under Army Regulation (AR) 165-1 for supervision and coordination by military chaplains. UMTs will make every effort to plan for adequate religious support resources with their unit such as literature and sacramental supplies to support the religious needs of contractors. UMTs will plan for area religious support coverage to include contractors. Coordination for additional ecclesiastical supplies or religious support personnel will be with the supervising UMT. Chaplains will provide support for death notifications, Red Cross notifications by command, liaison with CONUS clergy, or others for contractor casualties. Religious support is coordinated through the chain of command.

RELIGIOUS SUPPORT MISSION TASK

4-72. Before the battle, the supervisory UMT submits a religious support appendix that both gives clear guidance to subordinate UMTs and allows all unit commanders to be informed of the religious support provided. The chaplain assistant gathers information that will allow the UMT to battle track the operation and current unit positions, understand overlays and SOI information, and successfully navigate the battlespace

4-73. The brigade UMT also coordinates with combat support and combat service support units in the brigade AO. Units with organic UMTs may need additional faith group support. Some UMTs may provide religious support for units whose organic support is not available. Time, distance, and battlespace conditions may restrict the travel of the UMTs and make religious support difficult.

Preparation for Combat

4-74. The brigade UMT uses the religious support estimate to plan religious support for all units in the brigade AO. It assesses the faith group needs of the battalions and requests additional support from the division UMT. The brigade UMT coordinates times and locations for forward link up points with the maneuver battalions.

4-75. The brigade UMT coordinates also with combat support and combat service support units in the brigade AO. Units with organic UMTs may need additional faith group support. Some UMTs may provide religious support for units without organic religious support or for units whose organic support is not available. Time, distance, and battlespace conditions may restrict the travel of the UMTs and make religious support difficult.

4-76. During offensive operations, the brigade support area (BSA) moves frequently. The brigade UMT may establish an alternate operational base in the BSA, but it must know how to move its equipment if the BSA relocates. As an alternative, the team may choose to establish its operational base near the brigade tactical operations center (TOC), where the majority of planning and communications are found. In the final analysis, the BDE UMT must locate in a position that best allows it to facilitate RS across the brigade. Regardless of location, the brigade UMT must be prepared to displace quickly.

The brigade UMT attends rehearsals to ensure religious support is fully integrated with the scheme of maneuver and the CSS plan.

4-77. During rehearsals, the brigade commander ensures the BSB commander has chosen adequate main supply routes (MSRs), future locations, logistics release points (LRPs), and refueling sites. The brigade UMT monitors this part of the rehearsal closely to integrate religious support with the scheme of maneuver and CSS plan.

4-78. The brigade UMT monitors the tactical situation, the status of each battalion UMT, and plans for future operations.

Develop Intelligence through reconnaissance and intelligence assets thru Exercise Command and Control

4-79. The brigade UMT monitors the tactical situation, the status of each battalion UMT, and plans for future operations. If changes in METT-TC warrant adjustments in religious support, the brigade chaplain recommends changes to the brigade commander or executive officer. Reassigning a chaplain or chaplain assistant, or changing the mission of a battalion UMT, requires a FRAGO. The brigade UMT must be prepared to augment TF UMTs in the event of mass casualties.

Perform Logistics and Combat Service Support

4-80. When the brigade reaches the objective, casualty evacuation increases. The brigade UMT again assesses the requirements for additional religious support in the AO. If the number of casualties is high, the team may choose to move to the forward support medical company in the BSA.

If the number of casualties is high, the brigade UMT may choose to move to the forward support medical company in the BSA.

4-81. If a TF UMT has not submitted a current SITREP, the brigade UMT NCOIC contacts the team to determine its status. When feasible, the brigade UMT moves forward to the TF UMTs to provide pastoral care. Priority goes to the TF executing the main attack.

4-82. If the brigade continues the mission, time may be very limited. This phase may start with reconnaissance, or it may go directly into deployment for another assault.

4-83. The brigade UMT considers the spiritual condition of the soldiers in follow-on planning. The severity of the previous operation and the number of casualties suffered are key factors in planning.

4-84. Split Operations could include a combined Legal, Medical, UMT and Civil Affairs Cell as an active battle-tracking proponent in TOC operations, while a separate religious support cell, employing all BSA UMTs, operates in the BSA TOC. This type of operation promotes better use of RS assets, prepares the BSA Chaplain to assume the responsibility of the TF Chaplain in the event of loss, provides greater opportunities to minister to TF BOS systems, and produces two avenues for receiving TF UMT daily reports. The TF UMT maintains contact with the BSA cell through digital nonsecure voice telephone (DNVT) channels.

4-85. If the brigade is operating as part of a division mission, the brigade UMT of the follow-and-support brigade establishes liaison with the UMT of the lead brigade to facilitate pastoral care to casualties and other religious support. The follow-and-support brigade may provide evacuation and treatment for casualties of the lead brigade, so that the lead brigade can continue its tempo. This mission is of particular concern to the brigade UMT because of its complexity, intensity, and momentum.

Division Cavalry Squadron

4-86. The division cavalry squadron may operate across the entire division front or along one or both flanks. When the squadron operates forward of the brigades, the squadron UMT coordinates with the brigade UMTs for the care of casualties and for additional religious support.

DEPLOYMENT

4-87. BDE UMTs are prepared at all times to deploy with their respective unit. Their deployment plan is synchronized and coordinated with the Brigade command and staff, subordinate TF UMTs, and higher supervisory UMTs. The UMT trains throughout the year on the needed skills and tasks for deployment. Planning considerations include—

- Air/Ground flow schedule.
- Intelligence information.
- Subordinate UMT readiness.
- Unit spiritual fitness.
- Ecclesiastical resupply kits and other religious support material.

Pre-combat Planning Considerations

4-88. After completing the planning process, the UMT is not yet ready to begin religious support operations. Coordination with key people and precombat checks still need to be completed. Listed are examples of areas for coordination, inspection, and team member responsibility.

- Chaplain—
  - Worship/Counseling preparation.
  - Coordination with—
    - Executive Officer for religious support plan/appendix; staff meetings; daily activities.
    - Commander for his guidance and focus; give advice on spiritual fitness of command.
- Chaplain Assistant—
  - Coordination with headquarters and headquarters company (HHC)/headquarters and headquarters troop (HHT) commander for bivouac site; local security plans; food and fuel; planned moves of trains or TOC.
  - Checks and Activities—
    - Individual precombat inspection (PCI).
    - Weapon/ammo.
    - Mission oriented protective posture (MOPP) gear.
    - Survival gear.
    - Food.
    - Water.
    - Issue equipment.
  - Vehicle PCI—
    - Meals ready to eat (MREs) stored.
    - Operator's vehicle material (OVM) stored.
    - Fuel.
    - Water.
    - PMCS complete.
    - Radio and frequencies, crypto, fill.
  - Team PCI—
    - Map with overlays.
    - CEOI/SOI information.
    - Night Vision Goggles with spare batteries.
    - TSOP.
    - Compass.

        - M256 kit.
    - Movement Plan—
        - Map Recon.
        - Plan posted in Combat Trains Command Post (CTCP)/Field Trains Command Post (FTCP)/Tactical Operations Center (TOC).
    - Ecclesiastical Supplies—
        - 30/60/90 day supply.
        - Chaplain's Kit complete.
- Unit Ministry Team—
    - With subordinate unit commanders/CSMs/1SGs—
        - Plan subordinate unit worship services.
        - Identify times and locations for visiting soldiers.
        - Identify special spiritual needs of the command.
        - Identify other special needs and issues.
    - Higher and lower-level ministry teams—
        - Religious Support Plan (at least two levels up).
        - Location during phases of battle.
        - Religious support requirements.
        - Availability of and plans for denominational support.
        - Emergency ecclesiastical re-supply.
        - UMT Replacements.
    - Staff—
        - Casualty information.
        - Intelligence updates (critical for the UMT to keep themselves constantly updated).
        - METT-TC updates.
        - Medical Plan updates.

OPERATIONS

Staff Coordination

4-89. Before departing the operational base, the UMT leaves a copy of its travel plan in the command post (CP) or TOC. The chaplain coordinates with the S1/S4 representative to receive messages and requests for religious support. The brigade chaplain is normally expected to attend shift change and battle update briefings (BUB). The brigade chaplain should ensure a means whereby those who try to contact the UMT can leave messages and that the UMT location is known at all times.

4-90. During offensive operations, the brigade support area (BSA) moves frequently. The brigade UMT may establish an operational base at the brigade TOC. The BDE TOC provides the UMT—

- Current operational updates.
- Availability to the commander and key staff personnel.
- Battle tracking.

4-91. As an alternative, the team may choose to establish its operational base in the BSA. The TOC is preferable. Regardless of location, the brigade UMT must be prepared to displace quickly.

4-92. During rehearsals, the brigade commander ensures that the Brigade Support Battalion (BSB) commander has chosen adequate MSRs, future locations, re-supply points, and refueling sites. The brigade UMT monitors this part of the rehearsal closely to integrate religious support within the scheme of maneuver and CSS plan.

Note: When moving independently among task force and slice units, the brigade UMT must be aware of the potential for being left behind. To prevent this situation, UMTs must have dedicated communications equipment that is compatible with the unit's secure communication equipment.

Coordination with the CSS planners
1. Refer to FM 71-2, CSS chapter.
2. The BDE will issue its OPORD to the TF leadership early. The BDE OPORD contains the BDE UMT's Religious Support Appendix. Coordinate with your S4, and ensure his familiarity with the religious support appendix.
3. The TF Cdr will issue a warning order to the TF. A possible format is in FM 71-2, appendix
4. At this point one of two things might happen; either the S4 will return to the CTCP and convene a meeting with the battalion maintenance officer (BMO), medical platoon leader, S1, and the Chaplain to review the BDE order and receive feedback for CSS planning; or the S4 stays at the TOC and holds the CSS planning meeting there. If the Chaplain is occupied with ministry, the chaplain assistant attends the CSS planning meeting and provides UMT input.
5. The Cdr will issue the OPORD to the command at the TOC or other designated location. One member of the UMT should attend that meeting.
6. After the UMT has studied the OPORD, coordinate with the medical platoon leader. Find out first hand what the CSS plan is, especially the planned locations during each phase of the battle.
7. Attend the CSS rehearsal and listen for changes in the plan. During the rehearsal, let everyone know where the UMT will be during each phase of the battle.
8. Monitor the Admin/Log net to stay current. Often the Command puts out changes to the plan on short notice. The tactical situation changes rapidly and you might find the planned attack has changed into a hasty defense. Be alert, your life depends on it.

SITUATIONAL AWARENESS

4-93. Frequently the operation does not take place as planned. The UMT must know how to get information from a number of sources in the unit. There will be times when the back-channel information flow is more timely and accurate than the official flow.

4-94. Timely coordination with the CTCP leadership (S4 and S1) and their staffs before deployment makes it easier to coordinate and publish UMT FRAGOs. If the CTCP has not planned to track the UMT before it goes to the field, then in all likelihood it will not track the UMTs movements and priorities when in the field. Prior coordination makes it happen.

4-95. If the UMT announces that it will visit a particular company or section, then it needs to follow through on that plan. The UMT must carefully consider any information it places in the TF OPORD.

4-96. The Chaplain is a key player on the CSS staff. There are critical missions the chaplain must accomplish. The UMT must integrate with the CSS staff before deployment and understand all the forms, reports, and planning process used by the staff. The TF XO is responsible for all CSS within the TF and the chaplain must insure the XO is familiar with the critical religious support coordination factors that must be completed before deployment.

4-97. Finally, the CSS staff will produce estimates of anticipated casualties and expenditures for the upcoming battle. These estimates help the UMT to plan their religious support mission more effectively.

THE BRIGADE SUPPORT BATTALION UNIT MINISTRY TEAM IN THE BRIGADE SUPPORT AREA

4-98. The Brigade Support Area (BSA) is generally located behind the rear boundary of the TF sector. The BSA contains the support elements for the entire Brigade. The Engineer Sapper Battalion, the Brigade Support Battalion (BSB), and the field trains for all the task forces are just a few of the CSS assets in the BSA (FM 71-3). Normally, the TF UMT rarely visits this area because of the distance from the TF fight. There will be a supply route (SR) from the BSA into each TF sector. Each BDE and TF Chaplain should read the CSS chapter in FM 71-2 to gain an understanding of the composition and doctrine of the Field Trains in a BDE.

4-99. The HHC Commander is in charge of the Field Trains and will have a Field Trains Command Post (FTCP). The FTCP has the job of tracking the battle and coordinating the CSS actions in the BSA. The support platoon locates here (including mess and maintenance) and thus the LOGPAC system for the TF. The HHC Commander will also attend the "tenant's meetings" given by the BSB leadership to coordinate CSS issues.

4-100. Communication with the FTCP is difficult due to distance. The best method is to determine how the S4 communicates with the FTCP and use the same method. Possible pathways of communication are: the mobile subscriber equipment (MSE); frequency modulation (FM) radio communications with the TF A/L (Administrative / Logistics) net are possible if the distance is not too great. Sometimes a messenger can be used. Individuals to carry messengers can be found on the LOGPAC convoy. Link up with the LOGPAC convoy at the Logistics Release Point (RLP) in the TF and meet with the Support Platoon Leader or Platoon Sergeant.

4-101. The FTCP Command and Staff must understand that the BDE UMT or another UMT might work with them on various issues. For example, the BSB UMT might have the mission of working with the replacements going forward before their assignment to the individual companies. The BDE UMT might use the FTCP to pass messages/reports forward to the TF UMT and vice versa. One of the UMTs in the BSA normally provides religious support for the soldiers in the field trains.

4-102. The BSB is a critical asset for the BDE. The medical company will provide evacuation assets to the TF medical platoon as well as serve as a replacement treatment team if needed. The flow of ambulances between the BSB medical company and the TF BAS provides a readymade transportation/communications network. Companies evacuate wounded to a Patient Collection Point where an ambulance from the BAS/JAS collect the wounded. The wounded receive emergency treatment at the BAS/ JAS. A BSB medical ambulance will then evacuate them directly or through an ambulance exchange point (AXP). The soldier receives further treatment in the BSB before further evacuation. The BDE chaplain will assign a UMT to work with the wounded in the BSB (normally the BSB UMT). Communications can go by the evacuation route. If the TF UMT uses this method, make prior coordination so BSB ambulance personnel understand their role in passing messages, and the BDE UMT arranges to get the messages. This should be in the TSOP for both the TF and the BSB to insure no misunderstanding. Again, the TSOP is very important!

4-103. Tenant's meetings held at the BSA by the BSB leadership are for all elements of the BSA community. A representative from the BDE S1/S4 command post will normally attend as well as the TF HHC Commander from the TF Field Trains Command Post. The BDE UMT normally has its base of operation in the BDE S1/S4 command post area. Information and messages may go with the TF HHC Commander to the BDE S1/S4 representative.

THE ENGINEER BATTALION

4-104. The Engineer UMT is usually based in the BSA. Engineer activities are directed through the engineer Administrative/Logistics Operations Center (ALOC) which is located near the BDE TOC. The Engineer UMT should assume some of the area RS requirements in the BSA. Normally, the engineer assets in the TF are in known locations at a Class IV point. If the Engineer UMT elects to move forward into the TF sector to visit its scattered unit, it should first coordinate with the BDE UMT.

4-105. Religious support within the BSA and for the entire BDE is critical. The BSB UMT should have a definite RSP for their area of operations and for the BDE through the BDE UMT. The BDE UMT should provide specific guidance and tasks for the BSB UMT. Some tasks are as follows—

- Provide religious support to the medical staff.
- Read casualty feeder cards to determine if religious support has been given by forward TF UMTs.
- Receive/send messages from/to TF UMTs then relay messages to BDE UMT.
- Brief and prepare replacements for combat operations.
- Provide BDE UMT information on soldier replacement issues, medical trends, number and types of casualties, and the BSA situation.

RELIGIOUS SUPPORT TO CASUALTIES

4-106. Spiritual care to wounded soldiers is one of the core functions of the UMT. It contributes to the soldiers' well-being and recovery. Prayers, scripture readings, rites, ceremonies, and sacraments constitute religious support for wounded or dying soldiers. While the chaplain gives care to individual casualties, the chaplain assistant moves among the other soldiers to provide care and to identify those in greatest need of assistance.

4-107. During combat operations, maneuver battalion UMTs may locate with the battalion aid station. If the medical platoon divides the aid station into two or more treatment teams, the UMT should consider locating where the most casualties are expected. Based upon experience, the UMT may split for a short time to maintain contact with the separate treatment teams. However, this is based upon training, experience, and necessity. The chaplain must carefully weight all aspects.

4-108. The UMT considers the physical condition of the casualties and gives priority to the most seriously injured. The following medical categories help the team determine priorities for care:

- ***Expectant:*** casualties with very serious injuries and low chance of survival. The UMT provides pastoral and spiritual care for the dying as its first priority.
- ***Immediate:*** casualties with a high chance of survival if medical measures are accomplished quickly.
- ***Delayed:*** casualties needing major surgery, but who are able to tolerate a delay in surgical treatment.
- ***Minimal:*** casualties with relatively minor injuries that can be treated with self-care. Do not ignore these casualties as most will be treated and returned to combat with potentially high levels of anxiety.

RELIGIOUS SUPPORT TO CAREGIVERS

4-109. Religious support to caregivers is a vital part of the UMT's mission. Often the gravity of the situation, the need to give aid quickly, and the absorbing engagement of the moment will carry caregivers through the crisis. When the crisis is concluded, however, the impact of what has happened may strike the caregivers suddenly and dramatically. Medical personnel and mortuary affairs personnel may be in particular need of religious support at the conclusion of a traumatic event.

RELIGIOUS SUPPORT TO GRIEVING UNIT MEMBERS

4-110. Religious support to everyone involved in a military operation is critical. Often survivors are stunned and immobilized by the loss of fellow soldiers. The emotional and spiritual impact of loss requires the finest skills of the UMT to deal with the denial, anger, guilt, and fear, and to facilitate the continuation of life and service.

4-111. Care should begin as soon as possible after the loss occurs. The knowledge that soldiers have received religious support gives comfort and encourages grieving soldiers to move on with their lives. Following the immediate intervention, supportive follow-on care should be provided when soldiers re-deploy to their home station.

OFFENSE

4-112. The offense is the decisive form of combat operations. The main purpose of the offense is to defeat, destroy, or neutralize the enemy force. While strategic, operational, or tactical considerations may require defending for a time, defeating the enemy requires shifting to the offense. Rapid movement characterizes offensive operations over large areas. The types of offensive operations are movement to contact, attack, exploitation, and pursuit.

4-113. Offensive operations place great demands on UMTs. The constant movement of units and troops, and the operation tempo (OPTEMPO) of the battle limit the UMTs' access to soldiers. The religious support focus will vary from mission to mission. Once an offensive operation begins, UMTs in combat units will provide ministry on a different scale and at different times from those in combat support and combat service support units.

OFFENSIVE PLANNING CONSIDERATIONS

4-114. The UMT plans religious support for each phase of an operation. Offense planning considerations include—

- Develop Intelligence through reconnaissance and intelligence assets.
- Conduct Maneuver.
- Deployment.
- Employ Firepower.
- Protect the Force
- Exercise Command and Control (C2)
- Perform Logistics and CSS Consolidation and Reorganization or Continuation.

(See FM 100-40 for more detail.)

INFANTRY OR ARMOR BATTALION/TASK FORCE

Preparation for Combat

4-115. The Task Force (TF) UMT begins planning for religious support by considering METT-TC. The mission and the time available will determine how the UMT provides religious support. With little time, the team plans and prepares quickly not to miss opportunities for religious support to soldiers. For example, the time available for religious support before a hasty attack is much less than the time available for religious support before a movement to contact or a deliberate attack.

4-116. The UMT carefully establishes religious support priorities. It considers the order of march, the units conducting the main and supporting attacks, and the severity and number of casualties in previous missions. Priority is given to elements that have sustained the most casualties or to those that will be engaged at the earliest. Pastoral care to casualties is always the priority. Therefore, the UMT coordinates with the medical platoon when casualty evacuation plans are formulated.

Develop Intelligence through Reconnaissance and Intelligence assets.

4-117. This phase begins as soon as possible after the unit receives a mission. Reconnaissance elements seek to detect enemy activity and the physical characteristics of areas of interest to the commander.

> The scout platoon is an early priority for religious support.

4-118. The TF scout platoon begins its reconnaissance and surveillance as soon as the mission is received from the brigade. It may leave as much as 24 hours before the main body crosses the LD. The scout platoon is an early priority for religious support.

4-119. While the scout platoon conducts the reconnaissance, the rest of the unit continues preparation for combat. This preparation normally takes place in a tactical assembly area (TAA). Worship services, sacraments and ordinances, individual counseling, and visitation are possible and very important for soldiers during this phase. The UMT coordinates with company commanders for times and places for religious activities. An excellent opportunity for this coordination is the meeting at the LRP.

> The UMT coordinates with company commanders and first sergeants for time and place of religious activities.

4-120. The UMT utilizes the intelligence preparation of the battlefield (IPB) process by the S-2 staff section to identify any aspect within the AO that will affect the manner in which they accomplish the religious support mission.

Conduct Maneuver

4-121. When attacking from positions not in contact with the enemy, units normally stage in a TAA. The TF does not move all at once. While the line companies move toward the LD, the main command post, the combat trains, battalion aid station (BAS), support elements of the TOC, mortars, and other elements of the TF prepare for the operation. Religious support is given to each element of the TF before it begins movement. Soldiers who are preparing for the operation may not have time to gather for worship.

> The Task Force UMT normally travels with the combat trains or with the BAS.

4-122. The TF UMT normally travels with the combat trains or with the BAS. If the team is visiting soldiers in other locations before movement, it must determine ahead of time which routes will be clear of obstacles, how long it will take to return to the BAS, and the location of link up points. The UMT times its movement to precede or coincide with the scheduled start time of the BAS. All movement is coordinated with the S4.

4-123. If the medical platoon intends to split the BAS, the team must plan support for more than one site. The UMT must know where the TF surgeon will locate; where most of the casualties are expected; and, if the UMT needs to separate, how it will maintain two operations.

4-124. The displacement of critical CSS elements, such as the combat trains, BAS, Unit Maintenance Collection Point (UMCP), Logistic Release Points (LRPs), and Casualty Collection Points (CCPs) will be triggered by the TF crossing phase lines or by other control measures. This means of movement allows the CSS elements to remain responsive to the task force. The TF UMT monitors the tactical situation to know when the TF is approaching each phase line and thus when it may need to move or shift its religious support focus.

4-125. Because of distances and the OPTEMPO, most units will re-supply on the move. The TF UMT will plan accordingly for its own re-supply.

4-126. As the TF moves, communication becomes difficult. The UMT will find it hard to monitor the tactical situation. Good planning, rehearsals, and the ability to adapt on the battlespace enable the UMT to provide comprehensive religious support despite the confusion and chaos of battle. Coordination with key staff members and the executive officer is critical.

The UMT must employ sleep plans and staggered shifts when necessary to remain fully capable of night operations.

4-127. Task force operations are continuous. The UMT must be fully capable of night operations, employing sleep plans and staggered shifts when necessary. If the TF moves during the hours of darkness, the TF UMT must maintain contact with other elements with which it is traveling.

4-128. The attacking force moves to positions of advantage during this time. The basic forms of maneuver are envelopment, penetration, frontal attack, turning movement, and infiltration.

4-129. In the maneuver phase, religious support may be restricted to small numbers of soldiers during halts in the movement, or at casualty collection points along the route of march.

4-130. During the assault, casualties are likely to increase. The TF UMT will be heavily engaged with providing religious support to casualties and caregivers. If overwhelmed with casualties, the BAS declares a MASCAL situation. The TF UMT requests additional support during a MASCAL from the brigade UMT.

4-131. During a MASCAL the TF UMT requests additional support from the brigade unit ministry team.

Employ Firepower

4-132. The fire support coordinator (FSCOORD) integrates fire support into the unit's maneuver scheme. Religious support during this phase is similar to the maneuver operation.

Protect the Force

4-133. To protect the force a commander can employ any number of actions to include: maintaining a high tempo of operations; conducting security operations, which include organizing and employing covering, guard, or screen forces; conduct area security operations; employ operations security (OPSEC) procedures; execute deception operations; conduct defensive information operations; employ camouflage, cover, and concealment, and conduct active and passive air defense operations.

4-134. The UMTs movement in the battlespace, communication methods, and information management techniques should reflect the commander's security and force protection objectives. The UMT closely monitors force security issues that may affect religious support planning The UMT aggressively monitors the battle and the commander's use of command and control (C2) in allocating forces. The decision-making process and the format for plans and orders are found in FM 101-5. Below are the UMTs planning considerations within this staff process.

The Goal: Comprehensive/Synchronized Religious Support		
Staff Action	***UMT Action for input to staff***	***UMT Action for internal use***
Receive the Mission	-Mission Analysis: Can we support the mission and how? -Coordinate with S4 and S1. -Coordinate with higher UMTs.	-Begin rehearsals of known actions/drills. -Execute current RS plan. -Begin considering the unit's proposed COA's for support of the mission.
Conduct Estimates by each staff section/	-Prepare RS Estimate of current/projected unit condition. -Provide input to staff on AOR religions and culture.	-Communicate issues to subunit UMTs. -Continue Spiritual Care.
Restated mission to Cdr/ Receive Initial Cdr's Guidance	-Review religious, moral, and ethical issues.	-Continue prep for next mission.
Develop/compare/ recommend the 'best' COA to Cdr	-Input to COA's for tactical mission. -RS Implications of each COA. -Compare feasibility/ acceptability/ supportability of each COA that the staff produces.	-Further develop RS plan for support of the upcoming mission. -How will we synchronize RS? -Continue provide RS to unit and area personnel.
Receive Cdr's Estimate including DECISION and Cdr's CONCEPT	-Refine RS concept to comply with Cdr's concept.	-Communicate with subunit UMTs concerning their own preparation of subunit RSP needs and issues.
Prepare Plans and Orders	-Submit RS annex/paragraph if needed.	-Update RS Section NCOIC. -Continue RS.
Receive Cdr's approval of Plans and Orders	-Pray with Cdr, and give 1 min update of RS situation.	-Continue information gathering. -Continue RS.
Issue Plans and Orders	-Attend OPLAN meeting; be prepared to brief.	-Communicate RS plan and provide support to higher RMSTs and subunit UMTs.
Rehearsals and Troop Leading Procedures	-Attend rehearsals, insure RSP is integrated and understood.	-Aggressively execute RS plan.
Conduct mission	-Begin planning for follow on mission, sequels, branches.	-Provide prioritized RS. -Battle track subunit UMTs.

4-135. The UMT must remain flexible and support changes to the plan due to METT-TC considerations and the commander's priorities.

PERFORM LOGISTICS AND COMBAT SERVICE SUPPORT

4-136. On reaching the objective, the unit eliminates remaining resistance, consolidates its gains, and reorganizes as needed. The unit may immediately continue the attack to a follow-on objective, or it may hold in place and set up a hasty defense.

The UMT provides religious support to the force paying attention to leaders and those who show signs of battle stress.

4-137. The time the TF spends during this part of the offensive operation will depend upon subsequent missions. If there is sufficient time, the UMT moves forward to maneuver elements. The UMT provides religious support to the force paying special attention to leaders and those who show signs of battle stress. The team may also conduct memorial services or ceremonies as directed by command.

4-138. A successful assault may develop into an exploitation or pursuit. The effectiveness of an exploitation or pursuit depends on rapid transition and vigorous execution. The UMT must be prepared to make the transition quickly with the least disruption to its operations.

TRAVEL IN THE BATTLESPACE

4-139. A UMT that travels to forward elements by itself is at great risk. Travel with the TF LOGPAC convoy reduces the security risk and the chances for getting lost. The LOGPAC moves from the field trains along the MSR to an LRP where the company first sergeants take control of their company LOGPACs.

The UMT meets with company first sergeants at the LRP.

4-140. Based on the scheme of maneuver, the TF S4 selects LRPs well forward. For security reasons, they are only used for short periods of time. The UMT finds LRP times and locations in the OPORD.

4-141. The S4 and company first sergeants meet at the LRP to discuss future logistical requirements. At this meeting, the UMT discusses the religious support needs of soldiers and coordinates for services and visits to the line companies with the first sergeants. The UMT then may go to one of the companies with the first sergeant and company LOGPAC

4-142. When the company has completed its re-supply, the first sergeant returns his portion of the LOGPAC convoy to the LRP. The TF support platoon leader reforms the convoy and leads it back to the field trains. The TF UMT follows the first sergeant to the LRP and returns to the combat trains or, if time permits, travels to the field trains.

4-143. Engineers. The engineers provide an alternative to traveling with the LOGPAC. Because LOGPACs are often conducted during times of limited visibility, the UMT may find travel with elements of the engineer company more effective. The engineers travel to all the company locations to dig in hasty defensive positions, clear mines and obstacles, and prepare for the follow-on mission. The UMT coordinates with the S4 to link up with the engineers for travel to the companies.

RELIGIOUS SUPPORT OF DEFENSIVE OPERATIONS

4-144. Army forces conduct defensive operations as part of major operations and campaigns in combination with offensive operations. The basic task of the defense is to force the enemy to attack under unfavorable circumstances. Military forces defend only until they gain sufficient strength to attack. Though the outcome of decisive combat derives from offensive operations, it is often necessary, even advisable, to defend.

4-145. Defensive planning considerations are very similar to offensive considerations. In the defense, the UMT has more time to synchronize the religious support plan and apply maximum religious support assets to critical areas. Shifting religious support assets as the operation evolves will require prior coordination with commanders and UMTs. Prior planning regarding the synchronization of the religious support mission and plan during different phases increases the effectiveness of the religious support defense plan.

BATTALION/ TASK FORCE

Develop Intelligence

4-146. As soldiers prepare defensive positions, the TF executes the counter-reconnaissance mission. This mission is meant to deny the enemy information about friendly dispositions and friendly preparations for battle. The counter-recon force typically consists of the scout platoon, ground surveillance radars, and maneuver elements.

> Starting as far forward as possible, while minimizing risk to exposing soldiers to artillery fire, the UMT makes brief visits to soldiers in fighting positions.

4-147. Throughout this phase soldiers are digging in and preparing fighting positions for the coming enemy attack. The UMT keeps movement to a minimum to avoid disclosing fighting positions, paying close attention to cover and concealment. Starting, as far forward as possible, the team makes brief visits to soldiers in fighting positions, working rearward towards the combat trains. The UMT coordinates its travel with the TOC and the combat trains command post (CTCP).

> The UMT must carefully plan movement and religious support to avoid being caught in the confusion.

Deploy/Conduct Maneuver

4-148. The rearward passage of lines by the covering force causes considerable movement and confusion. The UMT must be cautious when planning movement and religious support to avoid being caught in the confusion. The UMT locates with the battalion aid station or LOGPAC, communicates its location to the TOC and CTCP, and deliberately avoids the passage lanes.

Employ Firepower

4-149. Enemy artillery poses a significant threat to the safety of the UMT. The UMT must be near adequate overhead cover for protection. Gathering soldiers for worship may involve undue risk.

4-150. If the enemy uses chemical or biological agents, the team will make adjustments in religious support. The team may need to postpone religious activities until the hazard is eliminated or reduced to a safe operating level. Several factors that the UMT considers include:

- Degree of engagement with enemy forces.
- Level of exposure and risk to soldiers and equipment.
- Priorities for religious support.
- Anticipated operations in response to enemy attack.

The chaplain offers sacraments and ordinances to soldiers before combat.

4-151. Sacraments and ordinances are not celebrated in contaminated areas. The UMT does not resume services until soldiers have moved to safe areas. Therefore, the chaplain takes advantage of every opportunity to administer the sacraments before soldiers engage in combat, or before soldiers are exposed to a contaminated environment.

4-152. The UMT carefully coordinates movement around and through obstacles set in place for the defense. Obstacles installed after the team has departed its base of operations can put the team at risk. The team should know the breach marking system and MSR. To avoid obstacles, it carries a copy of the obstacle overlay.

The team carries a copy of the obstacle overlay when traveling.

4-153. The UMT must be prepared to adapt to the changing tactical situation of a mobile defense that combines offensive, defensive, and retrograde actions. This combination results in a nonlinear front to create confusion for the attacking forces.

4-154. As the enemy begins the attack, the UMT shifts its attention to soldiers in the battalion aid station (BAS). From this point on, the UMT's main effort will be the care of dying and other casualties as they are brought to the BAS.

The UMT must be prepared to move on a moment's notice.

4-155. During the enemy attack, the BAS may move frequently, evacuation routes may be blocked by an enemy penetration, and defending elements may become temporarily encircled or bypassed by enemy forces. Rapidly moving enemy units may threaten or overrun the BAS. The UMT must be prepared to move on a moment's notice. Equipment and supplies not required immediately should be stowed in the team's vehicle. The vehicle should be parked in a manner that allows for rapid departure if the BAS is threatened.

Perform Logistics and Combat Service Support

4-156. As the TF consolidates and reorganizes after the battle, the companies evacuate the remaining casualties to the aid station. Once the UMT has ministered to the dying and other casualties, it assesses METT- TC to determine its next actions. If the unit is planning a counterattack, the UMT prepares to move with the unit. If not, the team moves forward to units on the forward line of own troops (FLOT) to visit soldiers. The UMT NCOIC ensures the team's supplies are adequate for follow-on operations.

Coordinate with Command

4-157. The UMT must have a clear understanding of the battlespace situation so they can prioritize tasks at the right time and location. The UMT takes advantage of the war-gaming that takes place in the MDMP to derive their decision points. The UMT coordinates with the commanders and other UMTs in the AO to ensure all religious support needs are met according to the operational plans.

BRIGADE

4-158. In the mobile defense, battalions engaged in offensive, defensive, and retrograde actions present a challenge for the brigade UMT. The UMT plans religious support for each of the units based on METT- TC. The area defense differs from the mobile defense in that all units are engaged in defensive operations. Within the defense planning considerations by the staff, the brigade UMTs responsibilities to the BDE UMTs include—

- Faith Group support.
- Operational information.
- Logistics of Faith Group specific supplies.
- Personal spiritual support to UMTs.

STABILITY OPERATIONS

4-159. Conducting religious support in stability operations is essentially the same as in offense, defense, and support operations.

Battalion

4-160. The UMT will need to reprioritize planning considerations in order to fully support this type of operation. The UMT planning emphasis will include—

- Joint and multinational religious support cooperation and communication.
- More planning for faith-based NGO/IO liaison tasks.
- Working more closely with the CMO personnel in mutual support tasks.
- Host nation religious leadership liaison responsibilities.

Brigade

4-161. The BDE or equivalent will be challenged with coordination and information management requirements. The information flow will require the proper assessment of information to allow for timely and adequate religious support. Planning considerations in addition to BN include—

- Civil Military Operations Center (CMOC) representation involvement.
- Staff integration needs.
- Media and public affairs command guidance.
- Remote site religious support coverage planning.
- Force protection.
- Religious support to Contractors.

SUPPORT OPERATIONS

4-162. The tasks for planning religious support for Support Operations are identical for offense, defense and stability operations. The support operation requires the UMT to know the restrictions, limitations, and proper responsibilities of the UMT in providing religious support in a support operation environment.

Battalion

4-163. The UMT will be directly involved with civilians and the trauma associated with the operation. This will have a stressful effect on the UMT members. The BN UMT must be prepared for the area of operation METT-TC considerations. Planning includes—

- Personal burnout and combat stress.
- A daily log (See DA Form 1594).
- Soldier and civilian debriefings.
- Continual spiritual and morale assessment of individuals in the unit.

Brigade

4-164. The BDE UMT will work closely with state agencies and joint chains of command. The religious support plan and operation must operate within the mission, concept of operation, and commander's intent. Legal restraints may hinder the free exercise of religion as it relates to the military providing religious support to civilians who request assistance and guidance. The UMTs battle book and TSOP must reflect the planning considerations. Planning considerations include—

- Mass Casualty training.
- Information on civilian and state agencies for liaison tasks.
- Critical Event Debriefing as a norm.
- Moral and ethical dimensions.

TERMINATION AND POST-CONFLICT OPERATIONS

4-165. Planning for termination of operations must be ongoing during all phases of COA development, deployment of forces, and execution of operations. It must be fully integrated within the orders process and synchronized and coordinated with the staff and UMTs. In most cases, operations will be terminated in their own right when stated objectives have been met. In some operations (e.g. stability and support), the command may have to recognize that the mission is not achievable without restructuring and/or commitment of additional assets. Termination of the mission may be the best COA. The following checklist may assist in termination planning:

- What is the planned timeline for termination?
- Will there be follow-on forces and who?
- What are the post-conflict requirements?
- Is there a force protection issue for safely withdrawing forces?
- What are the requirements for redeployment?
- What are the considerations for release of the Reserve Component forces?

- Will there be humanitarian, governmental, and infrastructure assistance required for post conflict activities? Is there coordination required?

(See JP 5-00.2 and JP 1-05 for more detail.)

TRANSITION

4-166. As with planning for conflict termination, planning for transition must extend throughout the planning process and into operations and redeployment. Transferring control of command and staff responsibilities to another organization requires detailed planning and execution. It is dependent on the situation and each transfer will have it's own characteristics and requirements. Transition planning by the UMT continually requires close coordination with higher and communication with the UMTs of the transitioning organization. Regeneration and reconstitution is a primary consideration in the transition plan. Transition planning by BDE should encompass the following:

- What is the new mission?
- What U.S. forces, equipment, and/or supplies will remain behind?
- Who will support UMTs that remain behind?
- What is the classification of information to be shared with incoming UMTs?
- What will be the new ROE?
- What are the building/Chapels requirements?
- What is the timeline for transfer?

The mission may dictate that additional planning may considerations need to be planned and coordinated for as well.

REDEPLOYMENT AND RECONSTITUTION

4-167. Reconstitution (see FM 100-9) is a method of replenishing units that have been decimated to the extent they are no longer combat-effective. Such units are withdrawn from contact to a rear area. Personnel and equipment as well as rest and recuperation are the focus. These units require special religious support. BDE UMTs must be proactive in supporting reconstitution not only for unit personnel, but the UMTs of that unit, its leadership, and caregivers as well. Religious support will encompass the following:

- Honoring the dead with memorial services and ceremonies.
- Nurture of the living with counseling, worship, and critical event debriefings.

4-168. Post-conflict requirements and impact have a direct bearing on the redeployment religious support mission. The impact of battle and the stress associated with personal loss, fear, physical and emotional fatigue demands religious support application. METT-TC is still utilized in this stage by UMTs as it was in deployment. The balancing of available assets is essential during this phase to ensure effective religious support. UMTs must factor in redeployment not only back to home station, but also to other missions. Both require extensive religious support planning. UMTs of the BN, BDE, and equivalents must not re-deploy prior to their completing their religious support mission to their respective units and other associated religious support requirements of the TF. Family support and other religious support missions at home station is the responsibility of the installation religious support mission. Planning considerations for both echelons include—

- After Action Report.
- Reunion Briefings program.
- Logistics and time table.
- Religious support coverage plan.
- Echelon of UMTs back to home station ensuring AO religious support coverage.

GARRISON OPERATIONS

4-169. UMTs participate in the religious support program of the installation in accordance with their orders of assignment by Headquarters Department of the Army. They coordinate with their chain of command and the installation chaplain for garrison religious support needs for their unit soldiers and families while non-deployed and during all phases of deployment. Planning considerations include—

- Family Readiness Group (FRG): Support by briefings/classes on family issues.
- Killed in action (KIA)/missing in action (MIA)/wounded in action (WIA) memorial ceremonies and religious services.
- Hospital visitation.
- Casualty notification support.
- Worship support.

OTHER BRIGADE AND BELOW UNITS AND SPECIAL RELIGIOUS SUPPORT CONSIDERATIONS

Special Forces

4-170. Special Forces (SF) combine at the lowest tactical level, the functions performed by conventional branches of the Army. United States Army Special Operations Command (Airborne) (USASOC (A)) is the Army component of the joint and unified command, United States Special Operations Command (USSOCOM). USASOC (A) is comprised of Special Forces, Rangers, a Special Operations Aviation Regiment, the Special Operations Support Command and Civil Affairs and Psychological Operations forces. Special Forces is commanded by the U.S Army Special Forces Command (USASFC), which has seven subordinate commands known as Special Forces Groups consisting of three Special Forces (SF) battalions each. USASFC (A) has five primary missions and related additional capabilities.

Primary Missions

4-171. The five primary missions are Unconventional Warfare (UW), Foreign Internal Defense (FID), Special Reconnaissance (SR), Direct Action (DA), and Counter Terrorism (CT). An SF unit can have elements deployed to several different areas within its assigned region. These elements range in size from the individual soldier to units of 12 to 150 soldiers. SF missions are directly influenced by politico-military considerations. A change in national security policy or strategy may change SF missions.

Additional Capabilities

4-172. Special Forces units are trained to conduct MOOTW missions. Language capabilities, area orientation, small unit, and multi-agency experience are invaluable in MOOTW. Additional capabilities include noncombatant evacuation operations (NEOs), arms control, support to domestic authorities, humanitarian assistance (HA), disaster relief, security assistance, nation assistance, counter-drug operations, combating terrorism, peacekeeping and peace enforcement, show of force, support to insurgencies and counterinsurgencies, personnel recovery, and special activities.

4-173. The Special Forces branch or related MOS is the only non-accessioning branch or MOS. A soldier selected for the Special Forces branch or MOS has successfully completed service in an accession branch or MOS and graduated from the Special Forces Qualification Course.

UMT Personnel Qualifications

4-174. Religious support in the Special Forces community requires unique qualifications. The most critical is the ability to provide sustained, independent religious support. Others include—

- Security clearance for all UMT members equal to the mission classification.
- Airborne qualification.
- Extensive orientation in special operations, or graduation from the Special Forces Qualification Course, or Special Forces Staff Course (SFSC), or an "S" enlisted MOS identifier.
- Training in the religious background of the region of deployment.

The Nature of Religious Support in Special Forces Units

4-175. The unconventional nature of special operations, the high OPTEMPO, and the multiple mission deployments create stress on units, soldiers, and families. These pressures place special demands for religious support on SF UMTs.

4-176. The USASFC (A) chaplain, under the guidance of the USASOC (A) Chaplain, gives staff supervision to the USASFC (A) religious program. Group chaplains supervise religious support operations and provide technical guidance and resources to subordinate UMTs. SF chaplains frequently deploy to advanced operational bases (AOB). If a Special Forces Operational Base (SFOB) is deployed, a Catholic chaplain, when available, is attached to perform religious support at the SFOB, as well as, at the AOB, and down to the Operational Detachment Alpha (ODA).

Ranger Regiment

4-177. The 75th Ranger Regiment is a major subordinate command (MSC) of the USASOC (A). The Ranger Regiment is capable of conducting both conventional and special operations across the full spectrum of conflict. As a potent, rapidly deployable direct action infantry force, it is particularly well suited to conduct short duration strike operations or other small-scale offensive actions against targets of high strategic or operational importance. Such operations include actions to seize, destroy, or capture enemy forces or facilities, or to recover designated personnel (noncombatant evacuation operations, liberate friendly prisoners of war, capture designated enemy personnel) or equipment from hostile, denied, or politically sensitive, high threat environments requiring exceptionally competent, mature and disciplined forces.

4-178. The Rangers may conduct operations independently, or in support of a campaign plan. While Ranger operations may be conducted in coordination with conventional forces, they differ from conventional operations in degree of risk, operational techniques, and modes of employment. Ranger operations will on occasion be conducted before political preparation of the battlespace has been completed or, in certain cases, before such preparation has even been initiated. Because of its rapid deployment ability and capabilities as a formidable strike force for forced entry operations; the Ranger Regiment maintains an exceptionally high level of operational readiness.

Organization

4-179. The Ranger Regiment's structure is similar to that of a conventional light infantry brigade, but with a robust regimental headquarters that includes organic reconnaissance, signal, military intelligence, and training detachments. Ranger units are not designed for sustained combat and therefore, they are limited by very austere organic logistic support, and by equally limited organic combat multipliers. During all phases of operations and training, Ranger units require responsive and adequate support, either from theater Army assets or USASOC (A).

Unit Ministry Team Personnel Qualifications

4-180. Religious support to Army Rangers is physically, emotionally and spiritually demanding. Unique qualifications include—

- Security clearance for all UMT members equal to mission classification.
- Airborne qualification.
- Ranger qualification.
- Completion of the Ranger Orientation Program (ROP) by chaplains and assistants E-5 and above.
- Completion of the Ranger Indoctrination Program (RIP) by chaplain assistants E-1 through E-4.

The Nature of Religious Support in Ranger Units

4-181. All Ranger elements, from the single platoon to the entire Ranger Regiment, are capable of conducting extremely decentralized training exercises and combat operations. Some additional religious support planning requirements include:

- A thorough understanding of the operational and strategic mission.
- Crisis action planning and execution of religious support

4-182. The Ranger Regimental chaplain exercises staff supervision over regimental religious support and subordinate Ranger battalion UMTs.

U.S. ARMY CIVIL AFFAIRS AND PSYCHOLOGICAL OPERATIONS COMMAND (AIRBORNE)

4-183. U.S. Army Civil Affairs and Psychological Operations Command (Airborne) (USACAPOC (A)) provides civil affairs (CA) and psychological operations (PSYOP) activities and information for both SF and conventional commanders. The CA/PSYOP mission is to inform commanders on the nature of foreign audiences to influence the behavior of foreign governments, organizations, groups, and individuals. To influence a target audience's thinking and behavior, the CA/PSYOP unit must be extremely well versed in the language, culture, thought patterns, and belief systems of the people in the AO.

PSYCHOLOGICAL OPERATIONS UNITS

Organization

4-184. Psychological Operations (PSYOP) personnel often deploy for extended periods as individuals and small teams to support SOF and conventional forces. Deployed personnel are required to work with high-level leaders of other armed forces, and with government officials from the U.S. and other nations.

Psychological Operations Unit Ministry Team Personnel Qualifications

4-185. Requirements for members of UMTs in PSYOPS varies between components to include:

- Security clearance for all UMT members equal to the mission classification.
- Airborne Qualification.

The Nature of Religious Support in Psychological Operations Units

4-186. Tailored PSYOP units in support of SF and conventional forces require flexible religious support. UMTs deploy with a battalion-size or larger PSYOP Task Force (POTF) or PSYOP Task Group (POTG). When POTF or POTGs do not deploy, the religious support is coordinated through the Supported unit.

- PSYOP UMTs frequently coordinate religious support with joint or combined forces.
- When a PSYOP UMT does not deploy, the supported unit provides religious support to PSYOP personnel.
- PSYOP UMTs participate extensively in family support programs during all stages of deployment.

CIVIL AFFAIRS UNITS

4-187. Civil Affairs (CA) units support the commander's relationship with civil authorities and the populace, promotes mission legitimacy, and enhances military effectiveness. CA units support all USASOC (A) missions with regional orientation briefings and information on political, social, cultural, religious, and economic aspects of the Joint Special Operations Area (JSOA).

Organization

4-188. CA units are task organized to support USASOC (A) and supported conventional units' mission requirements. Some CA units are organized as battalions with FID and UW support missions. Other units are organized into special teams, such as Cultural Affairs teams.

Unit Ministry Team Personnel Qualifications

4-189. The assignment of a Unit Ministry Team requires unique qualifications, to include—

- Security clearance for all UMT members equal to the mission classification.
- Airborne qualification.
- Extensive knowledge of religious and cultural issues and their effect on operations.

The Nature of Religious Support in Civil Affairs Units.

4-190. UMTs serving within USACAPOC (A) primary missions are to provide and perform religious support to the unit assigned in accordance with U.S. Code Title X. Civil Affairs UMTs may provide all UMTs in an AO a unique avenue of information and expertise on religious cultures in the AO. CA UMTs can also coordinate for command directed CA activities involving liaison between the command, local religious leaders, and UMTs. CA teams attached to supported commands are provided religious support through the supported command UMT. CA UMTs serve as another religious support asset for the combatant command chaplain's religious support plan.

MEDICAL UNITS

4-191. Deployable Army Medical units are in the process of transformation under the Medical Reengineering Initiative (MRI). MRI is expected to be complete by FY 2009.

4-192. Under MRI, deployable medicine includes C2 units, corps medical augmentation units, and hospitals. MRI C2 units include Medical Brigades, and Medical Commands (MEDCOMs) at Corps and Echelons Above Corps. MRI hospitals come in one type: the Combat Support Hospital, which can be located at Corps and Echelons Above Corps.

Organization

4-193. Unit Ministry Team personnel are organically assigned to MRI C2 units and to MRI hospitals. Corps medical augmentation units receive their religious support through area coverage.

4-194. Since health care skills are highly perishable and in great demand, the largest number of the Army's health care providers are assigned to TDA Medical Treatment Facilities (MTFs). When needed for operational requirements in deployable units, these health care providers fill necessary slots in deploying MTOE medical units under the Army Medical Department (AMEDD) Professional Officer Filler System (PROFIS).

4-195. Likewise, since medical ministry and pastoral care skills are also highly perishable and in great demand, Unit Ministry Team personnel may be assigned to TDA MTFs with PROFIS obligations to deployable MTOE medical units.

Unit Ministry Team Personnel Qualifications

4-196. Qualifications for Unit Ministry Team personnel serving medical units include:

- 7S, Combat Medical Ministry Specialist and
- 7R, Hospital Ministries Clinician for Chaplains;
- Emergency Medical Ministry Course (EMM) for chaplain assistants.

4-197. To meet the complex demands of combat medical ministry and pastoral care, the AMEDD Center and School's Chaplain Training and Operations Branch oversees an institutional training program to prepare Unit Ministry Team personnel providing religious support in medical unit. This program includes—

4-198. Emergency Medical Ministry Course (EMM) is the entry-level course, training chaplains and chaplain assistants in the basics of wartime emergency medical ministry. It is a two-week TDY course taught at AMEDD Center and School, with practical application at Brooke Army Medical Center. EMM is the basic qualifying course for Chaplain Assistants serving in medical units.

4-199. Combat Medical Ministry Course (CMM) is the intermediate course and requires one prerequisite unit of accredited Clinical Pastoral Education (CPE). Also a two-week TDY course taught at AMEDD Center and School, with practical application at Brooke Army Medical Center. CMM focuses upon six core competences: Crisis Ministry; Trauma Ministry; Mass Casualty Ministry; Biomedical and Battlespace Ethics; Grief, Loss, Death and Dying; and Spiritual Hope/Suicide Prevention. CMM is the qualifying course for award of the 7S, Combat Medical Specialist, ASI, the minimum qualification for Chaplains serving medical units.

4-200. Clinical Pastoral Education (CPE): Residency is the advanced level, accredited by the Association for Clinical Pastoral Education, Inc. and administered by the MEDCOM Army CPE System. The Army CPE system offers residencies at Walter Reed, Eisenhower, Madigan and Brooke Army Medical Centers. CPE prepares Chaplains for service within both institutional and deployed medical environments. CPE is required for award of the 7R ASI.

4-201. Clinical Pastoral Education: 7R Trainer (CPE 7R) is the fellowship level. Accredited by CPE, CPE 7R prepares select 7R Chaplains to supervise CPE residency training within the Army CPE System. CPE 7R is required for award of the 56D7R AOC.ASI.

The Nature of Religious Support in Medical Units

4-202. Unit Ministry Teams serving with medical units bring specialized training in spiritual assessment: mass casualty ministry; crisis ministry; grief and loss; spiritual hope; trauma ministry; and ethics to their provision of religious support. They have a dual religious support focus, in caring for the dying and other casualties and honoring the dead, while also nurturing the living in support of the medical staff.

SUMMARY

4-203. The UMT is assigned to a unit at either BN/Squadron or BDE/Regiment/ or Group. The unit is either operationally committed, about to be deployed, returning from deployment, or is in a garrison situation. In garrison, the priorities of religious support must fit in with the command's primary responsibilities:

- Training.
- Family Support.

4-204. Key to performing religious support in this environment is executing staff responsibilities to fulfill various tasks assigned to the UMT in garrison.

4-205. For the operationally committed unit, the UMT is involved in the highest calling for military support: supporting the soldier in combat situations, operations that fall into four categories: Offensive, Defensive, Stability and Support.

4-206. UMTs supporting these units may participate in joint command, and multi-national operations.

RESOURCES

4-207. The following is a list of main resources utilized for writing and updating this doctrine.

- AR 165-1.
- JP 1-05.
- JP 5-00.2.
- JP 3-05.
- FM 16-1.
- FM 3-0 (100-5).
- FM 3-90.3 (71-3).
- FM 4-0.
- FM 25-100.
- FM 25-101.
- FM 71-2.
- FM 100-9.
- FM 100-10-2.
- FM 100-21.
- FM 7-30.
- FM 7-20.
- FM 1-111.
- Lessons Learned from past and current operation
- Lessons Learned from the training centers

Chapter 5

DIVISION UNIT MINISTRY TEAM RELIGIOUS SUPPORT

TACTICAL ECHELON DESCRIPTION

5-1. The division is the largest fixed Army organization that trains and fights as a tactical team. The division echelon possesses great flexibility. The division is a unit of maneuver, organized with varying numbers and types of combat, combat support (CS), and combat service support (CSS) units. It may be armored, mechanized, medium, light infantry, airborne, or air assault. Each are capable of conducting operations over a wide range of environments. Their commanders tailor their organic brigades and attached forces for specific combat missions.

CONTENTS

5-2. The early 1990s brought significant changes to the world and the Army. The end of the Cold War presented threats that were more ambiguous and regionally focused. Force Projection has become our strategy for supporting the power-projection element of U.S. national security. Army divisions are prepared for force projection operations, as part of a joint, interagency, or multinational effort for full spectrum operations.

5-3. U.S. Army corps and divisions conduct Army operations that span the spectrum from peaceful engagement to actual combat. Army divisions train to fight and win battles and engagements as part of a joint force anywhere in the world. Their organization, capability, flexibility, and versatility allow them to conduct operations for stability and support multinational and interagency operations, as well as, in joint operations. The unique capabilities of armored, mechanized, infantry, light infantry, airborne, and air assault divisions provide the U.S. Army the ability to serve the national interests worldwide. (See FM 71-100 for further information.)

AS PART OF A CORPS

5-4. Traditionally, divisions have operated as part of a U.S. corps. In corps operations, divisions normally comprise 9 to 12 maneuver battalions, organic artillery battalions, and supporting CS and CSS units. Divisions perform a wide range of tactical missions and, for limited periods, are self-sustaining.

Corps augments divisions, as the mission requires. A corps or a numbered army may direct a division to perform tasks of operational importance.

AS PART OF A JOINT TASK FORCE

5-5. Division decision-making processes and capabilities make Army divisions effective in joint, multinational, and interagency operations. Army divisions focus on fighting and winning battles and engagements as part of a joint force anywhere in the world.

5-6. The division may deploy as part of a JTF without its traditional corps headquarters and corps support unit. In these types of operations, the division often works directly for the JTF commander. Therefore, divisions must know joint doctrine, tactics, techniques, and procedures. Divisions assigned to a JTF normally conduct traditional tactical operations and may be involved in nontraditional actions, such as interagency operations or operations with the host nation. Divisions are not normally designated as a JTF headquarters.

5-7. A division in a JTF requires experienced liaison personnel. The division may operate with or receive support from joint, multinational, interagency, or NGOs. The division may receive mission support such as joint fires, joint air defense, national level intelligence, and theater and national level communications. Competent liaison to coordinate with these agencies and other headquarters helps ensure properly executed operations.

JOINT TASK FORCE STAFF AUGMENTATION

5-8. Without significant augmentation, a division headquarters has a limited capability to perform as an Army force (ARFOR) or JTF.

DIVISION AS AN ARMY FORCE

5-9. Depending on the mission assigned to the JTF, the Army forces required for the operation may only be a division or a tailored force of units from a division. The division may find that it is the largest land component command group or staff organization assigned to the JTF. In this scenario, the JTF commander may designate a division under its C2 as the Army Forces (ARFOR) headquarters in small-scale contingency (SSC) operations, i.e., the senior Army headquarters within the JTF. Operations include foreign humanitarian assistance, peace operations, noncombatant evacuations, enforcing sanctions, reinforcing key allies, and limited strikes and interventions. Operations are conducted with and in support of government agencies, nongovernmental organizations, regional and international security organizations, and coalition partners. A division designated as an ARFOR executes these responsibilities unless they are specifically excluded. The Army Service Component Command (ASCC), in coordination with the division and DA staff, tailors the forces provided to the ARFOR to meet specific METT-TC conditions anticipated in the JOA. The ARFOR UMT must plan additional RS assets for unit, area, and denominational coverage.

5-10. Designation of the division as the Joint Forces Land Component (JFLC) or the ARFOR of the JTF incurs numerous additional staffing responsibilities. The division will have to coordinate with other JTF-level component or functional commands as well as the service component headquarters of the

JTF establishing authority for administrative and logistic support. The division as ARFOR plans and conducts operations in accordance with the CJTF guidance and detailed plans. The ARFOR will establish liaison personnel to the CJTF, other component commanders, and supporting commanders as necessary or as directed by the CJTF. The division staff will require augmentation to its staff to accomplish the tasks inherent to operating as the JFLC or ARFOR. In the event that a deployed division becomes the JFLC or ARFOR for the JTF, the commander must request staff, CS, and CSS augmentation. The division may need assistance in providing the necessary support units and sustainment activities for prolonged support of significant numbers of assigned or attached forces. (See FM 3-91 (71-100), Appendix F for further information.)

FUNDAMENTALS OF RELIGIOUS SUPPORT OPERATIONS

5-11. The mission of the Division UMT is to coordinate religious support to soldiers, families, and as the force structure evolves, to forward-deployed authorized civilians. Division Chaplains also serve as critical advisors to commanders at all levels of the division providing essential information on soldier and unit morale as affected by religion, quality of life matters, moral/ethical decision making, and the impact of indigenous religion on military operations. The Division Chaplain develops the CMRP and the division RSP on behalf of the Division Commander.

5-12. The battlespace of the 2lst century involves conditions of modularity, split-based operations, increased lethality, non-linearity, greater troop dispersion, increased agility, increased pace of operations, sustainment of tempo and simultaneity of combat operations.

5-13. Combat power in the 21st century will rely heavily upon technology. The cumulative affect of the numerous factors will increase the spiritual, psychological and interpersonal demands placed upon soldiers of the 21st century. Technology can provide additional means to plan, coordinate, and execute religious support to address these demands, but the personal delivery of religious support by the UMT will always be imperative.

ARMY FORCES

5-14. As an Army forces (ARFOR), the impact upon religious support planning and execution is significant. The role of the ARFOR UMT is significantly more diverse and broad than a division operating as a tactical headquarters in a corps. While the division plans, prepares for, and executes many of the same types of activities, its focus is always on the tactical operation in its AO. Unlike the division UMT focus, the ARFOR UMT not only focuses on tactical religious support tasks in its AO, but also provides critical operational-level religious support sustainment to U.S. Forces UMTs and others throughout the JOA. The ARFOR UMT may provide religious support management to tactical and sustainment activities in contiguous or noncontiguous areas, or a combination of both. The ARFOR UMT must focus its planning on multiple, simultaneous actions that accomplish its own, or sustains other's, religious support mission. The tasks performed include—

- Information management including processing, assessment, and dissemination for command and subordinate UMTs from strategic and operational sources.
- Managing religious support tactical missions and tasks to accomplish missions assigned by the Joint Force Commander or Joint Force Land Component Commander.
- Planning and executing self-deployment of religious support assets to ensure comprehensive and seamless religious support.
- Planning and integrating CMO mutual support tasks.
- Coordinating with Special Operations Forces (SOF) UMTs.
- Executing religious support sustainment operations.
- Providing logistical support.

JOINT TASK FORCE

5-15. The joint task force (JTF) chaplain is responsible for planning, coordinating, and implementing all religious support within the JTF for the commander of the JTF. In addition to the already established doctrinal tasks, the JTF UMT is responsible for the following:

- Developing joint religious ministry support plans in support of the operation. This includes preparation of the Chaplain Activities Appendix, under the Personnel Annex of the CJTF's OPLANs, OPORDs, or campaign plans, to include appropriate tabs.
- Identifying RMST requirements such as augmentation, logistics, and communication requirements.
- Coordinating religious ministry with MSC Chaplains.
- Maintain liaison with chaplains of international forces and appropriate host nation (HN) civilian religious leaders.
- Coordinating appropriate training for RMSTs.
- Coordinating with the other staff sections to ensure sufficient religious assets are available to support combat forces, medical services, civilian detainees, EPWs, and others.
- When appropriate and in coordination with the CMOC, the JTF may conduct liaison with faith-based NGOs and IOs.
- Advising distribution of humanitarian assistance supplies from religious organizations.

PLANNING CHECKLIST

- General
  1. Staff: communication, annex, integration, briefings
  2. Specialized:
- Nurture the living:
  - Spiritual Fitness
  - Worship
  - Counseling
  - Visitation to staff and subordinate UMTs
- Care for the Dying:
  - Spiritual triage
  - Leadership
  - Battle Stress
- Honor the Dead:
  - Memorial Services
  - Memorial Ceremonies
  - Rites
  - Sacraments
  - Prayers

Predeployment

5-16. The Division UMT translates the commander's intent and decisions into a fully coordinated and supported religious support operation. Division UMTs, in coordination with their commander and staff, can—

- Accurately determine the current situation.
- Meaningfully assess courses of action for the commander.
- Issue the necessary guidance and instructions, having received guidance from the chief of staff.
- Monitor how well the commander's religious support plan and instructions are being executed.
- Alter the plan in accordance with unit feedback, commander's guidance, commander's intent, and METT-TC.
- Plan future operations.
- Synchronize actions prior to combat operations.

5-17. The division UMT is expected to operate with a great deal of freedom and responsibility. From information available, it must discern what is essential for the commander's decision making. It must coordinate the entire division religious support effort to support the commander's intent and concept of operation. Normally, the commander issues guidance and makes tactical decisions from field locations, then moves to the point of the division's main effort. The UMT develops appendices and plans necessary coordination between subordinate and higher UMTs to execute the commander's intent.

Deployment

5-18. Religious support, like other battlespace support systems, requires C4I and leadership. The Division Chaplain provides UMT C4I and leadership. To perform these duties, the Division Chaplain, like the commander, must be positioned wherever the Division Chaplain needs to be to influence religious support coverage throughout the division area of operations. The Division Chaplain, therefore, cannot be tethered to a specific geographic location but must retain the flexibility to respond to changing conditions as they occur in the battlespace. The Division Chaplain must be able to provide C4I while at the same time provide direct religious support to the commander and staff.

5-19. The operational tempo of conflicts will demand 24-hour operations by the units engaged, religious support operations cannot be an exception to this demand if timely religious support is to be provided. Twenty-four hour operations allow for uninterrupted tracking of the battle. The purpose for tracking the battle is to monitor the fight so as to enable timely responses to emerging battlespace conditions requiring religious support.

5-20. In a division fight, 24-hour operations can be confined to a battle-tracking cell if that cell has the common relevant picture and can communicate to outlying UMTs. These outlying UMTs can then maintain 24-hour availability. Battlespace conditions will require 24-hour operations to monitor both current operations and future operational planning.

5-21. Religious support relies upon the availability of sufficient UMT assets to provide coverage. The capability to man 24-hour battle tracking operations cells and the flexibility to reallocate religious support personnel is dependent upon sufficient religious support resources. Flexibility must include the ability to tailor UMTs to match the force packages engaged in the operations.

Chaplain Detachments-Division

5-22. The Chaplain Detachments-Division (CD-D) allows the Division UMT to conduct 24-hour operations. The Chaplain Detachment will come from the Reserve Component and be organic to the Division it supports. This team consists of one chaplain (MAJ) and one chaplain assistant (SSG). The team will drill, train and deploy with the supported division. (See Appendix G and 6-69) Division Chaplains must incorporate the use of CD-Ds into their RSPs and request their deployment through proper staff actions.

5-23. Flexibility must include provisions to transition offense, defense, stability, and support missions.

5-24. Additional planning considerations include—

- The HQ may be structured for incremental deployment.
- Each deployment increment must be synchronized to ensure comprehensive religious support.
- TPFDD information on augmented religious support assets.
- Religious support assets and resource monitoring and management.
- Rules Of Engagement.

Pre-combat

5-25. This phase for the division UMT requires finalized preparation for combat operations. The UMT works closely with the staff, especially the G3/J3, monitoring the tactical situation for last minute changes to the religious support plan. The chaplain assistant NCOIC ensures that—

- Logistical requirements are finalized.
- Battle tracking requirements for the chaplain section are finalized and prepared.
- Appropriate support to Information Operations is coordinated.
- The NCO support channel is advised and monitored for support.
- Personnel issues for the division chaplain assistants are resolved and a contingency replacement plan is staffed with G-1/J-1.

5-26. UMT planning considerations include—

- Spiritual care to subordinate UMTs.
- Spiritual care to division command and staff.
- Managing the implementation of the religious support plan.
- Monitor the current operations situation applying METT-TC.
- Ensuring support for combat operations religious support plan.
- Division chaplain section personnel are spiritually, emotionally, and physically prepared for the operation.

Operations (See Figure 5-1.)

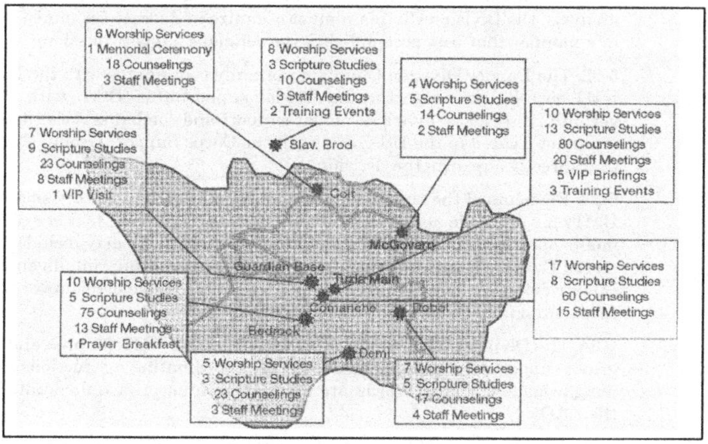

Figure 5-1. Week of Religious Support in Task Force Eagle

5-27. *Critical capabilities requirements.* The ability to provide religious support in the 21st century battlespace will depend upon five critical capabilities requirements: information access, situational awareness, communication connectivity, mobility, and force structure. A deficit in any of these five requirements will radically impair the ability to provide timely religious support.

5-28. Division UMTs must have real time operational awareness. They must be able to determine an appropriate response and expeditiously communicate that information to religious support personnel who will respond directly to the situation. To enable religious support throughout the battlespace, UMTs must be able to move and communicate seamlessly throughout the AO.

5-29. Battle tracking operations require sufficient personnel and the required technical training to conduct 24-hour operations.

Offense

5-30. The Division Chaplain coordinates and supervises religious support for the comprehensive support of all soldiers and units in the division. Based on the religious support estimate, the division chaplain determines area support requirements and where to position the division UMT to best accomplish its mission.

5-31. Normally, the Division Chaplain monitors the tactical situation, plans religious support, and provides care for soldiers from the main Command Post (CP). Prior to an operation, the Division Chaplain moves forward and provides pastoral care to brigade UMTs. Division UMT personnel, under the

Deputy Division Chaplain, provide religious support to soldiers in the rear CP; monitor the volume of casualties, replacements, logistics; and coordinate religious support in the division rear area. These are norms, but in the final analysis, the Division Chaplain must organize and locate his ministry team in a manner that best accomplishes the religious support mission.

5-32. The Deputy Division Chaplain normally coordinates with the DISCOM UMT for religious support in the division support area (DSA); with maneuver units in reserve, division combat support and combat service support units not located in the DSA; and with the Corps Support Group (CSG) UMT that directly supports the division.

5-33. Elements of the supporting CSG may be located in the division AO. Its UMTs may provide additional support to the division and to other corps elements operating in the division area. Other corps units may include the Armored Cavalry Regiment (ACR), artillery, signal, military intelligence, medical, air defense artillery, aviation, and engineers. Some elements of these corps units may be forward in brigade support areas.

5-34. The Division UMT coordinates with the CSG UMT for area and faith group support. The division RSA includes coordinating instructions authorizing brigade UMTs to coordinate with UMTs from corps units located in their AOs.

5-35. The Division UMT coordinates with the Corps UMT concerning area support of corps units in the division AO. The tactical situation may prevent corps UMTs from moving forward to their subordinate units. The teams may be separated from their forward elements by as much as 150 kilometers.

5-36. Area support is limited in light infantry divisions because of restricted transportation assets. Non-divisional UMTs with greater mobility may be tasked to provide more area support to a light division than they would to a heavy division.

Defense

5-37. Division religious support during the defense is similar to religious support during the offense. Access to subordinate UMTs is easier, but enemy reconnaissance and preparatory fires may complicate travel. As in the offense, the division UMT coordinates with the CSG UMT for mutual support.

5-38. The DISCOM UMT coordinates area support in the DSA.

Stability

5-39. Conducting religious support during stability operations remains fundamentally the same as in the offense and defense. The management of the religious support plan is challenged by the increased information management tasks required to properly respond to religious support issues. In stability operations, as in war, the division conducts simultaneous operations. For example, the division could conduct peacekeeping operations at the same time it is deploying additional forces and securing facilities. Stability operations are usually complex. Rarely will UMTs be given a single task to accomplish. Their capability to handle multiple tasks across a wide range will challenge them. When the division headquarters is deployed for stability operations, it is often assigned as the ARFOR and works for a joint task force commander.

5-40. Although often only part of the division deploys, as a controlling headquarters, it may receive attached or operational control (OPCON) units from many external sources. These sources include Army organizations, other U.S. military services, and multinational forces. Within these units are other UMTs that the division UMT must integrate into their religious support plan. Planning considerations for the division UMT—

- ROE understanding and religious support application.
- Command understanding of the religious culture and traditions.
- Media and Public Affairs Office/Officer (PAO) guidance.
- Religious support integration during Base Camp planning.
- Command and staff expansion for operations.
- Information management.
- Personnel issues for UMTs
- Increased liaison tasks with faith-based NGOs and IOs, and host nation religious leaders.

Support

5-41. Support operations consist of relief operations, support to WMD incidents, support to civil law enforcement, and community support. All are challenging for religious support planning and execution. Support operations can also be conducted simultaneously with offense, defense, and stability operations. Posse Comitatus (the legal guidelines defining the relationship between federal armed forces and the civilian populace) plays a critical part for planning. (See Annex K for more information.)

5-42. Religious support planning may include—

- Limits on deployed religious support personnel due to statutory agreements and other restrictions.
- Availability and type of transportation assets.
- Force protection requirements.
- Training requirements.
- SOF participation.
- Use of reserve component forces.
- Other U.S. government agencies' roles.
- Lodgment sites and responsibilities.
- Host nation requirements and expectations (OCONUS).
- Political and social implications of the division's actions.
- Funding and special equipment and supplies requirements.
- Impact of mass casualty upon the units.

Reconstitution, Redeployment, and Demobilization

5-43. Reconstitution is an extraordinary action that commanders plan and implement to restore units to a desired level of combat readiness. It transcends normal daily force sustainment actions. No resources exist solely to perform reconstitution; rather existing systems and units do it.

5-44. Reconstitution decisions belong to commanders. They control assets and decide whether to reorganize or regenerate a unit. Normally, the decision to reconstitute is made two echelons above the unit that must be rebuilt. The commander two echelons above, with his staff's assistance, are in the best position to assess a subordinate unit's effectiveness. The division UMT provides unit ministry support, particularly for cases of combat and operational stress, spiritual renewal, critical event debriefings, and worship opportunities. The chaplain coordinates worship and memorial services, sacramental acts, and pastoral counseling. Chaplains play a key role in assessing the spiritual morale and moral/ethical needs of the unit.

5-45. The OPLAN includes enough details to enable staffs and supporting units to prepare for rapid restoration of units within command priorities. The division UMT inputs the religious support plan in support of reconstitution operations after utilizing the estimation and analysis process. While the religious support plan cannot meet all the contingencies of Army operations, it must be one that the command and UMT can adapt to the situation. In addition to addressing reconstitution for a specific mission, the staff may have to write a separate OPLAN for reconstitution operations. The UMT is fully integrated within this process. The more fully developed a UMT's reconstitution SOP, the easier for it to develop a reconstitution plan.

5-46. Redeployment is the repositioning of forces after mission accomplishment or transfer of mission and authority to another unit. This may mean redeployment to another AO for subsequent missions, or return to home station. In either case, the division UMT should prepare subordinate UMTs to assist commanders, soldiers, and family members, in preparing emotionally and spiritually for the changes that come with redeployment and reunion.

5-47. Demobilization is the return of forces to a peacetime posture. Planning for demobilization is critical. The division UMT cannot wait for the operation to end to begin planning for demobilization. It must begin early in the process. The division UMT's role in demobilization may be much greater than in mobilization. It begins in the division's AO with reconstitution of attached Reserve Component forces and preparation for redeployment to CONUS or OCONUS demobilization stations. Demobilization planning begins concurrently with mobilization. Demobilization planning may impact on division UMTs in theater as reserve component units begin the tasks of reconstitution and demobilization, while under division or corps control. Religious support plans must address many issues including—

- UMT spiritual reconstitution.
- Decorations and awards.
- Evaluation reports.
- Memorial Services and Ceremonies.
- Critical event debriefings.
- Reunion briefings for command and staff, soldiers, and family members.
- The division UMT may also coordinate with the corps and installation chaplain at both the division's home station installation and the RC unit's demobilization station.

Garrison Operations

5-48. During garrison/installation operations, the division UMT plans and implements the CMRP in coordination with the unit training schedule to support the religious needs of the division's soldiers and their families. The division UMT manages and supervises the subordinate UMTs assigned to the division for the commander and coordinates ministry as applicable with the garrison/installation UMT. The UMT participates in all installation chaplains' training programs and other programs that enhance the religious quality of life of the installation. They also fully participate in the unit's training program. The division UMT will support the unit's training program by being available to conduct training for the command and staff, and subordinate UMTs. The division UMT is the mentor for the battalion/squadron UMTs of the division and will execute a mentorship program. Also, the senior division UMT supervises the personnel assigned or attached to the division chaplain's office.

ROLE OF THE DIVISION UNIT MINISTRY TEAM

5-49. The role of the division UMT is critical to the professional and tactical proficiency of the subordinate UMTs. Their leadership responsibilities shape and guide brigade, battalion, and squadron UMTs role in religious support execution process. They are the command and staff advisors for the officers and NCO chain of command.

PRINCIPLES

Table 5-1. Division/Task Force Tactical UMT

CHAPLAIN (O-5) ***CHAPLAIN (O-4)***	***CHAPLAIN ASSISTANT (E-8)*** ***CHAPLAIN ASSISTANT (E-5)*** ***CHAPLAIN ASSISTANT (E-3)***
Role Commissioned Officer/Staff Religious Professional Title 10 Protected	***Role*** Noncommissioned Officer Specialized Assistant Religious Support Specialist Team Security
Mission Advises commander on religious support and quality of life issues to support the unit's mission. Performs/provides/coordinates the religious support mission for the division. Division Team Leader for Religious Support.	***Mission*** Advises Chaplain and NCO Chain. Provides specialized assistance in areas of religious support to support the chaplain. Perform Battle Staff functions. Team NCOIC.
Parallel Tasks 1. Assess subordinate UMTs readiness. 2. Conducts staff assistance visits. 3. Determines Training Objectives. 4. Mentors subordinate UMTs. 5. Recommend personnel assignments.	***Parallel Tasks*** 1. Assess subordinate UMTs readiness. 2. Conducts staff assistance visits. 3. Conducts UMT religious support training. 4. Mentors subordinate UMTs. 5. Recommends Chaplain Assistant assignments.

5-50. The Division Chaplain is a personal staff officer, religious leader, and team leader. The Division Chaplain Assistant NCOIC must be a NCO leader and advisor, mentor, information manager for the team. They both must have a keen awareness of the world situation as it bears on potential military operations, know the role of military religious leader's and have a broad understanding of the impact of indigenous religions on military operations.

5-51. Unit Ministry Teams must use technology to enhance the personal delivery of religious support. They must be trained in joint and multi-national operations and skilled in synchronizing all aspects of religious support during military operations. They will operate in an environment of increasing volatility, complexity and ambiguity. The division UMT must be innovative and disciplined.

5-52. This means that the Division Chaplain section NCOIC will need to be a graduate of the Battle Staff NCO Course (with the 2S identifier). This will enable the UMT NCOIC to take part in the MDMP in the absence of the chaplain, track the battle analytically and better utilize the digital enablers that the division will share with other staff sections.

5-53. Division UMTs require the same level of technology as the command of the supported force. UMT automation, communication and transportation requirements must keep pace with the command's requirements in order to effectively coordinate religious support, communicate critical information, and remain physically flexible to meet requirements across the division battlespace. They are an integral part of the division.

RESPONSIBILITY AND ACCOUNTABILITY

Staff Responsibilities

5-54. The Division UMT works directly for the commander representing the religious support dimension of caring for soldiers and family members. The UMT executes the commander's plan for religious support initiatives and programs. Staff responsibilities include—

- Synchronizing religious support programs across the division AO.
- Coordinating denominational coverage.
- Performing UMT personnel tasks.
- Advising the Division Commander on matters of religion, ethics, morals, and morale as affected by religion, the impact of indigenous religions on military operations, and installation and community religious support issues.
- Preparing the staff religious support estimate and appendix for the division OPORDs.

Relationship to the Corps Unit Ministry Team

5-55. The Division UMT supports all of the Corps UMT programs including training, family support, and professional development. They work closely together implementing UMT personnel issues for the command. During combat operations, the division UMT utilizes the Corps UMT's religious support appendix to the corps OPORD for religious support planning. The Division UMT works closely with the corps to ensure the success of the religious support mission.

SUPERVISORY RESPONSIBILITIES

5-56. The Division UMT provides technical and professional supervision to subordinate UMTs in the division. As the senior UMT, the Division UMT provides leadership for UMTs in professional development, implementation of the CMRP, personnel issues, and other issues as directed by the commander. The Division UMT works closely with subordinate UMTs to mentor and provide direction.

TRAINING

5-57. The Division UMT assesses its current training strategies as well as the underlying assumptions regarding pre-training knowledge and skills in light of the Division Commander's training and mission requirements. The Division UMT will provide training in planning, preparation and execution of the 11 Religious Support Activities in accordance with the CMRP. Information literacy requirements for all UMTs are imperative in training.

5-58. Training will be profoundly affected since non-linear operations complicate the organization of the battlespace into close, deep and rear operations.

SUMMARY

5-59. Recent significant changes in international politics have created an ambiguous security environment in which threats to world stability and U.S. national interests are more regionally focused. Many crisis situations in which the United States exercises its power span the spectrum of operations from operations other than war to actual combat. Force Projection has become our chief strategy for supporting the power-projection element of U.S. national security. Army corps and divisions must be prepared for force projection operations, as part of a joint or multinational effort and in war as well as other than war environments.

5-60. The mission of the Division UMT is to coordinate religious support to soldiers, families, and as the force structure evolves, to forward-deployed DOD authorized civilians. Division Chaplains also serve as critical advisors to commanders at all levels of the division providing essential information on troop and unit morale, quality of life matters, and the impact of indigenous religion on military operations.

5-61. The Division UMT is expected to operate with a great deal of freedom and responsibility. From information available, it must discern what is essential for the commander's decision making. It must coordinate the entire division religious support effort to support the commander's intent and concept of operation. The UMT develops plans and conducts necessary coordination between subordinate and higher UMTs to execute the commander's intent.

5-62. Combat power in the 21st century will rely heavily upon technology. The cumulative affect of the numerous factors will increase the spiritual, psychological and interpersonal demands placed upon soldiers of the 21st century. Technology will provide additional means to plan, coordinate, and execute religious support to address these demands but the personal delivery of religious support by the UMT will always be imperative.

RESOURCES

5-63. The following is a list of main resources utilized for writing and updating this doctrine.

- JP 1-05.

- JP 5-00.2.
- JP 3-05.
- FM 16-1.
- FM 100-5.
- FM 3-0 (100-5).
- FM 25-100.
- FM 25-101.
- FM 71-100.
- Lessons Learned from past and current operation
- Lessons Learned from the training centers

Chapter 6

CORPS/ECHELONS ABOVE CORPS RELIGIOUS SUPPORT OPERATIONS

The corps is the Army's largest tactical organization. The corps organization is tailored for the assigned theater of operations and the assigned mission. Corps may be assigned divisions of any type required by the theater and the mission. They have a support command and will receive additional Combat Service (CS) and Combat Service Support (CSS) organizations based on the assigned mission.

Echelons above corps (EAC) consist of the Army at theater level and above. These units will be under the command of a regional combatant commander (unified command). This level is in the strategic context.

CONTENTS

OPERATIONAL ECHELON DESCRIPTION

6-1. Corps conducts operations with joint, interagency, or multinational forces, facilitated by information superiority across the full range of military operations, in a Major Theater War or small-scale contingency, against any threat, in any assigned AO. Corps operations may include a mixture of offense, defense, stability, and support operations executed sequentially or simultaneously.

6-2. Corps is usually the link between the operational and tactical levels of war. Corps may link the operational level of war to the strategic level of war. Corps plan and conduct major operations and battles. The corps headquarters synchronizes tactical operations including maneuver; fires, joint, deep, close, and Suppression of Enemy Air Defenses (SEAD); CS; and CSS. The corps is not a fixed standing organization, but a formation whose core characteristic is the ability to form task forces focused on mission requirements. It employs tailored forces that are optimized for specific functions or tasks. The corps employs divisions and applies force allocation rules of support to divisions based on mission. The corps normally has a wide range of Army capabilities to support and sustain any effort. The two critical requirements of corps operations are a robust C4 intelligence, surveillance and reconnaissance (C4ISR) and logistics support to plan, prepare, execute, and assess the full range of operations. Corps traditionally can employ multinational forces and operate with governmental and non-governmental agencies and capabilities. Corps operations are nested in political military dimensions of the combatant commander's theater effort, but are optimized for conducting military operations.

6-3. The corps is a much diversified organization that fulfills the synchronization of many functions that achieve tactical and operational advantage. The corps conducts planning together with joint forces to achieve campaign objectives and conducts simultaneous operations in depth across the corps battlespace. The corps integrates all joint combat, CS, and CSS efforts.

6-4. The corps plans and conducts campaigns and major operations. Corps campaign plans set broad long-term goals, such as control of a geographical area, defeat of an enemy in theater, or the reestablishment of international boundaries. The corps also plans and executes tactical-level battles that support the corps operational objectives, simultaneously conducting close, deep, and rear operations, as well as planning future operations.

6-5. Current U.S. collective security arrangements make it imperative that corps are ready to deploy in response to worldwide situations. In the event of a crisis situation, a corps headquarters considers the employment of light forces, armored maneuver forces, and special operations forces. (See Figure 6-1.)

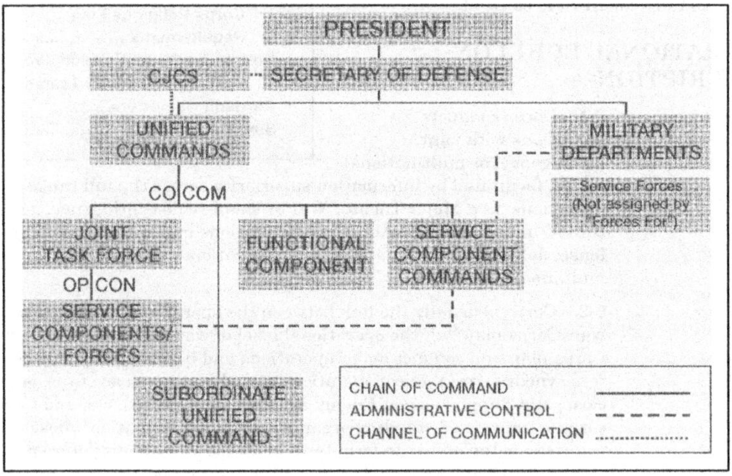

Figure 6-1. Chain of Command and Control

JOINT TASK FORCE CONFIGURATIONS

6-6. A corps headquarters may function as the Army component headquarters of a joint force, the Joint Forces Land Component Command (JFLCC) headquarters of a JTF or as the JTF headquarters itself. The corps' primary responsibility in the joint environment is to participate in the development of supporting joint campaign plans. Current U.S. collective security commitments and the very nature of the international political environment ensure that U.S. corps' will also operate as part of multinational coalition forces.

6-7. Corps with missions to respond to worldwide situations will frequently fight as the largest ground formation in the theater. Often this situation results in the corps commander filling the position of the commander JTF (CJTF). In such cases, the corps headquarters accepts the responsibility to plan operational and tactical level missions as well as oversee the tactical execution of campaigns or other extended operations.

6-8. As the CJTF, the corps commander will exercise OPCON or tactical control (TACON) of assigned forces. The CJTF must form a joint staff and establish the necessary joint boards and centers, as the mission requires. Joint service representation on the staff should be in relative proportion to the service composition of the JTF. It is imperative that all of the staff sections and agencies have joint service representation in the formation and execution of the joint campaign plan. The corps staff must transition to a joint staff structure to ensure unity of effort within the task force.

6-9. The JTF establishing authority is responsible for providing personnel and resources for the corps when the corps is designated the JTF headquarters. However, the corps commander, as the CJTF, must determine what augmentation requirements he needs for the task at hand and coordinate support through the establishing authority. This augmentation is essential in the transition of the corps to a JTF structure.

6-10. Augmentation must be tailored for the specific situation and will be different for every mission. The following areas usually require augmentation:

- Joint, special staff, and personal sections.
- Specific functional area augmentation, such as CA, movement control, and PSYOP.
- Headquarters life support functions.
- Communications support.
- Security support for the JTF headquarters

6-11. The combatant commanders usually organize augmentation for the joint staff into packages based on the nature and tempo of developments and the existing capabilities of the headquarters designated as the JTF. These packages fall into three broad categories:

- Quick response cells to rapidly augment the JTF headquarters with joint and operational area expertise during the early stages of a crisis action situation.
- Functional area augmentation to provide the JTF with expertise not organic to its staff or to enhance specific capabilities based on mission requirements

- Joint staff augmentation to provide the coordinating personal and special staff sections expertise in joint procedures and service-unique capabilities.

JOINT TASK FORCE STAFF AUGMENTATION

6-12. Augmentation facilitates coordination between the JTF staff and the combatant command joint staff, especially with regard to accessing information and capabilities available at the combatant command level.

6-13. Although the corps/ JTF commander applies the military element of power, he also considers those interagency, political, economic, and other resources that may be available and appropriate for the task at hand. The CJTF must understand the strategic and regional environment to include U.S. policy, treaty commitments, the status of forces agreements (SOFA); host nation concerns, religious and cultural considerations, and coalition members' interests. These influences will affect campaign and operational planning, and the establishment of ROE for the joint force. The CJTF must keep in mind the strategic end state through out the planning and execution of the operation.

THE CORPS AS AN ARMY FORCES/JOINT FORCE LAND COMPONENT COMMANDER HEADQUARTERS

6-14. A JTF-level service component command consists generally of forces from only one service of the JTF. The CJTF may designate the corps headquarters as the ARFOR headquarters giving it Title 10 responsibilities and, in most cases, OPCON of a number of assigned or attached forces.

6-15. The designation of Commander, ARFOR carries additional responsibilities for the Corps Commander. The corps as ARFOR must accomplish the same coordination as covered above. Corps headquarters are uniquely suited to be the ARFOR or Joint Force Land Component Commander (JFLCC) headquarters in a JTF. Joint Pub 1-02 defines the Joint Force Land Component Commander as "the commander within a unified command, subordinate unified command, or a JTF responsible to the establishing commander for making recommendations on the proper employment of land forces, planning and coordinating land operations, or accomplishing such operational missions as may be assigned. The joint force land component commander is given the authority necessary to accomplish missions and tasks assigned by the establishing commander. The joint force land component commander will normally be the commander with the preponderance of land forces and the requisite C2 capabilities." The primary purpose of a JFLCC is to provide unity of command for employing land power for the benefit of the joint force as a whole and to give the Joint Force Commander (JFC) a workable span of control.

6-16. The JFLCC must shape the battlespace for subordinate units by focusing on deep operations in space and time (more than 96 hours out).

6-17. Subordinate units are organized for executing the battle command functions required from the higher headquarters of ground maneuver forces. Corps headquarters can coordinate the sustainment activities for significant numbers of assigned or attached forces. Corps also has the staffing to do joint and multinational force coordination and liaison. In most cases the corps headquarters is capable of being the ARFOR or JFLCC headquarters in a JTF without significant personnel augmentation.

THE CORPS AS A FORCE PROVIDER

6-18. In day-to-day operations the corps is required to provide resources to support operations in which the corps headquarters is not otherwise involved. (As an example, each U.S. corps today has soldiers deployed around the globe.) As a force-provider, corps has considerable assets, which they may commit to other commands to support major and lesser regional contingencies around the world. Although these deployed organizations may be under the OPCON or TACON of another headquarters, the corps retains significant responsibilities as the parent organization.

6-19. Operations in stability and support environments are good examples in which corps' function as force providers. Stability operations and support operations often come with vague or ad hoc chains of command, requiring specific support. These missions are different. They require a variety of corps combat, CS, and CSS units to operate in concert with U.S. and foreign civil agencies of government, international, and private organizations.

ECHELONS ABOVE CORPS

6-20. EAC echelons include the theater command. Within this command are various commands that support theater operations. The Army Service Component Command (ASCC) is an Army component in a unified and subordinate unified command. Army forces, through the ASCCs, provide defenses to all joint force components. The ASCC establishes the link between ARFOR and the joint command. It also plans and executes operations in support of the joint campaign, plans and executes support operations to sustain subordinate ARFOR, and provides support to other services in accordance with executive agent responsibilities. The Theater Support Command (TSC) is an EAC-level support organization that is key to operations for CS and CSS execution. The TSC is structured and organized to deploy tailored functional modules. The ARFOR commander is provided key functional expertise and C2 to support joint forces.

FUNDAMENTALS OF RELIGIOUS SUPPORT OPERATIONS

6-21. The corps can rapidly tailor force packages formed from the full range of Army capabilities, both active and reserve. The corps C2 has the organic capabilities for split-based operations required for simultaneous functions of deploying and employing Army forces. The religious support planning and management mission considers force-tailoring UMTs to perform and provide religious support for all types and sizes of forces in all contingencies. Religious support delivery, coordination, synchronization, and synergy are dependent upon the communication capacity in both digital and analog environments to support decisive operations and to improve the synergy of joint and multinational operations. Superior situational understanding is available through information systems that facilitate the execution of decisive, shaping, and sustainment operations across all dimensions of the corps AO.

CORPS RELIGIOUS SUPPORT

6-22. The Corps Chaplain supports the Corps Commander's operational-level responsibilities and roles by engaging in three operational areas:

- Given guidance by the Corps Commander in coordination with G-5, PAO, CA, the Corps Chaplain establishes links with representatives of joint, multinational, interagency, faith-based NGOs and IOs, and religious leaders of the host nation.
- Plans and executes religious support for corps operations. Monitors religious support in major subordinate commands.
- Executes support operations to sustain subordinate Army forces. (See FM 100-7.)

6-23. From the main CP, the Corps UMT plans and synchronizes religious support in the corps area. The Corps Chaplain monitors the tactical situation and moves as far forward as possible to provide staff supervision. The Corps UMT at the rear CP recommends chaplain and chaplain assistant assignments. It coordinates religious support for displaced persons and also plans religious support for reconstitution. It provides an on-call UMT for the Corps Reconstitution Task Force (RTF).

6-24. The Corps UMT normally passes responsibility for religious support in the corps support areas (CSA) to the COSCOM UMT. The Corps UMT must be capable of simultaneous employment and deployment. Its flexible METT-TC planning speeds the process of force tailoring, while its capabilities allow it to provide religious support to the reception, staging, onward movement, and integration (RSOI) at the lodgment base or in an intermediate staging base (ISB) within a JOA. The Corps Chaplain's section will have secure, robust, and survivable communications networks for transmissions of voice, data, and video.

6-25. The senior echelons above division (EAD) Chaplain will tailor the staffing of religious support personnel to match force packages engaged in operations. Religious support relies upon the availability of sufficient religious support assets and the flexibility to reallocate those assets.

6-26. Flexibility must include provisions to transition between combat and other military operations. Planning considerations include—

- Provide interface/liaison between corps and local religious leaders and faith-based NGO/IOs.
- Support the resettlement of refugees and displaced civilians.
- Reconstitute and regenerate combat ineffective units.
- Manage religious support personnel assignments.
- Contractor religious support planning for subordinate UMTs.
- In coordination with the ASCC chaplain, establishes liaison with senior UMTs of other services for religious support to Army authorized personnel, as well as, religious support to service members of other services.
- Conduct specialized religious support assessment of the AO for the commander and staff.
- Coordinate all chaplains' fund activities within the corps and all budget and funding requirements for the corps UMTs.
- Provide spiritual care to subordinate UMTs.

- Coordinate with the ASCC chaplain for Chaplain Detachments (CD) and Special Medical Augmentation Response Team-Pastoral Care (SMART-PC) usage.
- Provide a UMT training program.
- Plan RS 96 hours to 5 days from execution for the corps fight.
- Utilization of "requests for information" (RFI) to respond to FRAGOs and mission RS implied tasks.

6-27. The Corps Chaplain's staff will be resourced to conduct continuous operations. The Corps Chaplains staff will consist of at least four (4) chaplains and five (5) chaplain assistants and will be augmented by the Chaplain Detachment – Corps. Further augmentation may be necessary if serving as the JTF or JFLCC UMT. The Corps UMT will execute sleep plans to minimize the degrading effects of sleep loss. The Chaplain's section must be equipped to minimize the stress imposed by extreme environmental conditions. The Corps UMT will have the capability to exercise religious support management for the commander from any point within the corps extended area of operations.

THE CORPS/EAC TACTICAL/OPERATIONAL/STRATEGIC UMT	
CHAPLAIN (O-6)	***CHAPLAIN ASSISTANT (E-7/8/9)***
Role Commissioned Officer / Staff member Religious Professional Title 10 Protected	***Role*** Noncommissioned Officer Specialized Assistant Religious Support Specialist Team Security
Mission Advises commander on religious support and quality of life issues to support the unit's mission. Performs/provides/coordinates the religious support mission for the command. Senior Team Leader	***Mission*** Advises Chaplain and NCO Chain. Provides specialized assistance in areas of religious support to support the chaplain. Perform Battle Staff functions. Senior Team NCOIC
Parallel Tasks 1. Conducts staff assistance visits. 2. Determines Training Objectives. 3. Mentors subordinate UMTs. 4. Recommends personnel assignments.	***Parallel Tasks*** 1. Conducts staff assistance visits. 2. Conducts religious support training. 3. Mentors subordinate UMTs. 4. Recommends Chaplain Assistant assignments.

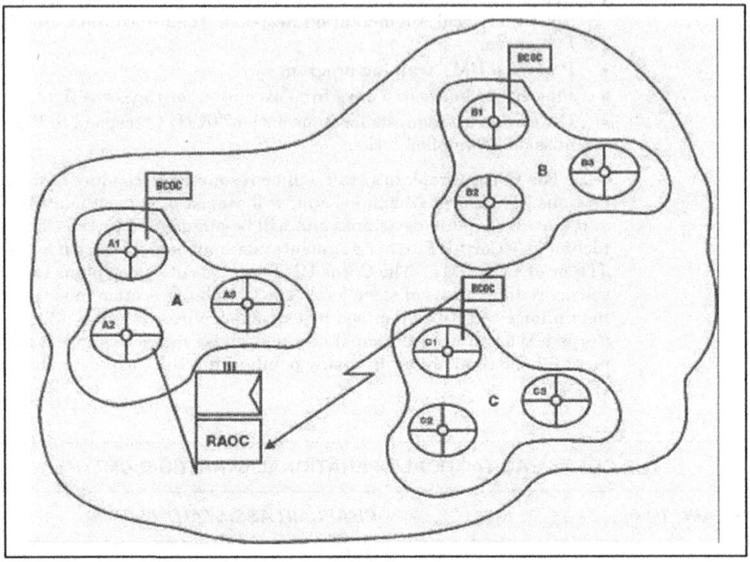

Figure 6-3. Religious Support Coordination in a Corps Support Area

CORPS SUPPORT COMMAND

6-28. The Corps Support Command (COSCOM) UMT coordinates religious support in the corps support area (CSA) for the Corps chaplain. The CSA constitutes the Corps Rear Area. The potential expansion of the CSA, the dispersion of units in bases and base clusters, and the movement of small units make the coordination of religious support difficult. The COSCOM UMT will manage and coordinate area religious support for the Base Cluster Commander to ensure synchronized RS to all units in the base footprint. As a measure of the magnitude of possible expansion, during Operation DESERT SHIELD/DESERT STORM, one COSCOM grew to over 60 UMTs. The area that the COSCOM may cover encompasses the entire battlespace, making authorized and assigned transportation and communication assets for all COSCOM UMTs essential.

6-29. The COSCOM UMT works closely with the Rear Tactical Operation Center (RTOC) and Rear Area Operation Center (RAOC) to track the locations and faith groups of the chaplains in the CSA.

6-30. The COSCOM UMT monitors the flow of religious support supplies with the Corps Material Management Center. The COSCOM UMT locates where communication, battle tracking, and religious support management are best executed. Twenty-four-hour operations are critical to ensure comprehensive religious support and personal staff support to the commander. Split operations at two locations may be considered.

CORPS SUPPORT GROUP

6-31. The COSCOM supports the corps with two or more Corps Support Groups (CSGs): forward and rear. CSGs are composed of multi-functional battalions that provide supply, services, and maintenance to supported units. Forward CSGs support the divisions and other corps units in their AOs. The Rear CSG supports the corps rear area, units in reserve, separate brigades, and units with special missions. The task organization of a CSG can change daily.

6-32. The Forward CSG UMTs monitor the rapidly changing situation through their assigned RAOCs. The tactical communication system found in the RAOC is the primary means of communication for the CSG UMTs. These teams are responsible for coordinating religious support throughout their AOs. They maintain a database with the locations of all UMTs by unit and faith group to process requests for area and faith group religious support.

6-33. The Rear CSG UMT provides religious support to the largest unit in the corps support area that includes the corps and COSCOM headquarters.

6-34. The Rear CSG UMT is often given the reconstitution mission.

RELIGIOUS SUPPORT IN REAR AREAS

6-35. The dispersion of units in rear areas presents a significant religious support challenge. Unit ministry teams have elements of units deployed far forward and out of reach. At the same time, they have elements of other units located in their areas. Of necessity, area support in the rear consumes a greater portion of a UMT's time and energy than does unit support. The Corps Support Chaplain has the responsibility for the rear. Timely religious support coordination is crucial and requires a clear understanding of rear operations.

REAR OPERATIONS

6-36. Rear operations support tactical forces in contact, guaranteeing freedom of maneuver, continuity of support, and uninterrupted C2. In the linear battlespace these actions occur behind forces engaged in active combat. In the non-linear battlespace they occur in the support base areas. Rear operations are characterized by a wide dispersion of multi-functional units conducting a variety of missions around-the-clock.

6-37. Rear areas may be small in a sparse theater, as in small-scale contingencies (SSC), or large in a mature theater, as during general war. In SSC, with tactical operations arrayed through 360 degrees, the "rear area" refers to the area designated as the support base.

6-38. The synchronization of rear operations is the responsibility of a Rear Tactical Operations Center (RTOC), a Rear Area Operations Centers (RAOC), or a Rear Command Post (RCP).

6-39. The Rear Tactical Operations Center (RTOC) is the primary staff element for planning and coordinating rear area security operations.

6-40. The Rear Area Operations Center (RAOC) is a subordinate command post in or near the rear tactical command post concerned with terrain management and security operations.

6-41. The RAOC maintains critical information for the UMT to accomplish its mission: a current situation map; continuous communications with the RCP, other RAOCs, and subordinate base clusters; the threat condition; and Base Defense Status Reports.

The RAOC maintains critical information for the UMT.

6-42. The most important RAOC function for religious support coordination is terrain management. The RAOC manages the use of terrain by assigning units to bases and base clusters.

Bases

6-43. Units are positioned in bases to share security responsibilities and capitalize on mutual strengths. Bases have clearly defined, defensible perimeters and established access controls.

6-44. Normally, the senior commander is the Base Commander. All forces assigned to the base are under his operational control for defense purposes. The commander forms a Base Defense Operations Center (BDOC) to assist in planning, coordinating, integrating, and controlling base defense efforts.

6-45. The base commander's UMT obtains information about the units (including those in transition) occupying the base from the BDOC. The UMT uses the BDOC's secure communication equipment to coordinate with higher echelons and to coordinate area religious support requirements. (See Figure 6-4.)

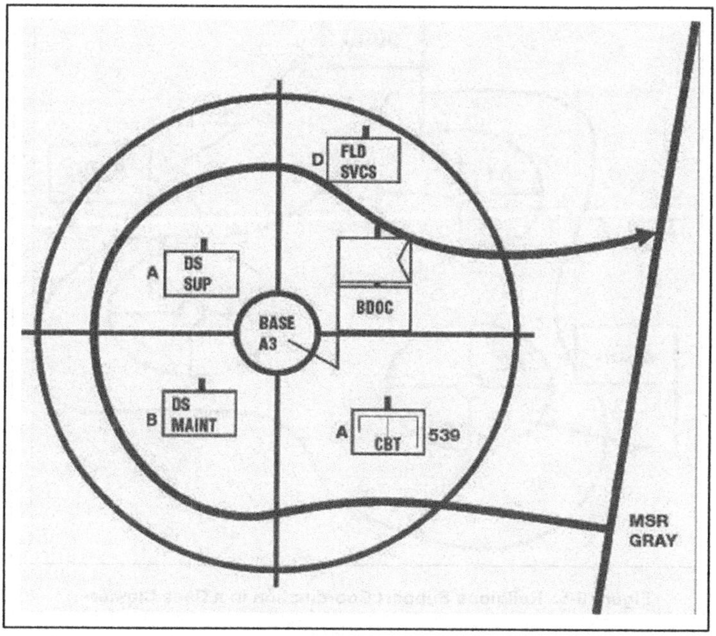

Figure 6-4. Religious Support Coordination in a Base

Base Clusters

6-46. A base cluster is a grouping of bases that are mutually supporting for defense purposes. The commander forms a Base Cluster Operations Center (BCOC) from his own staff and available base assets.

6-47. As in the base, the base cluster commander's UMT uses the BCOC as its source of information for coordinating area religious support in the base cluster and for communicating with higher and lower echelons. (See Figure 6-5.)

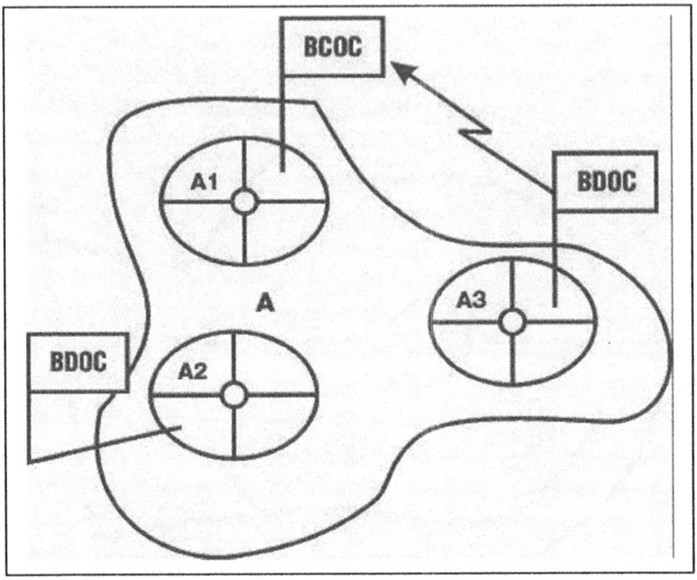

Figure 6-5. Religious Support Coordination in a Base Cluster

Religious Support Coordination In Rear Areas

6-48. The base or base cluster commander's unit ministry team (Base or Base Cluster UMT), regardless of the chaplain's rank, is responsible for area support planning and coordination. Tenant and transient units coordinate with the Base UMT. Base and Base Cluster UMTs coordinate scarce resources for optimum religious support.

6-49. Base and Base Cluster UMTs monitor unit movements, particularly company-size and smaller. They maintain a database of chaplains, their units, faith groups, locations, and means of communication.

6-50. When the commander does not have an assigned UMT, the senior chaplain in the base or base cluster coordinates area support. If there is no religious support available in the base or base cluster, the next higher echelon takes responsibility for area coordination. These relationships are specified in the Religious Support Appendix of the next higher headquarters.

Base Religious Support

6-51. Tenant units coordinate with the Base UMT for additional religious support. The Base UMT conducts a daily meeting to coordinate religious support for the soldiers in the base. Units without organic religious support contact the Base UMT through the BDOC for religious support.

6-52. The Base UMT reports tenant and transient chaplains by faith group to the Base Cluster UMT. This information is used to coordinate faith group support in the base cluster. Unit ministry teams provide area support as needed on a temporary basis until their unit's mission requires their movement from the base.

6-53. If there is no chaplain available in the base to meet a unique faith group requirement, the Base UMT requests support through the BDOC to the BCOC. For instance, if there is no Roman Catholic chaplain available, the Base UMT may request Catholic support using this procedure.

Base Cluster Support

6-54. The base cluster religious support is coordinated in the same way as in the base. The Base Cluster UMT monitors the movement of units into and out of the bases. As bases submit requests for additional faith group support, the Base Cluster UMT meets those requests with available UMTs. If the request cannot be met, it is forwarded to the next higher echelon. The Base Cluster UMT reports chaplains by faith group through the BCOC to the RAOC of the next higher echelon.

Higher Echelon Support

6-55. The procedure for coordination of faith group support described above is repeated up the chain of command. At higher echelons, as the area and the number of units increase, the maintenance of the chaplain database becomes more sophisticated. At these levels, a computerized database, using standard data base software, is required.

SPECIAL RELIGIOUS SUPPORT REQUIREMENTS IN REAR AREAS

RECEPTION, STATIONING, ONWARD MOVEMENT, AND INTEGRATION OPERATIONS

6-56. RSOI operations include the initial reception of units and individuals, the preparation of these units and individuals for combat, and their movement forward into the combat zone. Reception operations begin before hostilities start and continue after hostilities cease. RSOI may be conducted either forward in the theater of operation, in the rear of the area of operation at an Intermediate Staging Base (ISB), or at home station and/or en route to the AO. In the case of RSOI being conducted in the theater, the theater chaplain is responsible for coordinating religious support in the reception area. In the case of a RSOI being conducted closer to the AO, the COSCOM chaplain may be designated to provide area religious support.

MORTUARY AFFAIRS

6-57. The recovery, preparation, and transportation of the dead are one of the most difficult missions in the battlespace. Religious support is provided to soldiers in Mortuary Affairs units and to those detailed to perform this mission. The care for the dead requires sensitive religious support.

6-58. Planning Considerations:

- Religious support in sites where those killed in action are prepared for return to CONUS.

- Religious support for soldiers who are doing the work.
- Special faith group requirements for burial.

RECONSTITUTION AND REDEPLOYMENT

6-59. Reconstitution (both reorganization and regeneration) restores combat ineffective units to a specified level of effectiveness. The Army Service Component Command (ASCC) plans and conducts operational and tactical reconstitution operations.

6-60. Regeneration involves large-scale replacement of personnel and equipment and the rebuilding of units. This mission is usually assigned to the TSC, COSCOM, or DISCOM, depending on the size of the unit. Division, Corps, or ASCC UMTs assist in religious support to the units.

6-61. One or more UMTs may be assigned to a RTF to provide religious support to soldiers until the unit's assigned UMT has recuperated or been replaced. Religious support during reconstitution includes the ministry of presence, group and individual counseling and debriefing (see Appendix C, Critical Intervention Management and Critical Event Debriefings), worship and memorial services, and prayer. RTF UMTs play an important role in restoring the soldiers' spiritual fitness and rebuilding the unit's cohesion and morale.

6-62. Planning Considerations:

- RTF planning process and status.
- How many UMTs are required?
- Which UMTs are designated to join the RTF, how are they notified, and by whom?
- Anticipated duration of mission.
- Criteria for determining when RTF UMTs end the mission and return to their parent units.
- Logistical support for RTF UMTs.

THEATER RELIGIOUS SUPPORT

6-63. The Army Service Component Command (ASCC) serves as the senior army echelon in a theater and is the service component command of a unified command. It includes the service component commander and all Army personnel, organizations, units, and installations. The ASCC chaplain is the senior Army chaplain in a theater of operations and provides staff supervision over all Army religious support in the theater. The ASCC chaplain is responsible for recommending religious support policy to the ASCC commander. An example of an ASCC is Third U.S. Army, which is the ASCC of USCENTCOM.

OPERATIONS

6-64. The ability to provide religious support in the 21st century battlespace will depend upon five critical capabilities requirements: information access, situational awareness, communication, connectivity, mobility, and force structure. A deficit in any of these five requirements will radically impair the ability to provide timely religious support.

6-65. EAD UMTs must have real time operational awareness. They must be able to determine an appropriate response and expeditiously communicate that information to religious support personnel who will respond directly to the situation. To enable religious support throughout the battlespace, UMTs must be able to move and communicate seamlessly throughout the AO.

6-66. The complexity of coordinating religious support at EAD requires sufficient religious support personnel to provide flexible tailoring and augmentation to support the mission. The senior theater UMT estimates and plans campaigns with the staff and must focus on the different phases, determine priorities, and manage the coordination and execution of religious support tasks. When the theater commander establishes the combat zone (CZ) and communications zone (COMMZ), the theater UMT plans for simultaneous

6-66. activities across the full spectrum of operations throughout both AOs. Communication across both areas is critical for religious support synchronization; reach capabilities, reporting, and comprehensive planning. Division rear operations and Corps MSR operations are linked as a part of the battlespace management system, which includes tracking religious support assets from COMMZ through assembly areas (AA) to corps rear to division.

STRATEGIC RELIGIOUS SUPPORT PLANNING

6-67. The combatant commander's UMT's strategic religious support planning provides the framework for the religious support assets in peacetime and in crisis. During predeployment, the senior command UMT develops an assessment that transitions the focus to crisis or war. Force projection and assessment transitions religious support assets to combat operations. Religious support planning is integrated into the combatant commander's staff plans and estimation process. The result is expressed in terms of religious support to mission objectives, concepts, vision, resources, and provides guidance to subordinate command UMTs for a broad range of religious support activities. Planning should consider—

- The command's desired end state and how religious support can support full accomplishment of military objectives.
- Risks associated with METT-TC and the religious support assets.
- Religious support courses of action meeting mission objectives.
- Visualization of the religious support needs.

STRATEGIC RS ESTIMATE

- Assigned objectives from combatant commanders.
- Translation of combatant commanders objectives applicable to theater RS.
- Visualization of the strategic environment; how it relates to the accomplishment of the commander's RS objectives.
- Assessment of the threats of accomplishment of assigned RS objectives.
- Assessment of strategic alternatives to available, with accompanying analysis, risks, and the requirements for RS plans.
- Impact of indigenous religions on operations.
- Considerations of available RS resources, linked to accomplishment of assigned RS objectives.

ARMY SERVICE COMPONENT COMMAND RELIGIOUS SUPPORT

6-68. The ASCC chaplain manages UMT, chaplain and chaplain assistant, assignments throughout the theater for the commander. The senior chaplain assistant NCOIC advises the senior chaplain on all chaplain assistant assignments. The ASCC command UMT provides training, policy, logistics, and funding for all Army UMTs in the theater and serves as primary liaison with the next higher commander's staff chaplain. The ASCC UMT stays in close communication with senior UMTs in each sector. They participate in joint planning to ensure that the interest of the commander's religious support goals and plan are represented and implemented. They coordinate with the TSC UMT for movement, and transportation unit information and resources. The Personnel Command (PERSCOM) UMT in theater coordinates assignments for UMTs for the TSC command UMT. Functions of the ASCC UMT include—

- Establish liaison with host nations religious leaders and other Joint Religious Ministry Support Teams at the highest levels in the theater.
- Coordinate fund transitions for logistical support.
- Coordinate religious support for force arrival in transfer of authority (TOA) and RSOI sites.
- Coordinate religious support for contractors for the command as part of the commander's policy.
- Provide guidance to all senior Army UMTs to support the combatant commander's RSP.
- Plan and provide training for UMTs.
- Manage UMT assignments within the theater to ensure appropriate denominational balance to all areas of the theater.
- Provide guidance for the employment of all Chaplain Detachments (CDs) and SMART-PC in support of the religious support mission.

CHAPLAIN DETACHMENTS

6-69. When requested from a MACOM, the HQDA G3 will task USARC through FORSCOM to mobilize and deploy the requested chaplain detachments needed to support planned, current, or ongoing operations. The Chief of Chaplains may request that HQDA G3 task USARC to mobilize and deploy chaplain detachments in support of specific missions. (See Appendix G.)

ADDITIONAL CORPS/ECHELONS ABOVE CORPS RELIGIOUS PLAN REQUIREMENTS

SPECIAL OPERATIONS

6-70. The mission, duration and scope of operations, security considerations, signal capabilities, and the desired degree of control determine Special Operations Force (SOF) command relationships. SOF units above group or regimental level are Joint. Joint force commanders include unified, subordinate unified, JTF, functional component, and commanders of specific operational forces. USSOCOM organizes, equips, trains, and provides Special Operations Force for combatant commanders of regionally oriented unified commands (European, Atlantic, Southern, Pacific, and Central) and for the Commanding General, U.S. Forces, Korea. Combatant commanders delegate operational control of these SOF forces to their theater-level Special Operations Commands (SOCOMs). The six SOCOMs are the focal point for in-theater SOF, form nuclei for the Joint Special Operations Task Force (JSOTF), and furnish expertise needed to effectively employ SOF independently or in concert with conventional forces.

6-71. A unified command normally consists of the four service-specific conventional forces and a SOCOM. The combatant commander or the SOCOM may establish a JSOTF commanded by the Joint Force Special Operations Component Commander (JFSOCC). SOF units from the three services are normally under the OPCON of the JFSOCC.

6-72. ARSOF missions are often highly classified and sensitive. To provide counseling, debriefing, and other religious support to ARSOF units, the chaplain must have a Top Secret clearance. Chaplain assistants are required to have a clearance equal to mission classification. The USASOC UMT recommends all ARSOF UMT assignments.

6-73. Normally, the JSOTF chaplain is a member of the service having the majority of forces. (See JP 1-05, Chapter 2.) If the Army supplies the majority of forces, an ARSOF chaplain is normally designated as the JSOTF chaplain.

6-74. The JSOTF chaplain coordinates with the senior chaplain of the next higher headquarters to ensure complete religious support for SOF units. The JSOTF chaplain, in turn, supervises SOF religious support throughout the theater. The chaplain coordinates for additional religious support when other SOF chaplains are unable to provide religious support. SOF chaplains may provide limited area support for conventional forces when it does not conflict with SOF mission requirements

MEDICAL COMMAND SPECIAL MEDICAL AUGMENTATION RESPONSE TEAMS-PASTORAL CARE

6-75. The U.S. Army Surgeon General (DASG) and the U.S. Army Chief of Chaplains (DACH), in consultation with the Chief of Staff of the Army, the Director of Military Support, and applicable Federal and Department of Defense regulations, requires designated DASG/MEDCOM/DACH assigned, attached or associated individuals and organizations to develop and maintain special organized, trained and equipped SMART-PC.

6-76. Operations consist of—

- Alert, assemble, and issue warning order within 6 hours of no-notice notification.
- Deploy from home station within 6 hours of the warning order.
- Conduct mission: supporting commander without organic ministry team or needing additional ministry team support for mass casualty or devastating contingency operation; or provide professional pastoral care augmentation (to include Critical Event Debriefing, Trauma Ministry, Mass Casualty Ministry, and Spiritual Assessment) to local authorities in the management of disaster/mass casualty incidents.

6-77. SMART-PCs deploy with enough food, clothing, personal hygiene supplies, and equipment for self-sustainment in an austere environment for 72 hours. Team missions normally conclude in 72 hours. When extenuating or mitigating circumstances require DASG/MEDCOM or DACH approval for mission extension, SMART-PC members will procure locally or formally obtain logistical support from supported units, or local civil or Lead Federal Agency. SMART-PC members deploy on orders articulated on DD Form 1610.

6-78. At the direction of DACH and/or on order of DASG/MEDCOM, at request of commanders, and/or the request of legitimate civil, Federal, or Defense authorities, using appropriate, recognized, and approved channels designated units will deploy SMART-PC to provide short duration, medical ministry augmentation to:

- Regional, domestic, Federal and Defense agencies responding to disaster, civil-military cooperative action, humanitarian and emergency incidents.
- Support units without organic ministry teams or who require additional ministry team support for mass casualty or devastating contingency operation.

6-79. SMART-PC is an addition to the list of required teams, which includes:

- Trauma/Critical Care (SMART-TCC).
- Chemical/Biological (SMART-CB).
- Stress Management (SMART-SM). (This team has one CH assigned, no assistant.)
- Medical Command, Control, Communications, Telemedicine (SMART-MC3T).
- Preventive Medicine/Disease Surveillance (SMART-PM).
- Burn (SMART-B).
- Veterinary (SMART-V).
- Health Systems Assessment and Assistance (SMART-HS).

6-80. SMART-PC is modeled on the standard Unit Ministry Team of one chaplain and one chaplain assistant. Operationally, multiple teams are task organized based upon METT-TC and medical-ministry/risk analysis to provide the appropriate level of response and technical augmentation to meet mission needs.

MANAGEMENT:

DASG, ASSISTANT SURGEON GENERAL, FORCE PROJECTION
Policy, supervision, and oversight to equip, deploy and redeploy SMART-PC. Validate and resource special equipment requirements; periodically test and evaluate individual SMART-PC teams; develop, monitor, and report readiness reporting requirements. Establish and maintain officer liaison with DACH.

DASG / MEDCOM, ASSIGNED REGIONAL MEDICAL COMMANDS
Organize, train, and equip 2-SMART-PC teams. On order, deploy assigned SMART-PC within assigned regional medical boundaries. Be prepared to deploy outside of regional boundaries in support of other SMART-PC or to other national/international sites.

EQUIPMENT: Configuration of SMART-PC deployment sets, kits, and outfits will be man portable, and utilize commercial backpacks, suitcases, and rugged containers. Each equipment piece will meet commercial airframe size, weight, and cube requirements.

MEDICAL COMMAND SPECIAL MEDICAL AUGMENTATION RESPONSE TEAM-STRESS MANAGEMENT

6-81. The role of the chaplain assigned to a SMART-SM is to bring spiritual aspects of stress management, as a clinician, in support of the team's overall tasks, which include:

- Critical event debriefings.
- Spiritual counseling.

CONTINGENCY FORCE POOL (CFP)

6-82. The FORSCOM Chaplain's Office will coordinate with CONUSA Chaplains where CFP chaplains are attached for duty to ensure that chaplain personnel are integrated into MSCA contingency training and planning scenarios, and are included in contingency plans. The FORSCOM Chaplain's office will coordinate with the CONUSA Chaplains to de-conflict training and planning objectives. FORSCOM will ensure that CFP personnel support is placed in the Regional Planning Agents (RPA) MSCA handbook, in the DOD database to the Defense Coordinating Office (DCO) and integrate CFP training / certification with DOD Emergency Preparedness Training and CONUSA MSCA contingency exercises. Upon request, FORSCOM has the ability to cut orders for CFP personnel using ARPERCOM Orders and Resource System (AORS) and can make one or more CFP personnel immediately available depending on the crisis level. In a small-scale contingency operation (SSCO) OCONUS response, CFP personnel will report to the CONUS MACOM for contingency support, providing religious support and liaison / coordination. During MSCA or Homeland Defense / Security (HLD / HLS) CONUS contingencies, upon request by a DCO and with appropriate funding for RS to DOD personnel, CFP personnel become assets to the JTF Chaplain.

ROLE OF THE CORPS/ECHELON ABOVE CORPS UNIT MINISTRY TEAMS

6-83. The personal delivery of religious support by the UMT will be imperative in all operations. It is essential that the Corps/EAC HQs UMT is able to rapidly tailor and modularize UMTs for operations. Corps/EAC UMTs must organize to optimize and use information. Flexibility in applying METT-TC is critical to accommodate specific missions. Rank structure and composition of UMTs may change to support specific missions. RS planning must be robust and flexible enough to provide religious support across the full spectrum of military operations. Corps/EAC UMTs require the same level of technology as the supported force. As technology advances the UMT automation, communication and transportation requirements must keep pace with the command's overall requirements.

GARRISON OPERATIONS

6-84. During garrison operations, all EAC UMTs, the Corps UMT, and subordinate UMTs support the training program of the installation chaplain. They implement their CMRP to support the religious needs of the families and soldiers. The EAD UMTs manages and supervises subordinate UMTs assigned for the commander. They support the commander's training program through participation and planning and implementing UMT training. The Corps UMT is a mentor to the BDE UMTs and will execute a mentorship program. The Corps UMT supports the ASCC UMT RS plan.

SUMMARY

6-85. Corps and EAC UMT role is crucial to the warfighting religious support mission success. AC and RC are fully integrated to supply combat ready UMTs to provide religious support to the full spectrum force that meets the needs of the joint force commander in war, conflict and peace. Command UMTs are fully trained, ready, and staff integrated to meet the religious support needs of soldiers, family members, and authorized civilians in all operations. The UMTs are strategically and operationally responsive and prepared to conduct prompt and sustained religious support operations as part of joint, multinational, and interagency teams.

RESOURCES

6-86. The following is a list of main resources utilized for writing and updating this doctrine.

- AR 165-1, Chaplain Activities in the U.S. Army
- JP 1-05, Religious Ministry Support for Joint Operations
- JP 5-00.2, Joint Task Force Planning Guidance and Procedures
- JP 3-05, Doctrine for Joint Special Operations
- JP 3-0, Doctrine for Joint Operations
- JP 3-07.3, Joint Tactics, Techniques, and Procedures for Peace Operations
- JP 1-0, Doctrine for Personnel Support to Joint Operations
- JP 3-08, Interagency Coordination During Joint Operations

- JP 3-10, Joint Doctrine for Rear Area Operations
- FM 1 (100-1), The Army
- FM 16-1, May 1995, Religious Support
- FM 3-0 (100-5), Operations
- FM 25-100, Training the Force
- FM 25-101, Battle Focused Training
- FM 100-15, 29 OCT 1996, Corps Operations
- FM 100-11, 15 JAN 1998, Force Integration
- FM 100-9, 13 JAN 1992, Reconstitution
- FM 100-8, 24 NOV 1997, The Army In Multinational Operations
- FM 100-7, 31 MAY 1995, Decisive Force: The Army in Theater Operations
- FM 100-6, 27 AUG 1996, Information Operations
- FM 22-100, Army Leadership
- DA Pam 165-3, Chaplaincy Training Strategy
- National Military Strategy
- Joint Vision 2010
- Joint Vision 2020
- Army Vision 2010
- Concept for Future Joint Operations, May 1997
- U.S. Army Chaplaincy Strategic Plan FY 2000-FY 2005
- Chaplain Support Force XXI White Paper
- Lessons Learned from past and current operation
- Lessons Learned from the training centers

Chapter 7

INSTALLATION UNIT MINISTRY TEAM RELIGIOUS SUPPORT OPERATIONS

7-1. The Army, Active and Reserve components, is designed, equipped, and trained for force projection operations. Army installations function as force projection platforms, providing support for deployed tactical units throughout all stages of force projection. To ensure readiness, the Army continually plans and trains for the full range of military operations.

CONTENTS

7-2. Combat readiness also produces an Army whose organization, skills, leadership, and discipline can advance U.S. interests in a variety of operations not characterized as war. These operations include, but are not limited to, nation assistance; humanitarian and disaster relief; counter drug operations; peace operations; antiterrorism; shows of force; attacks and raids; noncombatant evacuation operations; insurgency and counterinsurgency support; civil disturbance and support to domestic authorities.

7-3. Force projection is the military component of power projection and is a critical component of U.S. strategy. AC and RC units, the mobilization base, DA civilians, and the industry are directly involved projecting the force anywhere in the world. Training, mobilization, deployment, employment, sustainment, and redeployment are the processes utilized for force projection. They are continuous, overlapping, and interactive throughout an operation. The process of force projection is an integral part of the Joint Operation Planning and Execution System (JOPES). JOPES is used to monitor, plan, and execute mobilization, deployment, employment, sustainment, and redeployment activities associated with joint operations. The Army force projection process is imbedded within this system.

7-4. Religious support is a vital part of the combat readiness mission of the installation in force projection. Force projection operations are inherently joint and religious support planning and synchronization are required.

INSTALLATION DESCRIPTION

MISSION

7-5. Army installations, which include STARCs and Reserve Support Command (RSCs), contribute to the Army's ability to support the National Military Strategy (NMS). Peacetime missions include: counter-drug operations on our national borders, nation building, counter-terrorism, WMD support, deployment of land forces for a conflict, or fighting and winning the nation's war. Each installation, STARC, and RSC has a primary mission that determines its assignment to a MACOM or Component.

ORGANIZATION

7-6. Installations may host dozens of organizations that are either part of the MACOM's mission or simply tenants assigned to the installation for support. Installations are organizationally characterized by the nature of the major activity they host. For force basing they would have Corps and Division(s). For unit training they would host such units, as Combat Training Centers (CTC), for example, the Joint Readiness Training Center (JRTC). For the logistic support they would provide depot and commodity management. Within the Army, an installation may be referred to as a post, camp, station, fort, barracks, subpost, depot, arsenal, proving ground, base, laboratory, port, or ammunition plant. For the Army National Guard and Reserve Component the term "installation" refers to the RSC for the RC and the STARC for the ARNG.

As Part Of A Major Army Command

7-7. Army installations are assigned to and operate under commanders of Major Army Commands (MACOMs) or MSCs where applicable, such as in the Army Materiel Command (AMC), and the U.S. Army Reserve Command (USARC). STARCs operate under the guidance of the National Guard Bureau. Installations consume, provide, maintain, and otherwise control significant resources while executing DOD, DA, and MACOM mandated policy programs. These programs are reviewed, analyzed, and resourced at DA level. Transformation Installation Management (TIM) will eventually transform the relationships between installations and MACOMs, to include, the relationships between MACOM Chaplains and Installation Chaplains. UMTs should keep abreast of such changes.

As Part Of Full Spectrum Operations

7-8. The entire installation management system is a key component of developing and sustaining a total force capable of meeting the full spectrum of missions required by the National Military Strategy (NMS). The Army's mission to expand and rapidly deploy trained and ready units is dependent upon installation capabilities. Within the installation support system is the Army's mobility infrastructure. Total force readiness is linked to installation operations.

KEY TERMS

- Joint Operation Planning Execution System: (JOPES)
- Area Support Group: (ASG)
- Port of Embarkation: (POE)
- Arial Port of Embarkation: (APOE)
- Seaport of Embarkation: (SPOE)
- Port of Debarkation: (POD)
- Aerial Port of Debarkation: (APOD)
- Seaport of Debarkation: (SPOD)

7-9. Army installations are standardized to ensure that operations can be effectively supported in peacetime. This effectiveness must continue throughout the transition from peacetime to wartime (mobilization) and during post conflict (post-deployment). Army installations provide the sustaining support necessary for units to accomplish their missions.

7-10. Installations provide this support by accomplishing both their primary and support missions. The primary mission of the installation includes, but is not limited to, readiness and preparation for operational deployment, training, supply, and/or maintenance. The support mission includes real property acquisition and upkeep, community service, and other personnel support.

7-11. Sustainment includes the primary and support missions of the installation. Throughout each phase of deployment, the installation manages the sustainment of units.

FUNDAMENTALS OF INSTALLATION RELIGIOUS SUPPORT OPERATIONS

7-12. Military chaplains and chaplain assistants are specially trained and assigned to provide for the government's interest to protect free exercise of religion rights in the military, which is a unique pluralistic and multi-faith environment. Military chaplains and chaplain assistants assigned in Table of Distribution and Allowances (TDA) positions provide direct support to the military commanders for the installation religious program, Family Life Centers, units, hospitals, initial entry training (IET) units, Chaplaincy resources management, Chaplaincy personnel management, Chaplaincy force structure and training. The content of the religious support mission of the chaplain and chaplain assistant is no different in the TDA than the TOE; only the command context changes. It must be understood that the Chaplaincy is not a social service under the city management context of the Installation, but a direct support function to the commander. Religious support is the military commander's mission. TDA positions are required to develop military religious support leaders and train religious support skills for the full spectrum of military operations, from the sustaining base to the forward edge of the battlespace.

7-13. Installation religious support has three major areas:

- To prepare and assist soldiers and families for military operations.
- To provide religious support for units and personnel on the installation.
- To sustain religious support operations during deployments and mobilizations.
- To train UMTs for comprehensive RS.

7-14. The installation chaplain is responsible for all religious support on the installation including all religious services and activities. The installation chaplain coordinates with tenant units to implement the CMRP. The installation's UMTs provide a broad religious support programs for units, soldiers, and their families. Worship services, pastoral care, religious education, and spiritual fitness training are provided for the religious support needs of the installation population.

READINESS

7-15. Military chaplains and chaplain assistants in the TDA are necessary to support peacetime deployments, as well as, to man operational units in mobilization and provide wartime rotation, and rapid casualty replacement. All TDA military chaplain and chaplain assistant positions are required to be trained and ready to use combat skills and training to implement the military religious support mission of the commander in any contingency.

7-16. The Reserve Component (RC) holds a significant portion of the mobilization assets for the Chaplaincy. The RC Chaplaincy is a military combat augmentation in direct support of the commander. RC chaplains and chaplain assistants in the TDA are necessary to support peacetime deployments, as well as, to man operational units in mobilization, provide wartime rotation and rapid casualty replacement. All RC TDA chaplains and chaplain assistants are required to be trained and ready to use combat skills and training to implement the religious support mission of the commander in any contingency. The RC Chaplaincy provides the recruitment base for both the AC and RC. The RC provides the training base for the Chaplain Candidate Program. The IMA and Active Guard Reserve (AGR) chaplains and chaplain assistants are in the TDA structure.

Assessment of Unit Ministry Team Readiness

7-17. As the commander's personal staff officer for religious support, the command installation chaplain is responsible for assessing the technical proficiency of UMTs in subordinate and tenant units. They conduct assessments utilizing the units training program such as during field training exercises (FTXs), external Army Training and Evaluation Program (ARTEP) events, unit training, Reserve component annual training and Command Inspections. Observer controllers may provide additional assessment assistance when the UMT is involved in Battle Command Training Program (BCTP) events or Combat Training Center (CTC) rotations. They acquire, in coordination with the senior unit UMT lessons learned to integrate into the installation UMT training program. (See DA PAM 165-3, The Chaplain Training Strategy.)

7-18. Chaplains and chaplain assistants participate in the installation's Command Master Religious Plan (CMRP) to help soldiers develop the personal spiritual readiness needed to sustain them during combat. Spiritual readiness is also essential to sustain soldier families while the unit is deployed. As part of the installation's CMRP, the UMTs conduct worship services, offer religious education, and provide spiritual fitness training and other activities in installation chapels and facilities.

Training

7-19. Training is the cornerstone of preparation for military operations. The Army's primary activity during peacetime is realistic, battle-focused training oriented on the unit's mission essential task list (METL). Installation chaplains are responsible for individual training, MOS 56A and 56M. The installation chaplain's training managers coordinate with subordinate UMTs and unit S3/G3s and ensure that UMTs have adequate technical and tactical training for deployment and combat. This is implemented through the development and execution of an annual Installation wide UMT training strategy and plan (the religious support METL provides training tasks along with the Chaplain Officer and Enlisted Critical Task Lists). This training is integrated within the implementation of the Chief of Chaplain Training Strategy (DA Pam 165-3) and higher headquarters training strategy.

7-20. Tactical proficiency is a demonstrated understanding of the Army's war-fighting doctrine and tactics. To provide effective religious support in combat, the UMT must know its unit's doctrine, tactics, techniques, and procedures. The UMT must also maintain the requisite skills of field craft essential for survival in the battlespace. Tactical training is primarily the responsibility of the unit to which the UMT is assigned or attached.

7-21. Technical proficiency is the demonstrated ability to perform all tasks related to one's functional area. Technical proficiency is required for both garrison and field missions. Technical proficiency is acquired from schools, self-study and training with other UMTs in performing the unit's and installation's religious support mission. Training and continuing education conferences, functional courses offered by USACHCS, and courses offered through the AMEDD Center and School, and the NCO Education System (NCOES) supplement the installation religious support training program.

7-22. UMTs must balance religious support training for combat and for garrison. The UMT must participate fully in both unit and religious support training to assure their tactical and technical proficiency. If training for either combat or garrison operations is neglected, ministry teams will not be prepared to provide comprehensive religious support to soldiers and their families.

7-23. Many installations serve as mobilization bases. (See Figure 7-1.) Mobilization can overwhelm the installation's resources. Thorough planning and realistic training reduces confusion. Mobilization training gives UMTs an opportunity to practice their mobilization religious support missions and an opportunity to assess the adequacy of mobilization plans. Installations assist with the following RC UMT training events:

- Annual training (AT).
- The Individual Mobilization Augmentee (IMA) training.
- Individual training events, Individual Ready Reserve (IRR), and chaplain candidate training.
- AC/RC Sustainment Training.
- Annual Chaplaincy Regional Sustainment Training (CREST) as mandated by the Chief of Chaplains.

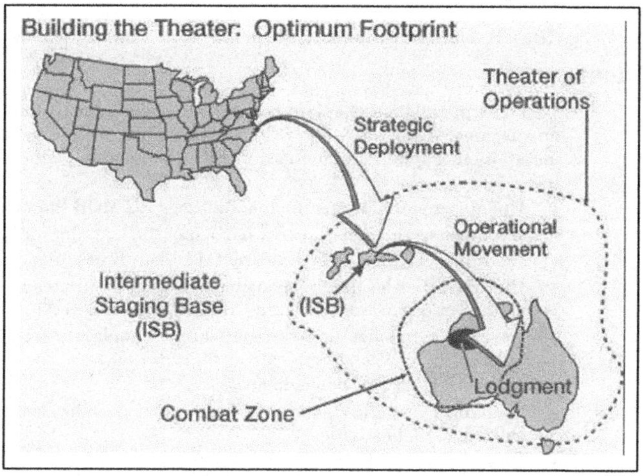

Figure 7-1. Building the Theater

FORCE PROJECTION

7-24. When the installation is serving as a force projection platform, the installation UMT must be prepared to provide as a staff the religious support mission. The installation chaplain's staff must be prepared to support all phases and sustainment requirements of the mission. Provision is made for religious support for all phases of deployment including pre-deployment, movement to the POE, strategic lift, theater reception, and theater onward movement. This includes counseling, training of UMTs, leadership training, reunion briefings and counseling services (and as a family readiness group resource). Installation UMTs will support peacetime deployments, as well as, to man operational units in mobilization, provide wartime rotation and rapid casualty replacement.

7-25. Force Projection planning includes—

- Mobilization, receiving, housing, processing, validating, deploying of soldiers including UMTs.
- Support the increased information communication requirements that will occur.
- Assist in reception as the force shifts MRC priority and re-deployment of forces.
- Maintain installation religious support requirements for the remaining military community after deployment.
- Training of UMTs as required.

MOBILIZATION PHASES

(See CH 3 for more information.)

PLANNING PHASE

7-26. The installation chaplain becomes the mobilization station chaplain and assumes a key role when RC UMTs mobilize. Before mobilization, the installation chaplain designates a mobilization planner who does the following:

- Updates and maintains the installation's RC MOB Plan book.
- Updates the religious support portion of the installation MOB PLAN.
- Identifies positions to be filled by IMAs or retirees.
- Clarifies cross-leveling responsibilities and coordinates with the G1/Director for Personnel and Civilian Affairs (DPCA).
- Develops a mobilization-training plan for chaplains and chaplain assistants.
- Prepares UMTs for deployment.
- Prepares for and participates in installation mobilization exercises (MOBEX) and activities.
- Plans for demobilization activities such as after action reviews, reunion briefings, awards, reports, etc.
- Maintains and improves combat readiness posture.
- Ensures all sites, such as RSOI, are fully manned by UMTs during all phases of mobilization, deployment, and redeployment.
- Family Readiness Group (FRG) religious support.
- Prepare the UMT MOBTDA.
- Coordinates with and tracks RC units with and without UMTs scheduled to report to the installation.
- Communicates with IMAs and retirees with orders to report to the installation upon mobilization.

7-27. This phase concerns all AC and RC UMTs efforts at all levels during peacetime to plan and train for assigned wartime tasks. The installation command UMT should ensure that the installation RS mobilization plan addresses civilian support for the mission. They must review the plan annually ensuring it is up-to-date and fully integrated into the installation commander's mobilization plan.

7-28. STARC UMT mobilization plans and SOPs clarify training and MOBTDA personnel to provide religious support to families at family assistance centers (FACs). Important consideration must be given to identifying the religious support requirements of the FACs and developing MOBTDAs with sufficient UMT personnel support.

7-29. The CONUSA UMT coordinates between the FORSCOM staff UMT and RC UMTs within its geographic area. CONUSA staff chaplains and NCOICs may redistribute UMT members in order to match anticipated UMT mobilization requirements.

ALERT PHASE

7-30. This phase begins when an AC or RC unit receives notice of a pending mobilization order. During this phase, the STARC and Army Reserve Command chaplains and NCOICs continue to cross-level individual members of UMTs within their state or region to bring alerted units to required religious support deployable status.

HOME STATION PHASE

7-31. This phase begins with the RC unit's entry on active federal duty or the AC's preparation for deployment. Actions are preformed to bring units to active status. Installation, STARC, and USARC Command UMTs continue cross-leveling, as well as, ensuring critical UMT logistical support is in place.

MOBILIZATION STATION PHASE

7-32. This phase begins when the AC/RC unit arrives at the MS or mobilization site. The unit UMT is validated to meet deployment unit validation criteria resulting in assuring religious support mission capability. Necessary chaplain and chaplain assistant individual and collective training are conducted. Timeliness is critical. Integration to the Soldier Readiness Program (SRP) for preparation of overseas movement (POM) activities for religious support input is conducted.

PORT OF EMBARKATION PHASE

7-33. This phase begins with the arrival at the APOE or SPOE. Religious support is critical to ensure the forces are spiritually prepared for deployment and combat. Worship, counseling, and other religious support tasks are initiated. UMTs departing are ensured spiritual support.

DEPLOYMENT PHASE

7-34. In addition to routine religious support operations, the installation's TDA UMTs have the responsibility to plan and to prepare to sustain religious support operations on the Installation throughout the mobilization, deployment, and redeployment and demobilization process.

7-35. The senior UMTs of the deploying units and the TDA UMTs coordinate plans for religious support during deployment. Planning considerations include the following:

- The care of families and authorized civilians.
- Procedures for consolidating chapel programs.
- The close-out of chapels (if necessary).
- The accounting, turning-in, or securing of unit and installation property.
- Survivor and casualty assistance.

7-36. FORSCOM Regulation 500-3-2 (FORMDEPS II), Deployment Guide, provides guidance and assigns responsibilities for support of the deployment of units.

7-37. UMTs of non-deployed units are included in installation plans to support deploying units. A garrison support unit (GSU) may be available to augment the installation's UMTs during deployment and mobilization.

Along with the installation's non-deployed UMTs, they assist with religious support to families until the deployed forces return.

7-38. A thorough understanding and appreciation for the Transfer of Authority of deploying units to the theater commander is critical to the installation chaplain's mission. At a time designated in the deployment order, the chain of command shifts to the gaining unit, usually the theater commander. When that TOA occurs, the relationship of the deploying UMTs to the installation's UMTs changes correspondingly. This new relationship remains in effect until the unit redeploys.

SUSTAINMENT PHASE

7-39. To sustain the force projection platform, the installation UMT planner must consider all sustainment requirements for the non-deployed units, units at all deployment stages, and family members of both military and authorized civilians. The continued training and readiness posture is maintained for sustaining the forward religious support mission.

Sustainment: Family Readiness Groups

7-40. Command sponsored Family Readiness Groups (FRGs) are a vital link between soldiers, their families, the command, and the unit. Volunteer family members lead FRGs. Each deploying unit establishes its own FRG. Establishing the FRG prior to the deployment alert is extremely helpful in developing support plans and operational procedures for providing assistance to the unit' families. Chaplains and chaplain assistants support these groups.

Sustainment: Casualty Notification

7-41. Notification of the next of kin (NOK) is the commander's responsibility. Chaplains are not notification officers, but they may be a part of the team that makes the notification. In the event of a disaster or mass casualty situation, they may serve on the Casualty Assistance Team. Religious support to grieving families is a priority for UMTs and appropriate training in such religious support is a necessity.

REDEPLOYMENT AND DEMOBILIZATION PHASE

7-42. Religious support during the redeployment and demobilization phases consists of the spiritual and emotional preparation of units from an area of operation to follow-on designated CONUS or OCONUS bases. Redeployment religious support must be planned and executed to insure the proper allocation of resources of an installation. Planning must encompass all phases of an AC force returning to its home station, mobilized RC forces, and individuals through DMS back to their home stations.

7-43. This phase begins with the arrival of forces at the mobilization station/installation. The installation UMT supports the commander in receiving the forces, receiving sustainment equipment and supplies, and helps develop a reception and reconstitution or onward movement plan.

7-44. Demobilization is a critical factor for strategic reconstitution. The installation staff UMT plans and executes the commander's demobilization plan. Reunion support to soldiers, authorized civilians, and families are given to ease the transition from mobilization/deployment to demobilization/redeployment. Counseling, both individual and group, workshops, and worship are critical for reconstitution. Demobilization of some units may occur at the same time other units are being mobilized, deployed or redeployed. Religious support planning must encompass a possible ongoing planning and execution process.

7-45. Religious support plans encompass—

- Reunion briefing for family members and friends.
- Reunion support for returning units.
- Back briefing from returning and TDA UMTs.
- Religious support to garrison command and staff.
- Critical event debriefing.

DOMESTIC SUPPORT

7-46. 7-46. A presidential declaration of a major disaster or emergency usually precedes Domestic Support Operations (DSO). The installation staff UMT plans and executes the commander's domestic support mission. As in other operations, the UMT mission does not change. The installation becomes the force projection for Army forces to the area of operation. DOD Directive 3025.15, Military Assistance to Civil Authorities (MACA) provides directives and guidance in responding to natural and manmade disasters and includes military assistance to civil disturbance, counter drug activities, combating terrorism, and law enforcement. Installation UMTs must know their role and limitations in this type of operation. Installation UMTs support local, state, and federal civil authorities (see JP 3-07.7; JP 1-05; FM 100-20).

ROLE OF THE INSTALLATION UNIT MINISTRY TEAM

7-47. The Installation Chaplain and staff provide seamless religious support to the non-linear battlespace extending from the CONUS deployment and sustaining base to the forward edges of the battlespace. The religious support mission extends throughout the full spectrum of military operations. The community religious support mission encompasses all military units, activities, organizations, and authorized personnel.

7-48. The importance and influence of the Chaplaincy to the spiritual and moral health of the unit (to include the sustaining base or garrison) and in spiritual matters has been valued throughout the history of the Army. Commanders recognize the value of the chaplain and the total UMT both in combat and garrison.

RESPONSIBILITIES AND ACCOUNTABILITY

Staff Responsibilities

7-49. The Installation Chaplain provides support and advice to the commander and staff and to all installation agencies on all matters of religion, ethics/morals, and morale as affected by religion. The Installation Commander is responsible for the religious support and spiritual fitness of soldiers, family members, and authorized civilians. This includes insuring comprehensive worship opportunities that are part of the CMRP.

7-50. Army chaplains and chaplain assistants are the agents of trust representing the Army's values and the needs of the military in the commander's religious support program. Chaplains and chaplain assistants are required to provide religious support in all military units, activities, organizations, and communities.

CHAPLAIN (O-6)	***CHAPLAIN ASSISTANT NCO (E-7/ 8)***
Role Commissioned Officer/Staff Religious Professional Title 10 Protected	***Role*** Noncommissioned Officer Specialized Assistant Religious Support Specialist
Mission Advises commander on religious support and quality of life issues to support the unit's mission. Performs/provides/coordinates the religious support mission for the command. Installation Director of Religious Support	***Mission*** Advises Chaplain and NCO Chain. Provides specialized assistance in areas of religious support to support the chaplain. Perform Garrison Staff functions. Installation Religious Support NCOIC.
Parallel Tasks 1. Supervise, mentor and assess subordinate Chaplains. 2. Recommend personnel assignments. 3. Supervises the contingency, mobilization plans. 4. Responsible for religious support training and professional development.	***Parallel Tasks*** 1. Supervise, mentor and assess subordinate chaplain assistants. 2. Recommend Chaplain Assistant assignments. 3. Prepares the contingency, mobilization plans. 4. Plans and conducts religious support training.

7-51. Installation Chaplains have staff responsibility for the following:

- Development and execution of the Command Master Religious Program under the direction of the commander; which includes all worship services, memorial services, religious activities, professional training, religious education, spiritual fitness and moral leadership training, mobilization plans and contingency operation planning, and pastoral support for soldiers, families, and authorized civilians.
- Conduct of officer personnel management and coordination enlisted personnel actions for Unit Ministry Team (UMT) assignments and professional development.
- Development of an annual training plan for chaplain and chaplain assistant training.
- Development and execution of the installation commander's suicide prevention program.
- Chaplain Family Life Centers which provide comprehensive programs for training UMT members, and family programs, such as, marriage and family counseling and family enrichment programs.

- Administration and management of the installation non-appropriated chaplains' fund.
- Supervision of the use of chapels and religious facilities to include planning for new and renovated religious use facilities.
- The Installation Chaplain is the personal staff officer to the commander for the religious support portions of planning, programming, force structure actions, budget input, crisis management, and force protection.
- Provide the commander unit spiritual fitness assessments.
- Support the commander's METL through implementing a religious support installation METL.

7-52. The TDA force structure is required in the active and reserve component to maintain adequate military rotation and surge capability for mobilization and deployments and for military career progression of chaplains and chaplain assistants. These TDA positions are required to train and develop military religious support leaders in religious support skills for the full spectrum of military operations.

Relationship To The Chief Of Chaplains

7-53. The Installation Chaplain section implements the Chief of Chaplains (CCH) Strategic Plan and guidance on providing religious support across the full spectrum of operations from the power projection base to the theater of operations and return. They plan and execute the Chaplaincy mobilization and contingency plans. The installation staff UMT implements the CCH training strategy for individual UMT members and collectively. They train and prepare UMTs in support of the CCH guidance for the Army.

Relationship To Major Army Commands

7-54. Installations are part of Major Army Commands (MACOMs) or MSCs. The installation will coordinate, review, and recommend assignment of chaplains and chaplain assistants. They will support the MACOM or MSC staff chaplain contingency plans. Also, they will implement the religious program as set forth through the command. Again, as Transformation Installation Management takes effect, the role between Installation Chaplains and MACOMs will likely change. UMTs should keep abreast of such changes.

SUPERVISORY RESPONSIBILITIES

7-55. The assignment instructions received by every chaplain at the time of his/her receipt of permanent change of station (PCS) orders, contains the following statement: "Chaplain will perform and/or provide direct and general religious support in accordance with Unit Religious Support Plan and Command Master Religious Program." The installation staff UMT establishes staff supervision and implements the CMRP. They supervise the training of UMTs for the CCH in cooperation and coordination with tenant command UMTs. They supervise garrison religious support force structure for the commander, both active and civilian. On an installation or in a community, the staff chaplain has overall staff responsibility for the religious support mission. In fulfilling that responsibility, the staff chaplain will coordinate and use total UMT assets from both the TDA and TOE units. The installation chaplain assistant NCOIC advises the chaplain on chaplain assistant assignments, training, and other critical support issues.

SPECIALIZED RELIGIOUS SUPPORT

7-56. The Army is committed to supporting its families during routine operations and during deployment. Commanders employ the full range of installation resources to support families.

7-57. The provision of religious support to families is the responsibility of the installation chaplain. The Family Life UMT and the hospital UMT assist in their related areas in providing support. When not deployed, each unit's UMTs also have a responsibility to support their unit's families.

7-58. When units are deployed the RC will provide UMTs to assists in sustaining support to families. UMTs support the commander's Family Readiness Group (FRG) initiatives and also provide a vital role in supporting families when the unit's soldiers are deployed. Though not specifically a UMT responsibility, FRGs have a traditional relationship with the UMTs.

FAMILY LIFE UMT

7-59. The Family Life Chaplain's primary mission is to train UMTs and to direct the Chaplain Family Life Center (CFLC). Family life ministry consists of three major elements: enrichment, prevention, and intervention. These elements are expressed through educational programs, family wellness development programs, and marriage and family counseling.

7-60. The Family Life Chaplain consults with senior chaplains to determine the specific training needs for subordinate units. Each Family Life Chaplain tailors the training program to meet the needs of the installation and the mission of the tactical units. Training could be in the form of technical assistance, educational resources, and specific training for UMTs. In addition, the Family Life Chaplain helps UMTs prepare families for the stress of mobilization and deployment. This support is directly related to the installation's mission as the force projection platform. Each Family Life UMT supports the UMTs of the tactical units on the installation. They will tailor the training program to meet the needs of the installation and the mission of the tactical units on the installation. For example, a Family Life UMT will consult with a Division UMT to determine the division's specific training needs in family life issues. By training the Division's UMTs in family issues, the Family Life Chaplain serves as a direct link between the installation chaplain and the UMTs in sustaining the force. Training topics include—

- Ministry to FRGs.
- Family stress.
- Communication skills.
- Family counseling skills.
- Reunion after deployment.
- Grief and loss.

7-61. A healthy home environment directly influences the soldier's performance of duty. With its support of the family, the family life ministry program sustains the soldier's spiritual fitness in peacetime and during combat.

HOSPITAL UMT

7-62. The Army Medical Department preserves the fighting strength of the Army, maintains the health of soldiers, and supports family members with medical services.

7-63. Chaplains and chaplain assistants assigned to hospitals help patients and families deal with medical problems and crises. Hospital chaplains at all levels have special training and skills that equip them to do the following:

- Assist people with feelings of fear, loneliness, anger, failure, and loss of faith.
- Provide pastoral care for those suffering from critical or chronic illnesses or injuries.
- Prepare religious support contingency plans for mass casualty situations.
- Facilitate support groups for patients, their families, and staff.
- Serve as members of interdisciplinary case management teams and hospital committees.
- Provide pastoral care to hospital staff.
- Provide training for staff members and UMTs.

CONFINEMENT UMT

7-64. Most Army installations do not have local installation confinement facilities. Soldiers are usually held in confinement by special arrangement with local authorities. The installation staff chaplain provides pastoral care for locally confined soldiers in keeping with the guidance of the installation commander. The United States Disciplinary Barracks (USDB) is provided a UMT for RS. UMTs provide religious support for both prisoners and facility staff members. This religious support includes performing or providing for the following:

- Worship
- Religious education
- Counseling
- Moral leadership
- Spiritual fitness training
- Family support

GARRISON SUPPORT UNITS

7-65. Garrison Support Units (GSU's) come from the RC to support base operations (BASOPS) during continuing operations, and assist in other installation missions during Annual Training (AT) periods. Each GSU has UMT assets within their TDA. (See The Chief of Chaplains Crisis Management UMT Assets Handbook, October 2000.)

7-66. When installations are called upon to act as projection platforms, they require augmentation to meet their force projection responsibilities. The installation's increased religious support mission also requires additional support and augmentation. The scope of the installation's religious support mission encompasses support to soldiers remaining on the installation, soldiers processing through the installation; families of soldiers assigned to the installation, families of soldiers processing through the installation, and increased religious support to the installation hospital. The requirement for religious support augmentation will increase as the level of mobilization increases.

Training.

7-67. The supported Installation Chaplain will provide training to the GSU UMT personnel.

Organization

7-68. UMTs will be embedded in the GSU. The number of UMTs required varies from installation to installation. The number of UMTs is determined by applying an allocation rule to the number of projected personnel associated with the each installation's mobilization mission.

SUMMARY

7-69. The installation staff UMT is a critical part of the overall contingency plans of the commander. They provide a trained ready force of UMTs to meet the full spectrum operational mission of the U.S. Army. To adequately support the future requirements of an Army capable to conducting prompt and sustained operations, installation UMTs must plan and execute the religious support mission in support of the National Military Strategy, the OCCH (Office of the Chief of Chaplains) strategic plan, and the installation commander's CMRP. The installation staff UMT must—

- Support the UMTs assigned to war-fighting units.
- Focus on Army and Chaplaincy core competencies.
- Implement the command CMRP for readiness, mobilization, and deployment.
- Provide quality and effective training for UMTs.
- Maintain stewardship of religious support assets.
- Provide the projection base for force projection and quality ministry for families and tenant units.

RESOURCES

7-70. The following is a list of main resources utilized for writing and updating this doctrine.

- AR 165-1.
- JP 1-05.
- JP 5-00.2.
- JP 3-0 (100-5).
- JP 3-10.
- FM 16-1.
- FM 25-100.
- FM 25-101.
- FM 100-11.
- FM 100-9.
- FM 22-100.
- FM 100-20.

- FM 100-19.
- DA Pam 165-3.
- National Military Strategy
- Joint Vision 2010
- Army Vision 2010
- U.S. Army Chaplaincy Strategic Plan FY 2000-FY 2005
- Lessons Learned from past and current operation
- Lessons Learned from the training centers
- The Chief of Chaplain's UMT Crisis Assets Handbook, October 2000

Appendix A

RELIGIOUS SUPPORT IN CIVIL MILITARY OPERATIONS

Throughout U.S. history, chaplains have supported commanders through advisement on matters pertaining to the interface between Religious Support Operations and Civil Military Operations (CMO). This appendix provides the necessary guidelines for chaplains to engage in this role as required by their commanders. (See Appendix G for discussion on CD-C) UMTs are reminded that CMO support is a secondary responsibility and that the personal delivery of religious support is always the UMT's imperative.

ADVISING THE COMMANDER

A-1. Chaplains will support the commander through advisement in the following areas that may influence CMO:

- Indigenous religious, ethnic, and cultural influences.
- Beliefs, practices, and customs of religious groups in the AO.
- Religious issues related to displaced civilians.
- Human welfare needs as requested by religious NGOs and IOs.
- Relations with indigenous religious leaders when directed by the commander.

A-2. Under Title X of the U.S. Code, Chaplains should not perform the following:

- Direct participation in negotiations or mediations as sole participant.
- Human intelligence (HUMINT) collection and/or target acquisition.

PROVIDE AND PERFORM RELIGIOUS SUPPORT

A-3. Chaplains will provide worship and pastoral care in Civil Military Operations in accordance within the following guidelines:

- The priority of responsibility for Army Chaplains is to provide worship opportunities and pastoral care to soldiers and authorized civilians in the AO.
- Chaplains will provide for pastoral counseling, visitation, religious education, spiritual fitness training, moral leadership training, critical event debriefing and the provision of religious literature and items.
- Chaplains will identify and notify the command of meaningful opportunities for soldiers to participate in Civil Military Operations to promote the morale and spiritual well being of deployed soldiers.

- In coordination with the Civil Military Operations Center, chaplains may provide certain religious support to refugees and displaced persons when directed by the Commander, and after consultation with Staff Judge Advocate (SJA). In such cases, it is critical to avoid any activities that can be construed as proselytizing among refugees or displaced persons.
- Every effort to include indigenous religious leadership to meet these needs should be taken.
- *CMO and RS tasks will need to be coordinated to ensure success.*

CMO Tasks	Religious Support Tasks
Advise commander of legal obligations and moral considerations.	Advise commander of moral considerations as affected by religion.
	Advise command on accommodating religious practices in AO.
Establish policy for the handling of dislocated civilians.	Monitor care of displaced civilians to ensure it is moral/ethical.
Plan displaced civilian operations.	
Coordinate displaced civilian operations.	Advise command on religious support plan for displaced civilians.
Provide displaced civilian support.	
Plan humanitarian/civic assistance activities.	Plan religious support activities. Assist CMO personnel in coordinating with local religious leaders for implementation of CMO activities.
Provide public welfare support.	Assist command in meeting human welfare needs in conjunction with religious NGOs & IOs –clothing, food shelter.
Coordinate health, welfare, and morale services.	
Provide civilian supply support.	
Provide food and agriculture support.	
Provide information on cultural considerations.	Advise leadership on beliefs, practices, and customs of religious groups in AO
Provide cultural affairs support.	Advise CMO personnel on the role/influence of religion in cultural affairs.
	Assess distinct religious, cultural, ethnic, and gender ministry needs
	Brief the commander of religious influences on contingency operations
Provide language support.	
Provide arts, monuments and archives support.	Advise command on impact of local religions, sites and facilities which affect the mission
	Advise command on the training of soldiers with respect to religious beliefs, practices, sites and facilities

CMO Tasks	Religious Support Tasks
Provide legal support.	
	Implement religious support in a Combined and/or Joint Area of Operation with Consideration to Host Nation Religious Background.
Prepare an area assessment.	Performs analysis of local religion as it affects mission.
Prepare an area study.	Prepares the RS annex to the OPORD, which includes impact of indigenous religions on AO and mission.
Implement the command map program.	Provide input to CMO personnel regarding local religions.
CMO must analyze the effect of FNS on the morale of U.S. soldiers and on the psychological condition of the local populace.	Advise leadership on negative changes in solder attitude on local population.
Plan disaster relief operations.	Provide religious support for humanitarian assistance survey team (HAST) operations.
Plan noncombatant evacuation operations.	
Commanders must segregate civilians from EPWs and civilian internees to protect them as required by international law.	Coordinate religious support for Enemy Prisoners of War.
Coordinate operations.	Coordinate religious support activities with CMO personnel and local religious leaders.
Perform command functions.	Provide religious support to CMO personnel.

Appendix B

COMBAT AND OPERATIONAL STRESS CONTROL AND RELIGIOUS SUPPORT

References: FM 8-51.
FM 22-51.
Textbook of Military Medicine: War Psychiatry, Part I; Warfare, Weaponry, and the Casualty Office of the Surgeon General of the United States Army, Falls Church, Virginia: 1995

INTRODUCTION

This appendix addresses the general role of the Unit Ministry Team in the commander's program for Combat and Operational Stress Control. Unit Ministry Teams, imbedded within units down to battalion level, provide immediate support to leaders in fulfilling their battle fatigue responsibilities. Unit Ministry Teams also assist in training leaders to recognize battle fatigue identification and intervention responsibilities. In cooperation with unit medical personnel, Unit Ministry Teams serve as a primary referral agency to mental health resources.

BACKGROUND

B-1. The negative impact of Combat and Operational Stress has been known by many names throughout the history of warfare: Nostalgia, Soldier's Heart, Shell Shock, War Neurosis, Combat Stress Reaction, and Battle Fatigue. Combat Stress Behaviors cover the full range of behavior in combat from the highly positive to the completely negative.

B-2. Positive Combat Stress Behaviors include heightened alertness, strength, endurance, and tolerance of discomfort. Examples of positive Combat Stress Behaviors would include the strong personal trust, loyalty, and cohesiveness, which develops among, peers in small units, as well as between leaders and subordinates.

B-3. Dysfunctional Combat Stress Behaviors come in two forms: Misconduct Stress Behaviors and Criminal Acts, and Battle Fatigue.

B-4. Misconduct Stress Behaviors and Criminal Acts range from simple UCMJ infractions to outright criminal acts such as killing enemy prisoners of war, threatening or killing superiors, or looting, pillaging, and rape.

B-5. Battle Fatigue, also referred to as Combat Stress Reaction or Combat Fatigue, is the distress and impaired performance resulting from the accumulated stressors of the combat or operational situation. Signs of Battle Fatigue would include hyper alertness; fear and anxiety; grief, self-doubt, and guilt; loss of confidence and hope; impaired duty performance; erratic actions and outbursts; freezing and immobility; terror, panic, and flight; and impaired speech, vision, touch, and hearing. In extreme cases Battle Fatigue may result in paralysis and hallucinations.

THE BASICS OF TREATMENT

B-6. The principles for stabilizing treatment of soldiers experiencing Battle Fatigue are summarized by the acronym **PIES**.

- **Proximity**: Battle Fatigue cases are treated as far forward as tactically possible and in close proximity with the soldier's unit with the goal of returning the soldier to duty after a period of several days.
- **Immediacy**: Immediate initiation of treatment.
- **Expectancy**: Expectation of rapid and full recovery and return to duty.
- **Simplicity**: Simplicity of approach, according to the four Rs:
    - *Reassurance*: Reassurance of normality.
    - *Rest*
    - *Replenishment*: Replenishment of nutrition, hydration, hygiene and sense of physical well being.
    - *Restoration*: Restoration of confidence through talk and activities.

B-7. It is important to note that the soldier's identity as a soldier is maintained and reinforced. Throughout treatment Battle Dress Uniforms are worn, not hospital pajamas. Rank distinctions and military courtesy are maintained, and these soldiers are responsible for self-care and helping others.

RELIGIOUS SUPPORT AND COMBAT/OPERATIONAL STRESS CONTROL

B-8. Soldiers' inner resources are generally rooted in their religious and spiritual values. In combat, soldiers often show more interest in their religious beliefs. When religious and spiritual values are challenged by the chaos of combat, soldiers may lose connection with the inner resources that have sustained them. Then they become targets of fear, and hopelessness, and eventually fall victim to Battle Fatigue. Further, they become at risk for Misconduct Stress Behaviors. The Unit Ministry Team is the primary resource available to soldiers experiencing these dilemmas and seeking to refocus their spiritual values.

The Unit Ministry Team's Role

B-9. Unit Ministry Teams provide preventative, immediate, and restorative spiritual and emotional support and care to soldiers experiencing Battle Fatigue.

B-10. Preventative Religious Support. The Unit Ministry Team assists in preventing Battle Fatigue and Misconduct Stress Disorders through Spiritual Fitness training. Ministry of Presence with soldiers, assigned DA civilian and contractor personnel is critical. The Unit Ministry Team provides a stabilizing influence on personnel; and assists them in strengthening and regaining personally held spiritual values. Preventative activities include—

- Worship opportunities.
- Private and group prayer opportunities.

- Religious literature and materials.
- Scripture readings.
- Sacraments and ordinances.
- Opportunities for working through frustration, fear, anxiety, and anger.
- Assistance to personnel and families prior to deployment, emphasizing family strengths.

B-11. Immediate Religious Support. The Unit Ministry Team assists commanders in the identification of personnel experiencing negative reactions to combat and operation stress: Battle Fatigue and Misconduct Stress Behaviors. The Unit Ministry Team works closely with the unit's leaders and medical personnel to care for Battle Fatigue cases through religious support and comfort. Immediate Religious Support activities may include—

- Presence with the soldier.
- Conversation focused upon fears, hopes, and other feelings.
- Conversation focused upon forgiveness.
- Prayer with personnel.
- Prayer for fallen comrades and Memorial Ceremonies and Services.
- Rites, sacraments, and ordinances, as appropriate.
- Sacred Scripture.

B-12. Restorative Religious Support. Following an operation, a unit may require reconstitution. Surviving soldiers may need to rebuild emotional, psychological, and spiritual strength. Depending upon the spiritual, emotional, and physical condition of the unit's soldiers, the organic Unit Ministry Team may need augmentation from higher echelons or other units. Restorative Religious Support activities may include—

- Worship, sacraments, rites, and ordinances.
- Memorial ceremonies and services.
- Religious literature and materials.
- Grief facilitation and counseling.
- Reinforcement of the soldiers' faith and hope.
- Opportunities for soldiers to talk about combat experiences and to integrate those experiences into their lives.

COMBAT STRESS BEHAVIORS

- POSITIVE COMBAT STRESS BEHAVIORS
- DYSFUNCTIONAL COMBAT STRESS BEHAVIORS
  - MISCONDUCT STRESS BEHAVIORS AND CRIMINAL ACTS
  - BATTLE FATIGUE

POSITIVE COMBAT STRESS BEHAVIORS	MISCONDUCT STRESS BEHAVIORS AND CRIMINAL ACTS	BATTLE FATIGUE
UNIT COHESION LOYALTY TO BUDDIES LOYALTY TO LEADERS IDENTIFICATION WITH UNIT TRADITION SENSE OF ELITENESS SENSE OF MISSION ALERTNESS AND VIGILANCE EXCEPTIONAL STRENGTH AND ENDURANCE INCREASED TOLERANCE TO HARDSHIP DISCOMFORT PAIN INJURY SENSE OF PURPOSE INCREASED FAITH HEROIC ACTS COURAGE SELF-SACRIFICE	MUTILATING ENEMY DEAD NOT TAKING PRISONERS KILLING ENEMY PRISONERS KILLING NONCOMBATANTS TORTURE, BRUTALITY KILLING ANIMALS FIGHTING WITH ALLIES ALCOHOL AND DRUG ABUSE RECKLESSNESS, INDISCIPLINE LOOTING, PILLAGE, RAPE FRATERNIZATION EXCESSIVELY ON SICK CALL NEGIGENT DISEASE, INJURY SHIRKING, MALINGERING COMBAT REFUSAL SELF0INFLICTED WOUNDS THREATENING/KILLING OWN LEADERS ("FRAGGING") ABSENCE WITHOUT LEAVE DISERTION	HYPERALERTNESS FEAR, ANXIETY IRRITABILITY, ANGER, RAGE GRIEF, SELF-DOUBT, GUILT PHYSICAL STRESS COMPLAINTS INATTENTION, CARELESSNESS LOSS OF CONFIDENCE LOSS OF HOPE AND FAITH DEPRESSION, INSOMNIA IMPAIRED DUTY PERFORMANCE ERRATIC ACTIONS, OUTBURSTS FREEZING, IMMOBILITY TERROR, PANIC RUNNING TOTAL EXHAUSTION APATHY LOSS OF SPEECH SKILLS AND HEARING WEAKNESS, PARALYSIS HALLUCINATIONS DELUSIONS

Appendix C

CRISIS INTERVENTION MANAGEMENT AND CRITICAL EVENT DEBRIEFINGS

BACKGROUND

There are currently three major schools of Crisis Intervention Management, of which Critical Event Debriefing (CED) by a variety of names, are a part. These are the International Critical Incident Stress Foundation's Critical Incident Stress Management (CISM), the National Organization for Victim Assistance (NOVA) Group Crisis Intervention (GCI), and the American Red Cross Disaster Response. In each case, CED is a component of the larger Crisis Intervention model, and training and certification requirements vary from model to model.

PURPOSE

C-1. Crisis Intervention helps small units soon after exceptionally traumatic events to—

- Quickly restore unit cohesion and effectiveness.
- Reduce short-term emotional and physical distress.
- Mitigate long-term distress and "burnout".
- Facilitate the transition from traumatic event (s) to a sense of normalcy and aids in future transitions.

REQUIREMENT

C-2. Leaders of small units should ordinarily conduct an after action review (AAR) when a mission is completed. After an exceptionally distressing mission or event, when METT-TC dependent, unit leaders may coordinate crisis intervention. Members of the Unit Ministry Team are equipped to provide this intervention.

C-3. Situations that may warrant crisis intervention include—

- Death of the unit member.
- Death or suffering of noncombatants (especially women and children).
- Handling of human remains, management of carnage, extensive disaster devastation.
- Friendly fire incident.
- Situation involving serious error, injustice, or atrocity.

C-4. Crisis intervention may be conducted by the Unit Ministry Team or may be coordinated with Mental Health/Combat Stress Control or other medical personnel.

PERSPECTIVE

C-5. Participants in crisis intervention are normal people who have survived an abnormal situation. CED is a generic term for a specific tool within the broader practice of crisis intervention management. CED is neither therapy nor clinical counseling. Rather, it is basic and wise interventive maintenance for the human spirit.

THE CRITICAL EVENT DEBRIEFING

C-6. WHO

- A functional organization: crew, team, squad, or platoon.
- Strangers, civilians, family members thrown together into a traumatic situation by chance.
- Normally includes only those directly involved in the event.
- Chain of Command included only if involved in the event.
- Includes members of identified Crisis Intervention Team.
- Media and outsiders excluded.

C-7. WHAT

Normally, the team consists of a debriefing leader and an assistant, with one additional assistant for every 10 people in a group of more than 20.

C-8. WHEN

- Most effective when conducted 12-72 hours after the event.
- Allow for enough rest and recovery for all to be alert and involved.
- After completing an operation with no expected intense follow-on operations for next 72 hours.
- Expected duration: 2-3 hours.
  When ongoing tactical operations prohibit CED, a briefer "defusing debriefing" focusing on cognitive reactions to the situation can be substituted

C-9. WHERE

- An emotionally neutral place, relatively safe from enemy action, distraction, observation.
- Reserve position or assembly area.
- With enough light to see all participants.
- Shelter from bad weather.

C-10. HOW

- **Introductory Phase**
  Purpose: To introduce the Crisis Intervention Team and explain the process.
  Chaplain or Leader, explains the grounds rules:
  - Personal information or feelings shared in the group during CED are not to be shared outside the group

---

Note: This does not override legal or moral responsibility to report violations of UCMJ or the Law of Land Warfare.

  - No notes or recordings.

  - No breaks are scheduled, but anyone may leave as needed, returning as soon as possible.
  - No one is expected to speak after the Fact Phase.
  - Each participant speaks for self, not others.
  - Everyone is equal during CED. All ranks speak frankly, with proper courtesy.
  - CED is not an AAR, but a discussion to clarify what happened and to restore sense of well-being.
  - Fact-finding, not Fault finding.
  - "Facts" include personal reactions the event.
  - The Team is available following the CED.
- **Fact Phase**
  Purpose: Reconstruct the event in detail, in chronological order, as an unbroken "historical time line" viewed from all sides and perspectives.
  Chaplain/Leader encourages all participants to start their stories before the event occurred and to work up to it.
  Participants—
  - The first person involved in the event is asked to tell his version of the story- how it started, his role (duty position), and what he saw, heard, smelled, and did-step by step.
  - One by one, participants are drawn into the first person's story and then asked to describe their observations and actions.
  - The Chaplain/Leader encourages participation. Everyone is asked to speak.
  - When disagreements over what occurred arise, the Chaplain/leader elicits group observations to clarify memories.
- Thought Phase
  Purpose: To personalize the event.
  Chaplain/Leader facilitates transition from facts to the personal.
  Participants are asked to share—
  - "What were you thinking as the event started?"
  - "What went through your mind when you saw, smelled, or did...?"
  - "What was your first thought when you came off 'automatic' or 'autopilot'?"
- **Reaction Phase**
  Purpose: Identify and ventilates feelings (emotions) raised by the event.
  Chaplain/Leader emphasizes that all emotional reactions deserve to be expressed, respected, and listened to.
  Participants are encouraged to share "reactions":
  - "What was the worst thing about the event?"
  - "How did you react when it happened?"
  - "How are you feeling about that now?"
  - "If you could change one thing about the event, what would it be?"

  Chaplain/Leader listens for common themes, feelings, and misperceptions—
  - Feelings of anger at others for not helping.
  - Blaming self or others for things beyond one's control.

- Feeling changed, different, worse than everyone else, cut off from others.

Participants are asked to describe physical reactions to the event.
Common symptoms include—

- Gastro-intestinal distress
- Frequent urination
- Loss of bowel and bladder control
- Loss of sexual interest
- Heart pounding
- Shortness of breath
- Muscle, back, neck, and head ache
- Trembling, jumpiness, and startle reactions

• **Reframe Phase**
Purpose: To transition from emotional reaction to the cognitive.
Chaplain/Leader facilitates the search for meaning in the event.
Participants are encouraged to derive personal meaning in the event.
  - "What lessons could be learned from this event?"
  - "What is something positive that you will take away from this experience?"
  - "What good can be found in this tragic situation?
  - "What are you proud of in this event?"

• **Teaching Phase**
Purpose: To reassure by teaching the participants that feelings and stress symptoms are normal reactions to abnormal conditions. Symptoms may last a while, but can be expected to resolve in time. If they don't resolve in time, seeking professional help is advised.
Chaplain/Leader summarizes the thoughts, feeling, and symptoms expressed by the group, reemphasize normality; and reduce feelings of uniqueness, weakness, or injury.
No predictions or glamorization of long-term disability should be given.
The Unit Ministry Team may schedule additional training in stress management, coping strategies, grief process, and anger management at later dates.

• **Reentry Phase**
Purpose: Complete and close the CED.
Chaplain/Leader gives final invitation for comments and makes a summary statement. Follow-up resources are discussed.
It is important for Unit Ministry Team personnel to make themselves available for follow-on conversations with individual participants following the CED.

Appendix D

INSTRUCTIONS FOR USING THE MODEL RELIGIOUS SUPPORT MATRIX

GENERAL INFORMATION (ADAPTED FROM FM 101-5)

This appendix provides a sample Religious Support Matrix. The matrix provides a highly visible, clear method to ensure all religious support providers are synchronized. It shows the relationships between activities, units, support functions and key events. Although the unit order will not appear in matrix format, the staff can write their annex as an execution matrix. The religious support appendix as part of the Service Support annex can be done in matrix format. An execution matrix depicts when and where specific supporting actions must occur.

HEADER INFORMATION

D-1. *Classification.* Army Regulation 380-5 contains the detailed description of classification procedures. If the entire plan is unclassified, no classification is required on the appendix.

D-2. *Annexes, Appendixes, Tabs, and Enclosures.* FM 101-5, 31 May 1997, p. H-7ff gives details on how annexes, appendixes, tabs, and enclosures are used. Follow unit SOP and include a matrix as shown in the sample in Figure D-1. The matrix should be included as an appendix to the Service Support Annex. An Arabic number designates the appendix; a capital letter designates the Annex . The Service Support Annex should include the Religious Support Matrix in the list of appendixes along with the other appendixes, such as Personnel and Legal.

D-3. *References.* References list information related to the matrix. The information should help the intended users understand the matrix. Map sheet series numbers may be included. The unit SOP does not need to be referenced. (Cf. FM 101-5, 31 May 1997 p. H-14)

D-4. *Time Zone.* The time zone used throughout the order is placed here. Operations across several zones use ZULU time. If the local time is chosen, the appropriate time zone suffix is used. The time zone is not based on geographical location but rather on the relationship to ZULU time. For example, Central Standard Time would normally be SIERRA. However, during daylight savings time ROMEO would be the appropriate suffix.

D-5. *DTG.* Date time group (DTG) refers to the time of signature in day, hour with suffix, month, and year format. E.g. 260900S JAN04 would be 26 January 2004 at 0900 hr in the SIERRA time zone.

D-6. *Task Organization*. Task Organization is found in Annex A of the OPORD/OPLAN. The UMT coverage plan is made by matching the UMTs with the units listed in the Task Organization. The unit designator like "Iron Horse" goes under the Unit name e.g. (2BCT). List the chaplain and assistant who provide the coverage. When the units are Task Force organized the chaplain of the Task Force is automatically responsible for all the units that would show up in their level Task Organization. If the coordinated Religious Support Plan calls for chaplains to cover units that are not a part of their unit's Task Organization for the entire operation, note those exceptions under the units listed to the right. If the exceptions change by phase, list the coverage in the concept of operation under units. List the chaplain and assistant who provide coverage and indicate the denomination or faith group of chaplains to more completely track coverage needs in the unit AO.

- Enemy Situation. Normally a reference to the Intelligence Annex will suffice. This presupposes that the chaplain has coordinated with the S2/G2 during the Religious Area Analysis (RAA) phase of planning. If there is anything specific about the enemy related to the RS mission that is not listed in the Intelligence annex, then it would go here.
- Friendly Situation. Normally, the basic order can be referenced here. Significant religious issues for the friendly forces are also added here if necessary. Note Religious Preference Profiles (RPPs) that differ from the norm. National Guard units can at times take local flavors. For instance, a National Guard unit in Utah may have a high percentage of soldiers who are members of the Church of Jesus Christ of Latter Day Saints. The Hawaii National Guard might have a unique local flavor as well. Anything that affects your religious support mission that is not the norm is noted here.
- Mission. State the Religious Support tasks and their purpose. Mission is to ___________ in order to ______________.
- Concept of Operations.
    - Location. Location answers the question, "Where will people go for services." During the different phases as the senior chaplain changes location people may have to go different places in order to receive or coordinate for Religious Support Services. Grid designators, Area of Operation, ISB chapel, are types of information that you can use.
    - Units. Note here exceptions to the coverage plan from what is listed under Task Organization. The purpose of the block is to state what UMT is providing which unit service. If the aviation chaplain cover the smoke platoon in the first phase but the field artillery chaplain covers the smoke platoon in the second phase, then list it here.
    - Time Frame. The time that is referred to here is the time schedule for services rendered. This is not the beginning and the ending of the phase. The start and ending of a phase can be listed under the Title of the Phase. Are there any special services or special religious holidays that happen during the operation? These can be listed here. At battalion level you may be able to state when you are going to hold a service for a particular company.

- UMT Priority of Support. UMT Priority of support refers to the services rendered by the UMT. While the unit is in the ISB the priority might be generally nurture the living. Then you would want to spell out just what that means: religious services, counseling opportunities, classes on stress, etc. During the battle the priority might be caring for casualties in general. Then what does that mean specifically? You could list praying for casualties, ministering to care givers, etc. During the post-battle phase honoring the dead might be the priority. During redeployment nurturing the living might include reunion briefings.
- Priority of Support. Priority of Support refers to the units that will receive the services offered by the UMT. While in the ISB, you may give priority to the scouts because of the danger of their mission and because they will be hard to reach later. Often, the religious support priority matches the commander's priority of support as found in the order. Normally, the main effort receives the priority of support.

D-7. *Coordinating Instructions*. Coordinating Instructions tells the unit how they will coordinate for services. How do people coordinate for religious services? How does the unit coordinate for Roman Catholic coverage or some other minority faith need? Before chaplains go on a civil affairs type of mission what coordination needs to take place. How are extraordinary Ministers of the Eucharist being used? How will denominational service leaders (DSLs) function or will they?

D-8. *Service Support*. State how many days of supply each UMT will need. State how re-supply is conducted. Normally, re-supply is through S4 channels.

D-9. *Command*. Report times and formats can be listed here. You can list the specific chain of succession for this operation. You should have the general procedures for the chain of succession in your TSOP.

D-10. *Signal*. List primary and alternate means of communication. List specific information that is needed to contact the chaplain who is responsible for the appendix. If it is a Brigade appendix it is the brigade chaplain. If it is a battalion appendix, then it should be the battalion chaplain.

D-11. *Safety*. List specific safety concerns. Could state that the UMTs need to do a Risk Assessment before certain types of missions (e.g. Civil Affairs or movement to new locations).

D-12. Extraneous Information.

- Name and position example Dolinger, TF CH
- Time Zone: The time zone is the same one used throughout the order.
- OFFICIAL: If the commander signs the original this is not used. The signed copy is a historical record and should remain in headquarters files. The preparing officer signs under OFFICIAL when the commander does not sign it.
- Types of services available
- Designation and location of unit or place providing the service.
- Schedules for services

- Specific missions for units (Special religious groups or holidays)
- Concept of operations.
  - Priority of effort
  - Location
  - Procedures
  - Policies
  - Requirements
  - Phases are specific parts of an operation that differ from those that precede or follow. Phases are indicated by time, distance, terrain, or event occurrences. These will be identified in the order.

CLASSIFICATION

APPENDIX 4 (RELIGIOUS SUPPORT MATRIX) TO ANNEX I (SERVICE SUPPORT) to OPLAN 10-01 OPERATION OVERLORD

REF: AR 165-1, JP 1-05, FM 1-05, RB 16-100				TIME ZONE: ZULU		DTG: 060300JUN44		
TASK ORGANIZATION: See Annex A (Task Org) for each phase	**Unit Designator** CH Goode Assistants SSG Nock	**Subordinate Unit Designator** CH Poling Assistants SSG Bizarri	**Subordinate Unit Designator** CH Fox Assistants SGT Woods	**Subordinate Unit Designator** CH Washington Assistants SPC Duff	**Subordinate Unit Designator** CH Watters Assistants SGT Titus	**Subordinate Unit Designator** CH Rayburn Assistants SGT Calvin	**Subordinate Unit Designator** CH Spurgeon Assistants SGT Luther	**Subordinate Unit Designator** CH Augstine Assistants PVT Delta

1. SITUATION	**1a. ENEMY SITUATION:** Annex B (Intelligence)	**1b. FRIENDLY SITUATION:** Basic Order

2. MISSION: Provide RS to dispersed elements of BCT NLT 081100JUN44 and provide RS to MASCAL sites as directed.

3. EXECUTION: *(intent)*

3a. Concept of Operations	**Phase I** Title of Phase	**Phase II** Title of Phase	**Phase III** Title of Phases	**Phase IV** Title of Phase
Location	UMTs cover units starting w/ those closest to the front	UMTs locate at Casualty Collection Points	UMTs cover CCPs and units first to attack	TBD
Units	EN UMT covers smoke platoon	FA UMT covers smoke platoon	TBD	TBD
Time Frame	Average 5 services daily per UMT	Average 1 service daily plus personal casualty ministry	Average 2 services daily	TBD
UMT Priority of Effort	Religious services, Prayer, Counseling	Prayer for casualties and dying, ministry to care givers	Prayer for casualties, religious services	TBD
Priority of Support	IN TF, EN BN	IN TF, AR TF	EN BN, FA BN	TBD

3b. COORDINATING INSTRUCTIONS.

(1) All UMTs going on CMO missions must coordinate w/ the BDE UMT.

(2) Units desiring Roman Catholic Coverage must coordinate w/ the BDE UMT.

(3) UMTs will report service times IAW SOP.

4. SERVICE SUPPORT

(1) UMTs will carry 30 days of supply.

(2) Resupply is through unit S4 channels.

5. COMMAND AND SIGNAL

5a. COMMAND.

(1) Chain of succession is FSB, EN, FA UMTs

(2) BDE UMT will use TOC as primary location.

(3) Send in reports IAW TACSOP

5b. SIGNAL.

(1) Primary means of communication is FBCB2, DVNT, VIASAT, BDE Admin/log FM net, Messanger

(2) Radio suffix for all Chaplains is 9, expander for Chaplain Assistants is 0.

5c. SAFETY.

UMTs will conduct Risk Assessment prior to all convoy movement and CMO missions.

ACKNOWLEDGE:
OFFICIAL: MOSES, BDE Chaplain

NAME (HYATT)
Rank (COL)

Figure D-1. Sample Model Religious Support Matrix

Appendix E

UNITED STATES CODES RELEVANT TO RELIGIOUS SUPPORT OPERATIONS

This appendix contains extracts from the USC relevant to RS. The extracts range from general military law to laws governing conscientious objectors, accommodation of religious practices and Title 18, Posse Comitatus Act.

E-1. Section 3073. Chaplains. There are chaplains in the Army. The Chaplains include—

- The Chief of Chaplains.
- Commissioned officers of the Regular Army appointed as chaplains.
- Other officers of the Army appointed as chaplains in the Army.

E-2. Section 3547. Duties: chaplains; assistance required of commanding officers—

- Each chaplain shall, when practicable, hold appropriate religious services at least once on each Sunday for the command to which he is assigned, and shall perform appropriate religious burial services for members of the Army who die while in that command.
- Each commanding officer shall furnish facilities, including necessary transportation, to any chaplain assigned to his command, to assist the chaplain in performing his duties.

E-3. Section 3581. Command: chaplains. A chaplain has rank without command.

E-4. Section 774. Religious apparel: wearing while in uniform—

- General Rule. - Except as provided under subsection (b), a member of the armed forces may wear an item of religious apparel while wearing the uniform of the member's armed force.
- Exceptions. - The Secretary concerned may prohibit the wearing of an item of religious apparel.
    - In circumstances with respect to which the Secretary determines that the wearing of the item would interfere with the performance of the member's military duties; or
    - If the Secretary determines, under regulations under subsection (c), that the item of apparel is not neat and conservative.

- Regulations. The Secretary concerned shall prescribe regulations concerning the wearing of religious apparel by members of the armed forces under the Secretary's jurisdiction while the members are wearing the uniform. Such regulations shall be consistent with subsections (a) and (b).
- Religious Apparel Defined. In this section, the term "religious apparel" means apparel the wearing of which is part of the observance of the religious faith practiced by the member.

E-5. Section 6031. Chaplains: divine services—

- An officer in the Chaplain Corps may conduct public worship according to the manner and forms of the church of which he is a member.
- The commanders of vessels and naval activities to which chaplains are attached shall cause divine service to be performed on Sunday, whenever the weather and other circumstances allow it to be done; and it is earnestly recommended to all officers, seamen, and others in the naval service diligently to attend at every performance of the worship of Almighty God.
- All persons in the Navy and in the Marine Corps are enjoined to behave themselves in a reverent and becoming manner during divine service.

E-6. Section 8547. Duties: chaplains; assistance required of commanding officers—

- Each chaplain shall, when practicable, hold appropriate religious services at least once on each Sunday for the command to which he is assigned, and shall perform appropriate religious burial services for members of the Air Force who die while in that command.
- Each commanding officer shall furnish facilities, including necessary transportation, to any chaplain assigned to his command, to assist the chaplain in performing his duties.

DEPARTMENT OF DEFENSE DIRECTIVES

DODD 1300.6 – Conscientious Objectors

E-7. References:

- DOD Directive 1300.6, subject as above, May 10, 1968 (hereby canceled)
- DOD Directive 1332.14, "Administrative Discharges, " December 20, 1965
- Section 6(j) of the Universal Military Training and Service Act, as amended (50 U.S.C. App. 456(j))
- Section 3103, title 38, United States Code

Definitions

E-8. **3.1. Conscientious Objective 1-0**. A firm, fixed and sincere objection to participation in war in any form or the bearing of arms, by reason of religious training

E-9. **3.1.1. Class1-O Conscientious Objector**. A member who, by reason of conscientious objection, sincerely objects to participation of any kind in war in any form.

E-10. **<u>3.1.2. 1-A-O Conscientious Objector</u>**. A member who, by reason of conscientious objection, sincerely objects to participation as a combatant in war in any form, but whose convictions are such as to permit a Military Service in a non-combatant status.

E-11. **3.3.** Non-combatant service or non-combatant duties (1-A-O) used interchangeably herein)

E-12. **3.3.1.** Service in any unit of the Armed Forces which is unarmed at all times.

E-13. **3.3.2.** Service in the medical department of any of type Armed Forces, wherever performed.

E-14. **3.3.3.** Any other assignment the primary function of which does not require the use of arms in com bat provided that such other assignment is acceptable to the individual concerned and does, not require him to bear arms or to be trained in their use.

E-15. **3.3.4.** Service aboard an armed ship or aircraft or in a combat zone shall not be considered to be combatant duty unless the individual concerned is personally and directly involved in the operation of weapons

E-16. **<u>3.4. Non-combatant Training</u>**. Any training which is not concerned with the study, use or handling of arms or weapons.

E-17. **5.3.1.** In order to find that an applicant's moral and ethical beliefs are against participation in war in any form and are held with the strength of traditional religious convictions, the applicant must show that these moral and ethical convictions, once acquired, have directed his life in the way traditional religious convictions of equal strength, depth and duration have directed the lives of those whose beliefs are clearly found in traditional religious convictions. In other words, the beliefs are clearly found in traditional religious convictions. In other words, the beliefs upon which conscientious objection is based must be the primary controlling force in the applicant's life.

E-18. **5.3.2.** A primary factor to be considered is the sincerity with which the belief is held. Great care must be exercised in seeking to determine whether asserted belief is honestly and genuinely held. Sincerity is determined by an impartial evaluation of the applicant's thinking and living in its totality, past and present. Care must be exercised in determining the integrity of belief and the consistency of application. Information presented by the claimant should be sufficient to convince that the claimant's personal history reveals views and actions strong enough to demonstrate that expediency or avoidance of military service is not the basis of his claim.

E-19. **5.3.2.1.** Therefore, in evaluating applications the conduct of applicants, in particular their outward manifestation of the beliefs asserted, will be carefully examined and given substantial weight.

E-20. **5.3.2.2.** Relevant factors that should be considered in determining an applicant's claim of conscientious objection in the home and church; general demeanor and pattern of conduct; participation in religious activities; whether ethical or moral convictions were gained through training, study, contemplation, or other activity comparable in rigor and dedication to the processes by which traditional religious convictions are formulated; credibility of the applicant; and credibility of persons supporting the claim.

E-21. **5.3.3.1.** Church membership or adherence to particular theological tenets are not required to warrant separation or assignment to non-combatant training and service for conscientious objectors.

E-22. **5.3.3.2.** Mere affiliation with church or other group which advocates conscientious objection as a tenet of its creed is not necessarily determinative of an applicant's position or belief.

E-23. **5.3.3.3.** Conversely, affiliation with a church or group which does not teach conscientious objection does not necessarily rule out adherence to conscientious objection beliefs in any given case.

E-24. **5.3.3.4.** Where an applicant is or has been a member of a church, religious organization, or religious sect, and where his claim of conscientious objection is related to such membership, inquiry may properly be made as to the fact of membership, and the teaching of the church, religious organization, or religious sect, as well as the applicant's religious activity. However, the fact that the applicant's may disagree with, or not subscribe to, some of the tenets of his church does not necessarily discredit his claim. The personal convictions of each individual will be controlling so long as they derive from moral, ethical or religious beliefs.

E-25. **5.4.** The burden of establishing a claim of conscientious objection as grounds for separation or assignment to non- combatant training and service is on the applicant. To this end, he must establish clear and convincing evidence (1) that the nature or basis of his claim comes within the definition of and criteria prescribed herein for conscientious objection, and (2) that his belief in connection therewith is honest, sincere and deeply held. The claimant has the burden of determining and setting forth the exact nature of his request, i.e., whether for separation based on conscientious objection (1-O), or, for assignment to non-combatant training and service based on conscientious objection (1-A-O).

E-26. **6.3.** The applicant shall be personally interviewed by a chaplain who shall submit a written opinion as to the nature and basis of the applicant's claim, and as to the applicant's sincerity and depth of conviction. The chaplain's report shall include the reasons for his conclusions. This opinion and report will become part of the "case file." If the applicant refuses to participate or is uncooperative or unresponsive in the course of the interviews, this fact will be included in the statement and report filed by the chaplain and psychiatrist or medical officer.

E-27. **6.4.3.6.** The investigating officer's report, along with the individual's application, all interviews with chaplains or doctors, evidence received as a result of the investigating officer's hearing, and any other items submitted by the applicant in support of his case will constitute the record. The investigating officer's conclusions and recommended dispositions will be based on the entire record and not merely on the evidence produced at the hearings. A copy of the record will be furnished to the applicant at the time it is forwarded to the commander who appointed the investigating officer, and the applicant will be informed that he has the right to submit a rebuttal to the report within the time prescribed by the Military Service concerned.

E-28. **6.6.** The Secretary of a Military Service may delegate authority to approve applications to the commander exercising general court-martial jurisdiction (or equivalent level command for Reserve organizations) over the applicant. The completed record of a case approved in the field will be forwarded to the Headquarters of the Military Service concerned for appropriate disposition.

E-29. **6.9.** To the extent practicable under the circumstances, during the period applications are being processed and until a decision is made, every effort will be made to assign applicants to duties, which will conflict as little as possible with their asserted beliefs. Unless the Military Service concerned provides otherwise, an applicant shall be required to comply with active duty or transfer orders in effect at the time of his application or subsequently issued and received. During the period applications are being processed, applicants will be expected to conform to the normal requirement of military service and to perform such duties as are assigned. Applicants may be disciplined for violations of the Uniform Code of Military Justice while awaiting action on their applications.

DODD 1300.17 – Accommodation of Religious Practices within the Military Services

E-30. References:

- DOD Directives 1300.17, subject as above, June 18, 1985 (hereby canceled)\
- Public Law 98-525, section 554(d), DOD Authorization Act, 1985, October 19, 1984
- Public Law 100-180, section 508, DOD Authorization Act, 1988-89, December 4, 1987
- Title 10, United States Code, Chapter 47, Uniform Code of Military Justice

E-31. **3.1.** A basic principle of our nation is free exercise of religion. The Department of Defense places a high value on the rights of members of the Armed Forces to observe the tenets of their respective religions. It is DOD policy that requests for accommodation of religious practices should be approved by commanders when accommodation will not have an adverse impact on military readiness, unit cohesion, standards, or discipline.

E-32. **3.2.1.** Worship services, holy days, and Sabbath observances should be accommodated, except when precluded by military necessity.

E-33. **3.2.2.** The Military Department should include religious belief as one factor for consideration when granting separate rations, and permit commanders to authorize individuals to provide their own supplemental food rations in a field or "at sea" environment to accommodate their religious beliefs.

E-34. **3.2.3.** The Military Department should consider religious beliefs as a factor for waiver of immunizations, subject to medical risks to the unit and military.

E-35. **3.2.6.** Religious items or articles not visible or otherwise apparent may be worn with the uniform, provided they shall not interfere with the performance of the member's military duties, as discussed in subparagraph

3.2.7.5. below, or interfere with the proper wearing of any authorized article of the uniform.

E-36. **3.2.7.** Under Public Law 100-180, section 508 (reference (C)), members of the Armed Forces may wear visible items of religious apparel while in uniform, except under circumstances in which an item is not neat and conservative or its wearing shall interfere with the performance of the member's military duties.

E-37. **3.2.7.1.** Under this Directive, "religious apparel" is defined as articles of clothing worn as part of the doctrinal or traditional observance of the religious faith practiced by the member. Hair and grooming practices required or observed by religious groups are not included within the meaning of religious apparel, Jewelry bearing religious inscriptions or otherwise indicating religious affiliation or belief is subject to existing Service uniform regulation just as jewelry that is not of a religious nature.

E-38. **3.2.7.2.** In the context of the wearing of a military uniform, "neat and conservative" items of religious apparel are those that:

E-39. **3.2.7.2.1.** Are discreet, tidy, and not dissonant or showy in style, size, design brightness, or color.

E-40. **3.2.7.2.2.** Do not replace or interfere with the proper wearing of any authorized article of the uniform

E-41. **3.2.7.2.3.** Are not temporarily or permanently affixed or appended to any authorized article of the uniform.

E-42. **3.2.7.3.** The standards in subparagraph 3.2.7.2., above, are intended to serve as a basis for determining a member's entitlement under Public Law 100-80, section 508 (reference ©), to wear religious apparel with the uniform. For example, unless prohibited by subparagraph 3.2.7.6., below, a Jewish yarmulke may be worn with uniform whenever a military cap, hat or other headgear is not prescribed. A yarmulke may also be worn underneath military headgear as long as it does not interfere with the proper wearing, functioning, or appearance of the prescribed headgear.

E-43. **3.2.7.4.** Exceptions to the standards in subparagraph 3.2.7.2., above, and other special accommodations for members of particular religious groups may be granted by the Military Departments under section 4., below.

E-44. **3.2.7.5.** Whether an item of religious apparel interferes with the performance of the member's military duties depends on the characteristics of the items, the circumstances of its intended wear, and the particular nature of the member's duties. Factors in determining if an item of religious apparel interferes with military duties include, but are not limited to, whether the item may.

E-45. **3.2.7.5.1.** Impair the safe and effective operation of weapons, military equipment, or machinery.

E-46. **3.2.7.5.2.** Pose a health or safety hazard to the wearer or others.

E-47. **3.2.7.5.3.** Interfere with the wearing or proper functioning of special or protective clothing or equipment (e.g., helmets, flack jackets, flight suits, camouflaged uniforms, gas masks, wet suits, and crash and rescue equipment).

E-48. **3.2.7.5.4.** Otherwise impair the accomplishment of the military mission.

E-49. **3.2.7.6.** A complete prohibition on the wearing of any visible items of religious apparel may be appropriate under unique circumstances in which the member's duties, the military mission, or the maintenance of discipline require absolute uniformity. For example, members may be prohibited from wearing visible religious apparel while wearing historical or ceremonial uniforms; participating in review formations parades honor or color guards, and similar ceremonial details and functions.

E-50. **3.2.7.7.** The authority to approve the wearing of an item of religious apparel with the uniform, under the guidelines of this paragraph, shall be exercised at the command level specified by each Military Department. Denials of requests to wear religious apparel shall be subject to review oat the Service Headquarters level. Final review shall occur within 30 days following the date of initial denial for cases arising in the United States, and within 60 days for all other cases. Exceptions to these deadlines shall be limited to exigent circumstances, such as extended deployment. Service. Service members shall be obliged to comply with orders prohibiting the wearing of questionable items of religious apparel pending review of such order under regulations issued by the Secretaries of the Military Departments.

E-51. **3.2.7.8.** Notwithstanding paragraphs 3.2.6. and 3.2.7. above, chaplains may wear any required religious apparel or accouterments and the uniforms while conducting worship services and during the performance of rites and rituals distinct to their faith groups.

PROCEDURES 4.

E-52. **4.1.** Under rules prescribed by the Secretary of the Military Department concerned, military commanders should consider the following factors along with any other factors deemed appropriate in determining whether to grant a request for accommodation religious practices addressed in section 3., above.

E-53. **4.1.1.** The importance of military requirements in terms of individual and unit readiness, health and safety, discipline, morale, and cohesion.

E-54. **4.1.2.** The religious importance of the accommodation to the requester.

E-55. **4.1.3.** The cumulative impact of repeated accommodations of a similar nature.

E-56. **4.1.4.** Alternative means available to meet the requested accommodation.

E-57. **4.1.5.** Previous treatment of the same or similar requests, including treatment of similar requests made for other than religious reasons.

E-58. **4.2.** The factors in subsection 4.1. above are intended to promote standard procedure for resolving difficult questions involving accommodation of religious practices. In view of the different mission requirements of each command, individual consideration of specific requests for accommodation is necessary. With the exception of requests involving the wearing of visible items of religious apparel with the uniform, denials of which must be reviewed at the Service Headquarters level, the appropriate level of command for resolution of these issues shall be determined by each of the Military Departments, based on its particular requirements and circumstances.

E-59. **4.3.** When requests for accommodation are not in the best interest of the unit and continued tension between the unit's requirements and the individual's religious belief is apparent, administrative actions should be considered. These actions may include, but are not limited to, assignment, reclassification, or separation. Nothing in this Directive precludes action under the Uniform Code of Military Justice (reference (d) in appropriate circumstances.

USC, TITLE 18, POSSE COMITATUS ACT, SECTION 1385

E-60. The Posse Comitatus Act prohibits the use of the Army and Air Force to enforce civilian law (state or federal) except in cases and under circumstances expressly authorized by the constitution or act of congress. DOD policy extends this prohibition to include the Navy and Marine Corps.

E-61. This act restricts federal military law enforcement activities as follows—

- DOD military personnel do not participate directly—
  - In arrest, search and seizure, or stopping and frisking of persons, or interdiction of vessels, aircraft, or vehicles.
  - In surveillance or pursuit.
  - As informants, undercover agents, or investigators in civilian legal cases or in any other civilian law enforcement activity.
- The Posse Comitatus Act does not apply to —
  - Members of the National Guard when in Title 32 and state active duty (SAD).
  - Members of a Reserve Component when not on active duty or active duty for training.
  - DOD civilians, unless under the direct C2 of a Title 10 active duty officer.
  - Military personnel when off duty and acting solely in a private capacity.
  - Military personnel taking action for the primary purpose of furthering a military or foreign affairs function of the U.S. For example, enforcing military justice, maintaining law and order on military installations, or protecting classified materials.
  - The U.S. Coast Guard.

Appendix F

GUIDE FOR RELIGIOUS AREA/IMPACT ASSESSMENT

As the Army deploys to more areas of the world not familiar to Americans, it becomes increasingly important for chaplains to advise commanders and soldiers on the religious/spiritual aspects of the people they encounter. The term "religious/spiritual" refers to the system of beliefs and practices that give meaning and purpose to people's lives. The Unit Ministry Team (UMT) task is to understand and assess and to remain objective.

This guide gives general parameters to help the UMT fulfill its mission. Not all religious/spiritual systems will contain all of these categories. Rather they are a general matrix that will serve as a starting point to help the UMT in its assessment. The chaplain needs to first consider which of the categories are applicable to the particular society to which he has deployed. This can be done through consulting relevant literature or local subject matter experts. Having determined which categories are relevant, the chaplain can then proceed to assess how the local population understands these categories. The final step would be to covey this knowledge to commanders and soldiers as appropriate. The chaplain's answer to the questions below must address in detail how the chaplain's answer is of issue to the commander as he prepares to conduct operations.

F-1. Holy Day, Rituals and Customs—

- What Holy days and/or festivals may impact military operations?
- What are important religious "do's and don'ts" to observe?
- What are dietary restrictions?
- What are the group's distinctive symbols?
- What are the nature, frequency and traits of worship?
- What are appropriate protocols for issues related to birth/marriage/death?

F-2. Sites and Shrines—

- What are the places of worship, pilgrimage and memorial sites? Why?
- Where are the cemeteries and what is the character of their make-up?
- What are distinctive identifying characteristics (architecture, symbols, etc.)?

F-3. Primary Values—

- What are they willing to die for?

- What subjects incite an emotional response?
- What behaviors does the group reward? What are punished?
- What value is placed on women, children, ancestors, certain animals or objects?

F-4. Leadership—

- Who are the religious leaders (official and unofficial)?
- Do the religious leaders have an impact on the armed forces?
- What political influence do religious leaders have?
- What is the relationship of religious leaders to government officials?
- What do religious leaders wear to symbolize their position?
- How many leaders are there and where are they located?
- What are the leaders titles?
- How are leaders selected and trained?

F-5. Tolerance/Religious Intensity—

- What is the religious commitment in this group?
  - Nominal (in name only)
  - Mild
  - Strong
  - Radical / Fanatical
- How tolerant is the group to other parties?
- How accepting are they of conversion of their members to other groups?
- How are members disciplined?
- Can others join and quit the group easily?
- How are competing groups viewed and received?

F-6. Relationship to Society—

- How does this society relate to the religious group?
  - Stamp out the group
  - Contain the group
  - Assimilate (absorb) the group
  - Share power with the group
  - Promote pluralism with this group
- How is this group viewed?
  - Religious group
  - Secret society
  - Protest movement
  - Political party
- Does the group have a distinct subculture or communal life?
- How does the group seek to influence society?
- How do they use media resources?

F-7. Organization—

- What cell-like groups are present?
- What is the nature of hierarchy within the movement?
- What are the centers of learning?

F-8. Doctrines/Myths—

- What is/are the sources of doctrinal authority?
- What is/are the sources of ethics?
- What are their concepts of justice?
- What are their concepts of reward?
- Who are the heroes and villains? Rivalries (past, present, future)?

F-9. History/Background—

- What larger group is this group related to?
- What makes this group distinctive from the larger group?
- When did this group come into the area?

F-10. References/Sources

Appendix G

CHAPLAIN DETACHMENTS

Chaplain Detachments (CDs) provide flexibility, robustness and specific liaison functions for religious support. CDs represent a supplemental capability for conducting religious support operations and for providing/performing religious support across the full spectrum of military operations. When mobilized, these detachments increase the commander's ability to meet the religious support needs of soldiers.

There are four types of detachments:

- Chaplain Detachment–Alpha (CD-A), General Support (GS)
- Chaplain Detachment–Bravo (CD-B), (GS)
- Chaplain Detachment–Corps (CD-C), Direct Support (DS)
- Chaplain Detachment–Division (CD-D), (DS).

When mobilized, the CDs will normally come from the Reserve Components. It is important to note that these detachments are augmentation resources, not replacements for active component unit ministry teams (UMTs).

CD-A and CD-B UMTs offer maximum religious support flexibility for utilization primarily in major combat operations, but may also be utilized in homeland security and non-major combat operations. If directed by the Chief of Chaplains or FORSCOM, CD-A and CD-B UMTs may coordinate and/or provide religious support to first-responders, units without organic UMTs, DOD civilians, federal agencies, and other personnel as directed during MACA operations, WMD or other HLS incident. However, the CD-C and CD-D UMTs address specific critical religious support operational requirement shortfalls needed for the accomplishment of religious missions in the corps and division area of operations. These detachments will be aligned to support a specific corps or division. The CD-C and CD-D will have a habitual relationship for training and operational missions.

All CDs will rely on the unit to which assigned or attached for all life support and logistical support. Support may also be provided by other appropriate elements within the area of operation. This support includes, but is not limited to: legal, combat health support, finance, personnel and administrative services, field feeding, unit maintenance, and logistical support.

CHAPLAIN DETACHMENT- ALPHA (GENERAL SUPPORT)

G-1. *Mission.* The primary mission of the Chaplain Detachment – Alpha (CD-A) is to provide religious area support across the full spectrum of military operations, and spiritual reconstitution/recollection (SR2) in a Port of Debarkation (POD), Army, Theater, Corps, or equivalent separate task force. The CD-A also provides C2, operational planning and coordination support of Chaplain Detachment – Bravo (CD-B). The CD-A has the following capabilities:

- Planning, supervision, coordination, and provision of religious support for units operating in major ports of debarkation.
- Provide religious support for replacements and Force Reception, Onward movement operations in marshaling areas and ports of debarkation (PODs).
- Plan for and provide operational, specialized religious support and SR2 activities in rest, stand down, and convalescent/recovery areas. This detachment can conduct up to 2 activities per day.
- Plan for and provide area and denominational support for unit concentrations.
- Provide religious, indoctrination, and acclimatization support to incoming personnel, UMT replacements and UMTs during the reception, training, and acclimatization phase of the mobilization.
- Plan and provide religious support activities for units undergoing reconstitution.
- Provide religious support during NEOs.
- Conduct religious support operations in a JTF headquarters during operations other than war.
- Provide technical command and control (C2), supervision, control, and coordination of up to 5 CD-B UMTs.
- Plan and coordinate CD-B religious support operational missions.
- Provide the ASCC, Theater Support Command (TSC) or Corps chaplain with plans and recommendations for integrating specialized religious support and SR2 into contingency plans (COPLANs), operational plans (OPLANs) and operation orders (OPORDs) to enable support to unit/task force personnel.
- Perform other area and denominational religious support missions as required.

G-2. *Personnel—*

- Two chaplains (56A).
- Three chaplain assistants (56M).

G-3. *Assignment.* A CD-A will be assigned to an Army Service Component Command (ASCC), TSC, corps or equivalent separate task force headquarters. The unit will also be assigned as part of the port opening package of major PODs for religious support coverage of port elements. It may also be assigned to a task force or JTF headquarters.

G-4. *Employment.* A CD-A is used to augment religious support capabilities in rear areas from the division rear boundary through the communications zone in units and geographic areas where UMTs are not available or are unable to handle the religious support requirement, particularly areas such as ports of debarkation, marshaling areas, logistic support areas, combat support hospitals and other locations where there is a high concentration of soldiers. CD-A will be deployed as part of port opening packages to conduct religious support operations in major ports for port operational elements without organic UMTs. Additionally, it can be used with Force Provider or the Containerized Chapel to provide religious support to units in transit and to support personnel replacements arriving in theater. A CD-A can perform "spiritual reconstitution/recollection" activities for soldiers and members of other UMTs. If required, the CD-A can divide into two teams to provide expanded area support.

G-5. *Mobility*. The CD-A requires 100% mobility. It must have all required TOE equipment, supplies and personnel when mobilized in order to have 100% operational capability once deployed. This detachment will be required to operate independently throughout the theater, from the port(s) of debarkation (POD) to the division rear boundary while maintaining communication with the UMT of the unit to which assigned/attached. It must be able to provide technical control and coordination of up to 5 CD-Bs throughout the theaters area of operation. This detachment must be able to move all personnel and equipment in a single lift with its authorized organic vehicles.

G-6. Communications Equipment. The CD-A has a requirement to be fielded with the latest technology in both secured voice and digital data communications equipment. The CD-A will be required to communicate digitally and by voice with the unit(s) they are supporting (ASCC, TSC, or Corps or JTF HQs), CD-B UMTs, and back to home station from anywhere on the battlefield. In order for this detachment to meet its minimum mission essential wartime requirements, it must have communications equipment that is compatible and interoperable with the equipment used by the command UMT element of the unit to which assigned and units supported.

CHAPLAIN DETACHMENT-BRAVO (GENERAL SUPPORT)

G-7. *Mission*. The primary mission of the Chaplain Detachment – Bravo (CD-B) is to provide religious support by augmenting the ASCC, TSC, or Corps chaplain sections ability to provide unit and denominational religious support across the full spectrum of military operations at PODs, LSAs, base clusters, and for units without organic religious support assets or those units that deploy under concepts of split-based operations or modularity. It also provides religious support during NEOs, to units undergoing reconstitution, and mass casualties (MASCALs). The CD-B has the following capabilities:

- Provide and perform religious support activities to small units that deploy under the concepts of split-based operations or modularity and to units that do not have organic religious support assets.
- Provide and perform religious support activities for units from other services provided that, the unit to which the CD-B is assigned/attached has Army Support of other Services (ASOS) responsibility and that the unit does not have organic religious support assets.
- Provide and perform comprehensive religious support to family members prior to and during NEOs in forward deployed theatres.

- Provide and perform religious support activities for units undergoing reconstitution.
- Provide religious support to DOD civilians, contractors, and other authorized personnel, as dictated by the commander, that are in the area of operations.
- Provide SR2 when augmenting the CD-A.
- Deploy forward for immediate temporary UMT casualty replacement in divisional maneuver units as required for religious support.
- Provide other area and denominational religious support missions as required.

G-8. *Personnel—*

- One chaplain (56A).
- One chaplain assistant (56M).

G-9. *Assignment.* One CD-B will be assigned to each Corps Support Command (COSCOM) and Corps Support Group (CSG) to augment the chaplain section's ability to provide religious support within the AO. This unit can also be assigned to a TSC, ASG, or BSB.

G-10. *Employment.* Chaplain Detachment-Bravo's are used to augment religious support capabilities throughout the Theatre Army's and Corps area of operations. CD-Bs will normally be employed with a COSCOM or CSG. They can also be employed with, assigned to, or attached to Area Support Groups (ASG) or Base Support Battalions (BSB) to augment the religious support capabilities of those UMTs. Augmented UMTs will employ the CD-B to fill gaps in religious support coverage, specifically to units without organic UMTs within the assigned units area of responsibility. This would also include small units deployed under the concept of split-based operations, specifically at LSAs, and base clusters. This unit can also be utilized to assist Combat Support Hospitals (CSH) UMTs in the event of mass casualties. They can further be pushed down to the Division area of operations to support mass casualty operations when available UMT assets are unable to provide the required religious support. CD-B UMTs will not be employed as permanent UMT casualty replacements in maneuver units, but can be used as immediate temporary (usually 72 hours or less) backfill.

G-11. *Mobility.* The CD-B requires 100% mobility. It must have all required TOE equipment, supplies and personnel when mobilized in order to have 100% operational capability once deployed. This detachment will be required to operate independently throughout the theater, from the PODs to the division rear boundary. At times this detachment will also operate within the division battlespace to provide area and denominational religious support. This detachment must remain prepared to relocate on short notice (3 hours or less). It must also be able to move all personnel and equipment in a single lift with its authorized organic vehicles.

G-12. *Communications Equipment.* The CD-B has a requirement to be fielded with the latest technology in both secured voice and digital data communications equipment. The CD-B will be required to communicate digitally and by voice with the unit(s) they are supporting (ASCC, TSC, Corps, COSCOM, CSG, ASG, and BSB) and CD-A UMTs from anywhere on the battlefield. When providing temporary UMT backfill to maneuver brigades and battalions, the CD-B will be required to communicate with the division and/or maneuver brigade UMT to which they are operationally controlled to. In order for this detachment to meet its minimum mission essential wartime requirements, it must have communications equipment that is compatible with the equipment used by the command UMT element to which assigned and the unit supported.

CHAPLAIN DETACHMENT – CORPS (DIRECT SUPPORT)

G-13. *Mission*. The primary mission of the Chaplain Detachment – Corps (CD-C) is to augment the Corps Chaplain Section across the full spectrum of military operations to provide liaison with the G5/Civil Military Operation Center to facilitate the UMT supported Civil Military Operations activities within the corps on behalf of the corps chaplain. The CD-C is a religious support asset. The CD-C has the following capabilities:

- Plan for and coordinate religious support missions in support of area civil/military operations.
- Provide liaison with government, faith-based NGOs and International Organizations (IOs).
- Coordinate and monitor faith-based NGO and IO requirements and issues.
- Establish links with faith-based NGOs and IOs and with religious leaders of the host nation.
- Plan for and provide operational, specialized religious support and SR2 activities in rest, stand down, and convalescent/recovery areas.
- Provide the Corps Chaplain with plans and recommendations for contingency and reconstitution plans and support to the task force.
- Conduct specialized ministry/SR2 in the corps area at sites selected by using commanders.
- Perform other religious support missions as required.

G-14. *Personnel—*

- One chaplain (56A).
- One chaplain assistant (56M).
  - The chaplain and the chaplain assistant will have a need for specialized training in civil affairs, civil-military operations, and world religions to more effectively perform their mission. They should also have a foundational understanding and be familiar with the operations, missions, and goals of known faith-based NGOs and IOs.
  - The chaplain assistant will be required to have a 2S ASI (Battle Staff Qualified). This skill is required in order for the CD-C to work with the CMOC and the corps chaplain section in developing plans and operations. This skill is also important because it will allow the CD-C to maintain situational awareness and an understanding of the operational environment.

G-15. *Assignment*. The CD-C is assigned to all Heavy and Airborne Corps. They can further be attached or assigned to a JTF HQs or division when the division is operating as an ARFOR or JTF headquarters. The CD-C will habitually train and deploy with the corps to which assigned.

G-16. *Employment*. The CD-C will deploy with and provide habitual augmentation to the Corps chaplain section UMT. The CD-C will act as the liaison between faith-based NGO/IOs and the corps chaplain in an area of operation. The CD-C will be the principle point of contact between the Corps Chaplain and the CMOC concerning indigenous religious issues and religious support impact during Civil Military Operations missions. The CD-C may provide general religious support as the primary mission permits. The CD-C will normally co-locate with the Corps Chaplain section at the corps rear CP and operate out of either the corps chaplain cell or the CMOC. If required the CD-C can be deployed in support of a division that has ARFOR/JTF responsibility in situations where the corps HQs does not deploy. It may also be deployed forward into a division AO to conduct CA/HA and religious support liaison functions.

G-17. *Mobility*. The CD-C requires 100% mobility. It must have all required TOE equipment, supplies and personnel when mobilized in order to have 100% operational capability once deployed. This detachment will be required to operate independently throughout the corps area of operations. At times it will also be required to operate within the division area to conduct support to CMO and religious support liaison functions. This detachment will prepare to be separated from the Corps Chaplain Section for up to 72 hours. It must also be able to move all personnel and equipment in a single lift with its authorized organic vehicles.

G-18. *Communications Equipment*. The CD-C has a requirement to be fielded with the latest technology in both secured voice and digital data communications equipment. The CD-C will be required to communicate digitally and by voice with the corps and division chaplain sections from anywhere within the area of operations. They will also be required to communicate with faith-based NGO/IO representatives, host nation religious leaders, and other governmental organizations operating in the area of operations by voice communications equipment and face-to-face using either military linguists or a machine translation devices. The CD-C will also have a CONUS reach-back requirement in order to acquire information on indigenous religions and faith-based NGO/IOs from subject matter experts, databases, and the World Wide Web. In order for this detachment to meet its minimum mission essential wartime requirements, it must have communications equipment that is compatible with the equipment used by the command UMT element to which assigned and other units supported.

CHAPLAIN DETACHMENT –DIVISION (DIRECT SUPPORT)

G-19. *Mission*. The primary mission of the Chaplain Detachment – Division (CD-D) is to provide direct habitual augmentation to the Division Chaplain section. Provides assets required to enable the division chaplain section to conduct 24-hour and split-based religious support planning, coordination, and execution operations across the full spectrum of military operations. The CD-D is capable of the following:

- Provides required assets to the division chaplain section to enable sustained 24-hour and split-based operations between the DREAR and DMAIN.

- Battle track and monitor current division operations and subordinate unit and UMT locations.
- Operates battlefield-automated systems (BAS) to receive and analyze data for recommendations to the division chaplain for use in current operations and future planning of religious support.
- Collects SITREPs, reports, and requests for information (RFIs) from division CP sections/elements, corps chaplain section, brigade and separate battalion UMTs.
- Monitors locations and religious support needs and coordinates religious support for authorized civilians during operations to include: contractors, DA civilians, and interagency representatives operating in the division AO.
- Monitors location of faith-based NGO/IOs operating in the division AO and assesses religious support impact on operations.
- Perform other religious support missions as required.

G-20. *Personnel—*

- One chaplain (56A).
- One chaplain assistant (56M). The chaplain assistant will be required to have a 2S ASI (Battle Staff Qualified). This skill is required to enable the CD-D to be able to battle track and analyze current operations as a means to effectively anticipate religious support needs and develop input into current and future operational plans. It is also important because it will allow the CD-D to maintain situational awareness and an understanding of the operational environment; allowing for more accurate information exchanges to the division chaplain and staff regarding religious support operations.

G-21. *Assignment*. A CD-D will be assigned to each division.

G-22. *Employment*. A critical task for the division chaplain section is 24-hour religious support operations. The CD-D will be aligned to support and train with a specific division in order to develop a habitual relationship and operational knowledge of the division supported. Upon deployment notification of the division, the CD-D that has been assigned to the deploying division will be mobilized to deploy with that division. The CD-D will join the division chaplain section during the pre-deployment phase of the operation. This will permit the division chaplain section to conduct 24-hour and/or split-based religious support operations throughout the course of the mission and across the full spectrum of military operations. The CD-D will normally operate out of the DMAIN along with the division chaplain and NCOIC. While the CD-D can be used to perform general religious support functions, it is important to remember that the primary mission of the CD-D is to assist in performing 24-hour operations.

G-23. *Mobility*. The CD-D requires 100% mobility. It must have all required TOE equipment, supplies and personnel when mobilized in order to have 100% operational capability once deployed. This detachment will be required to have the same mobility as the division it supports. It must also be able to move all personnel and equipment in a single lift with its authorized organic vehicles.

G-24. *Communications Equipment*. The CD-D has a requirement to be fielded with the latest technology in both secured voice and digital data communications equipment. The CD-D will be required to communicate digitally and by voice with the division UMT, deputy division UMT, BDE and BN UMTs throughout the division AO. In order for this detachment to meet its minimum mission essential wartime requirements, it must have communications equipment compatible with the equipment used by the division chaplain section to which assigned and other units supported.

GLOSSARY

AA	assembly area
AAR	After Action Review
ABCA	American, British, Canadian, and Australian
AC	Active component
ACR	Armored Cavalry Regiment
ADA	air defense artillery
AGR	Active Guard Reserve
ALOC	Administrative/Logistics Operations Center
AO	area of operations
AOB	advanced operational bases
AOR	area of responsibility
AMC	Army Materiel Command
AMEDD	Army Medical Department
APOE	aerial port of embarkation
AR	Army regulation
ARFOR	Army forces
ARNG	Army National Guard
ARSOA	Army Special Operations Aviation
ARSOF	Army Special Operations Forces
ARTEP	Army Training and Evaluation Program
ASCC	Army Service Component Command
ASG	area support group
AT	annual training
AUTL	Army Universal Task List
AXP	ambulance exchange point
BAS	battalion aid station
BASOPS	base operations
BCOC	Base Cluster Operations Center
BCTP	Battle Command Training Program
BDE	brigade
BDOC	Base Defense Operations Center
BII	Basic issue items
BMO	battalion maintenance officer
BN	battalion
BOS	battlefield operating system
BSA	brigade support area
BSB	base support battalion

BUB	battle update briefings
C2	command and control
CCC (C3)	chaplain career course
C4ISR	C4 Intelligence, surveillance and Reconnaissance
CA	chaplain assistant
CA	Civil Affairs
CCH	Chief of Chaplains
CCP	Casualty Collection Point
CD	Chaplain Detachment
CD-A	Chaplain Detachment-Alpha (General Support)
CD-B	Chaplain Detachment-Bravo (General Support)
CD-C	Chaplain Detachment-Corps (Direct Support)
CD-D	Chaplain Detachment-Division (Direct Support)
CED	Critical Event Debriefing
CEOI	Communication electronic operating instructions
CFLC	Chaplain Family Life Center
CFP	Contingency Force Pool
CH	Chaplain
CHOBC	Chaplain Officer Basic Course
CISM	Critical Incident Stress Management
CJTF	combined-joint task force
CJTF	commander joint task force
CMM	Combat Medical Ministry Course
CMO	civil-military operations/officer
CMOC	Civil Military Operations Center
CMRP	Command Master Religious Plan
CMTC	Combat Maneuver Training Center
COA	course of action
CofS	Chief of Staff
COMMZ	communications zone
CONUS	Continental United States
CONUSA	Continental United States Army
COPLAN	contingency plan
COSCOM	Corps Support Command
COSR	combat operations stress response
CP	command post
CPE	Clinical Pastoral Education
CRC	Continental United States Replacement Center
CRP	Commander's Religious Program
CS	combat support
CSA	corps support area

CSB	Corps Support Battalion
CSG	Corps Support Group
CSM	Command Sergeant Major
CSS	combat service support
CSSCS	Combat Service Support Control System
CSU	Chaplain Support Unit
CT	Counter Terrorism
CTC	Combat Training Center
CTCP	Combat Trains Command Post
CZ	combat zone
DA	Department of the Army
DA	Direct Action
DACH	U.S. Army Chief of Chaplains
DASG	U.S. Army Surgeon General
DCO	Defense Coordinating Office
DCSOPS	Deputy Chief of Staff for Military Operations and Plans
DFG	Distinctive Faith Group
DFGL	Distinctive Faith Group Leaders
DISCOM	Division Support Command
DIV (IT)	Institutional Training Divisions
DIVARTY	division artillery
DLA	Defense Logistics Agency
DNVT	digital nonsecure voice telephone
DOD	Department of Defense
DOS	Days of Supply
DS	direct support
DSA	division support area
DSCP	Defense Supply Center Philadelphia
DSL	denominational service leader
DSO	domestic support operations
DTG	date time group
DTLOMS	doctrine, training, leader development, organization material, and soldier support
EAC	echelons above corps
EAD	echelons above division
EMM	Emergency Medical Ministry Course
EPW	enemy prisoner of war
FA	field artillery
FAC	family assistance center
FEMA	Federal Emergency Management Agency
FHA	foreign humanitarian assistance
FID	Foreign Internal Defense

FLOT	forward line of own troops
FORSCOM	Forces Command
FM	frequency modulation
FRAGO	fragmentary order
FRG	Family Readiness Group
FSCOORD	fire support coordinator
FSMC	forward support medical company
FTCP	Field Trains Command Post
FTX	field training exercises
GCI	Group Crisis Intervention
GS	general support
GSU	Garrison Support Unit
HA	humanitarian assistance
HHC	headquarters and headquarters company
HHT	headquarters and headquarters troop
HN	host nation
HQDA	Headquarters, Department of the Army
HSSN	home station support node
HUMINT	human intelligence
IET	initial entry training
IMA	Individual Mobilization Augmentee
IMT	Installation Ministry Team
IO	Information Operations
IPB	intelligence preparation of the battlefield
IRR	Individual Ready Reserve
ISB	intermediate staging base
JAG	Judge Advocate General
JAS	Jump Aid Station
JFC	Joint Force Commander
JFCOM	Joint Forces Command
JFLC	Joint Forces Land Component
JFLCC	Joint Forces Land Component Command
JFSOCC	Joint Force Special Operations Component Commander
JSOTF	Joint Special Operations Task Force
JOA	joint operational area
JOPES	Joint Operation Planning and Execution System
JRTC	Joint Readiness Training Center
JTF	joint task force
JSOA	Joint Special Operations Area
KIA	killed in action
LD	line of departure

LDRSHIP	loyalty, duty, respect, selfless service, honor, integrity, and personal courage
LFA	Lead Federal Agency
LL	Lessons Learned
LOGPAC	Logistics Package
LRP	logistics release point
MACA	military assistance to civil authorities
MACDIS	military assistance to civil disturbances
MACOM	Major Army Command
MASCAL	mass casualty
MDMP	Military Decision Making Process
MEDCOM	U.S. Army Medical Command
METL	mission essential task list
METT-TC	mission, enemy, terrain and weather, troops and support available, time available, and civil considerations
MIA	Missing in Action
MOB	Mobilization
MOBEX	mobilization exercises
MOOTW	military operations other than war
MOPP	mission oriented protective posture
MOS	military occupational specialty
MRI	Medical Reengineering Initiative
MRE	meal ready to eat
MS	mobilization station
MSC	Major Subordinate Command
MSCA	military support to civilian authorities
MSE	mobile subscriber equipment
MSR	main supply route
MTF	Medical Treatment Facility
MTOE	Modified Table of Organization and Equipment
NATO	North Atlantic Treaty Organization
NCO	noncommissioned officer
NCOES	Noncommissioned Officer Education System
NCOIC	noncommissioned officer in charge
NDMS	National Disaster Medical System
NEO	noncombatant evacuation operation
NG	National Guard
NGO	non-governmental organization
NMS	National Military Strategy
NOK	next of kin
NOVA	National Organization for Victim Assistance
NSN	national stock number

OCCH	Office of the Chief of Chaplains
OCOKA	Observation and fields of fire, cover and concealment, obstacles, Key terrain, and avenues of approach
OCONUS	Outside the Continental United States
ODA	Operational Detachment Alpha
ODSS	Offensive, Defensive, Stability and Support
OES	Officer Education System
OOTW	operations other than war
OPCON	operational control
OPLAN	operations plan
OPORD	operations order
OPSEC	operations security
OPTEMPO	operation tempo
OVM	operator's vehicle material
PAO	Public Affairs Office/Officer
PBO	Property Book Officer
PCI	precombat inspection
PCS	permanent change of station
PERSCOM	Personnel Command
PHA	Personnel Holding Area
PIES	proximity, immediacy, expectancy, simplicity
PMCS	preventive maintenance checks and services
POD	port of debarkation
POE	port of embarkation
POM	preparation of overseas movement
POTF	PSYOP Task Force
POTG	PSYOP Task Group
POW	prisoner of war
PROFIS	Professional Officer Filler System
PSYOP	psychological operations
RAA	Religious Area Analysis
RAOC	Rear Area Operation Center
RC	Reserve component
RCP	Rear Command Post
RIP	Ranger Indoctrination Program
RFI	request for information
RMST	Religious Ministry Support Team
ROE	rules of engagement
ROP	Ranger Orientation Program
RPP	Religious Preference Profile
RS	religious support
RSC	Reserve Support Command

RSE	religious support estimate
RSOI	reception, staging, onward movement, and integration
RSP	religious support plan
RTF	Reconstitution Task Force
RTOC	Rear Tactical Operation Center
S1/G1	Adjutant/Personnel Officer
S2/G2	Intelligence Officer
S3/G3	Operations and Training Officer
S4/G4	Logistics Officer
S5/G5	Civil Military Operations officer
S6/G6	Signal operations officer
SACRED	spirituality, accountability, compassion, religious leadership, excellence, and diversity
SAD	state active duty
SEAD	Suppression of Enemy Air Defenses
SF	Special Forces
SFOB	Special Forces Operational Base
SFSC	Special Forces Staff Course
SIDPERS	Standard Installation/Division Personnel System
SIGO	signal officer
SITREP	situation report
SJA	Staff Judge Advocate
SMART-B	Special Medical Augmentation Response Team-Burn
SMART-CB	Special Medical Augmentation Response Team-Chemical/Biological
SMART-HS	Special Medical Augmentation Response Team-Health Systems Assessments and Assistance
SMART-MC3T	Special Medical Augmentation Response Team- Medical Command, Control, Communications, Telemedicine
SMART-PC	Special Medical Augmentation Response Team-Pastoral Care
SMART-PM	Special Medical Augmentation Response Team-Preventive Medicine/Disease Surveillance
SMART-SM	Special Medical Augmentation Response Team-Stress Management
SMART-TCC	Special Medical Augmentation Response Team-Trauma/Critical Care
SMART-V	Special Medical Augmentation Response Team-Veterinary
SO	special operations
SOCOM	Special Operations Command
SOF	Special Operation Forces
SOFA	status of forces agreement
SOFCOM	United States Special Operations and Forces Command
SOI	special operating instruction
SOP	standard operating procedures
SPOE	seaport of embarkation

SR	Special Reconnaissance
SR	supply route
SR2	spiritual reconstitution/recollection
SRP	Soldier Readiness Program
SSC	small-scale contingency
SSCO	small-scale contingency operations
STARC	State Area Command, ARNG
TAA	tactical assembly area
TACON	tactical control
TDA	Table of Distribution and Allowances
TF	task force
TF A/L	task force (Administrative/Logistics)
TOA	transfer of authority
TOC	tactical operations center
TSC	Theater Support Command
TSOP	tactical standard operating procedure
TTP	tactics, techniques and procedures
UCMJ	uniformed code of military justice
UMCP	Unit Maintenance Collection Point
UMT	Unit Ministry Team
USACAPOC (A)	United States Army Civil Affairs and Psychological Operations Command (Airborne)
USAR	U.S. Army Reserve
USARC	United States Army Reserve Command
USASOC	United States Army Special Operations Command
USC	United States Code
USDB	United States Disciplinary Barracks
USSOCOM	United States Special Operations Command
UW	unconventional warfare
WIA	wounded in action
WMD	weapons of mass destruction
XO	executive officer

Bibliography

DOCUMENTS NEEDED

These documents should be available to the intended users of this publication.

Except where otherwise indicated, the following DA forms are available on the Army Electronic Library (AEL) CD-ROM (EM 0001) and the USAPA web site (www.usapa.army.mil); DD forms are available from the OSD web site http://www.dior.whs.mil/icdhome/ddeforms.htm

DA Form 7462, Religious Support Matrix, September 2002.

DD Form 1610, Request and Authorization for TDY Travel of DOD Personnel, January 2001.

JOINT AND MULTISERVICE PUBLICATIONS

Most joint publications are online at http://www.dtic.mil/doctrine/

DOD Directive 1300.6, *Conscientious Objectors*, 20 August 1971.

DOD Directive 1300.17, *Accommodation of Religious Practices Within the Military Services*, 3 February 1988.

DOD Directive 1332.14, *Enlisted Administrative Separations*, 21 December 1993.

JP 1-0, *Doctrine for Personnel Support to Joint Operations*, 19 November 1998.

JP 1-05, *Religious Ministry Support for Joint Operations*, 26 August 1996.

JP 3-0, *Doctrine for Joint Operations*, 10 September 2001.

JP 3-05, *Doctrine for Joint Special Operations*, 17 April 1998.

JP 3-07.3, *Joint Tactics, Techniques, and Procedures for Joint Special Operations Task Force Operations*, 19 December 2001.

JP 3-08, *Interagency Coordination During Joint Operations*, Volumes I and II, 9 October 1996.

JP 3-10, *Joint Doctrine for Rear Area Operations*, 28 May 1996.

JP 5-00.2, *Joint Task Force (JTF) Planning Guidance and Procedures*, 13 January 1999.

ARMY PUBLICATIONS

Most Army doctrinal publications are online at http:www.adtdl.army.mil/atdls.html or at http://www.lia.army.mil/dpmsh/pubs_rev_dt.html

AR 165-1, *Chaplain Activities in the United States Army*, 27 February 1998.

AR 380-5, *Department of the Army Information Security Program*, 29 September 2000.

AR 600-20, *Army Command Policy*, 13 May 2002.

AR 680-29, *Military Personnel – Organization and Type of Transactions Codes*, 1 March 1989.

DA PAM 165-3, *Chaplain Training Strategy*, 1 September 1998.

FM 1 (100-1), *The Army*, 14 June 2001.

FM 1-111, *Aviation Brigades*, 27 October 1997.

FM 3-0 (100-5), *Operations*, 14 June 2001.

FM 3-90 (100-40), *Tactics*, 4 July 2001.

FM 3-90.3 (71-3), *The Mounted Brigade Combat Team*, 1 November 2001.

FM 7-0 (25-100), *Train the Force*, 22 October 2002.

FM 7-20, *The Infantry Battalion*, 6 April 1992.

FM 7-30, *The Infantry Brigade*, 3 October 1995.

FM 8-51, *Combat Stress Control in a Theater of Operations Tactics, Techniques and Procedures*, 29 September 1994.

FM 16-1, *Religious Support*, 26 May 1995.

FM 22-51, *Leaders' Manual for Combat Stress Control*, 29 September 1994.

FM 22-100, *Army Leadership*, 31 August 1999.

FM 25-101, *Battle Focused Training*, 30 September 1990.

FM 55-30, *Army Motor Transport Units and Operations*, 27 June 1997.

FM 71-2, *The Tank and Mechanized Infantry Battalion Task Force*, 27 September 1988.

FM 71-100, *Division Operations*, 28 August 1996.

FM 100-6, *Information Operations*, 27 August 1996.

FM 100-7, *Decisive Force: The Army in Theater Operations*, 31 May 1995.

FM 100-8, *The Army in Multinational Operations*, 24 November 1997.

FM 100-9, *Reconstitution*, 13 January 1992.

FM 100-10-2, *Contracting Support on the Battlefield*, 4 August 1999.

FM 100-11, *Force Integration*, 15 January 1998.

FM 100-15, *Corps Operations*, 29 October 1996.

FM 100-17-3, *Reception, Staging, Onward Movement, and Integration*, 17 March 1999.

FM 100-19, *Domestic Support Operations*, 1 July 1993.

FM 100-20, *Military Operations in Low Intensity Conflict*, 5 December 1990.

FM 100-21, *Contractors on the Battlefield*, 26 March 2000.

FM 100-23, *Peace Operations*, 30 December 1994.

FM 101-5, *Staff Organization and Operations*, 31 May 1997.

FM 101-5-2, U.S. Army Report and Message Formats, 29 June 1999

OTHER PUBLICATIONS

Army Vision 2010

Chaplain Assistant Concept Paper, 2000

Chaplain Support Force XXI White Paper

Chief of Chaplains MSCA Concept Paper, September 2000

The Chief of Chaplain's UMT Crisis Assets Handbook, October 2000

Concept for Future Joint Operations, May 1997

Joint Vision 2010

Joint Vision 2020 –http://www.dtic.mil/jv2020

Military Rules of Evidence 503 UCMJ

National Military Strategy

A National Security Strategy for a New Century, May 1997

The Quadrennial Defense Review Report, September 2001.

Public Law 98-525

Public Law 100-180

Textbook of Military Medicine: War Psychiatry, Part I, Warfare, Weaponry and the Casualty Office of the Surgeon General of the United States Army, Falls Church, VA, 1995.

United States Code, Title 10, Armed Forces http://uscode.house.gov/title_10.htm

United States Code, Title 18, Crimes and Criminal Procedures: http://uscode.house.gov/title_18.htm

U.S. Army Chaplaincy Strategic Plan FY 2000-FY 2005

Preface

The mission of Army Public Affairs (PA) is to fulfill the Army's obligation to keep the American people and the Army informed, and to help establish the conditions that lead to confidence in America's Army and its readiness to conduct operations in peacetime, conflict and war. PA is a critical battlefield function in today's global information environment since every aspect of an Army operation is subject to instantaneous scrutiny.

This field manual (FM) sets forth tactics, techniques and procedures (TTPs) for conducting PA operations in accordance with the doctrinal principles contained in FM 3-0 (100-5), *Operations* and FM 3-61 (46-1), *Public Affairs Operations*. Although this manual is primarily designed to be used by public affairs officers, noncommissioned officers and civilians, it presents TTPs that all leaders conducting operations in the information age should be familiar with.

FM 3-61-1 (46-1-1) is applicable to units and individuals in both the active and reserve components. It serves as a foundation for integrating PA into Army doctrine, training, leader development, organization, materiel and soldier initiatives. In conjunction with the Army Training and Evaluation Program (ARTEP) and other training guidance, it should also be used to plan, integrate and execute individual and collective PA training in units throughout the Army.

The proponent for this manual is the Office of the Chief of Public Affairs. Send comments and recommendations on DA Form 2028 to Director, Army Public Affairs Center, ATTN: SAPA-PA, Fort Meade, MD 20755-5650.

Unless this publication states otherwise, masculine nouns or pronouns do not refer exclusively to men.

This document is approved for public release; distribution is unlimited.

Introduction

With the extremely sophisticated communication technologies of the global information environment (GIE), the nature of media coverage has a significant impact on the conduct of war and stability and support operations at the strategic, operational and tactical levels. Civilian and military news media coverage influences the perceptions of soldiers, family members, the public and political leaders, and affects the direction, range and duration of operations. It has a direct relation to the confidence these key audiences have in the Army and its execution of operations.

Effective PA operations are critical to successful Army operations in the information age. They assist the commander in monitoring and understanding public opinion, explaining the situational context of events and communicating the Army's perspective clearly and without filters. They enable the commander to interpret the perceptions of external and internal audiences and influence the way in which discussion of the operation is framed.

Synchronized, well-planned and actively executed PA tactics, techniques and procedures significantly clear the fog of war and impact the morale and effectiveness of the force. They reduce distractions, misinformation, uncertainty, confusion and other factors that cause stress and undermine efficient operations. They enhance understanding, acceptance and support. Effective PA operations contribute to soldier confidence, discipline, will to win, and unit cohesion.

FM 3-61-1 (46-1-1), *Public Affairs Tactics, Techniques and Procedures*, builds on the doctrinal foundation of FM 46-1. It translates the PA fundamentals and principles into detailed guidance for the planning, coordination and execution of PA operations. It provides what is required for the information age -- a sophisticated approach to conducting PA operations. It is the TTP that brings Army Public Affairs into the "information age."

FM 3-61 (46-1), *Public Affairs Operations*, addresses fundamental PA concepts in depth, and provides the linkage between PA and the Army's keystone doctrine, FM 3-0 (100-5), *Operations*. It recognizes that a refocused, restructured military will conduct operations in an information environment in which detailed, graphic, and live coverage of events are transmitted around the world. It builds from the understanding that information availability will influence strategic decisions and the direction, range and duration of operations.

FM 3-61 (46-1) also examines PA operations at the different levels of war and across the range of operations. It discusses PA operations with respect to the Principles of War and the Tenets of Army Operations. It analyzes the PA contributions to build and sustain combat power, defines the PA mission, and establishes strategic PA goals, fundamental PA principles, and underlying considerations for planning integrated information strategies.

Public affairs frequently deals in intangibles -- perceptions and implications -- that are not easily quantifiable or qualifiable, but are essential to commanders. The PA objectives, processes and methods presented in FM 3-61-1 (46-1-1) will assist Army leaders and PA professionals to develop solutions to the complex PA issues they will confront.

FM 3-61.1

Field Manual
No. 3-61.1

Headquarters
Department of the Army
Washington, DC, 1 October 2000

Public Affairs Tactics, Techniques and Procedures

Contents

Chapter 1

Public Affairs Fundamentals

Public Affairs fulfills the Army's obligation to keep the American people and the Army informed and helps establish the conditions that lead to confidence in America's Army and its readiness to conduct operations in peacetime, conflict and war.

FM 3-61 (46-1), Public Affairs Operations

THE PA MISSION

1-1. The American public, internal Army audiences, allies, adversaries and other critical audiences have access to an ever expanding array of public and military media. Newspapers, magazines, radio, television and electronic media are independent conduits of information. They provide news, analysis, interpretation and commentary and serve as a forum for ideas, opinions and public debate. What appears in the media, both civilian and military, shapes perceptions, attitudes and opinions, and can have a direct impact on mission success.

1-2. The vast majority of both civilian and military media representatives are committed to providing responsible, accurate, balanced coverage. Although there are exceptions, most media representatives, even when editorializing, are focused on achieving a credible presentation. To accomplish this, media representatives investigate issues, ask tough, challenging questions, and pursue verifiable answers. They seek information, interpretation and perspective. Army leaders at all levels need to educate media representatives and facilitate their efforts to provide an accurate, balanced and credible presentation of timely information.

1-3. Army leaders do this by integrating public affairs into the planning process and synchronizing PA operations with other facets of the operation. Integrating and synchronizing public affairs issues allows commanders to communicate their perspective and achieve a balanced, accurate, credible information presentation.

1-4. The PA mission identifies the essential contribution that PA makes to America's Army. The mission and the strategic goals derived from it provide the foundation on which public affairs operations are built. Strategies, which are developed from the perspective that every aspect of every operation could become an issue of interest in the global information environment, are the most successful. Developing such strategies requires that PA personnel access, analyze and anticipate potential issues by conducting a thorough mission analysis.

1-5. The challenge for commanders, and personnel supporting them, is to plan and execute operations, which accomplish this mission and support these goals. To do this, PA must be integrated into the planning and decision-making process from receipt of the mission.

1-6. The need to integrate and synchronize PA early derives from the fact that in most situations media representatives will be present in an area of operations before the arrival of Army forces. They will know the area of operations and because they are covering the story as it evolves, will have an understanding of, and opinion about, the situation.

1-7. Media interest will normally be the most intense at the onset of operations. Media representatives will cover the deployment of Army forces, their arrival in the area and their initial conduct. To support the commander and the force in their interactions with media representatives during these early stages, public affairs personnel should be deployed in the first days, if not hours, of the operation. Dealing with a large international press corps constitutes the most immediate public affairs challenge facing the commander during contingency.

PUBLIC AFFAIRS FORCE

1-8. The changing information environment in which the Army conducts war and stability and support operations makes it necessary for PA officers, NCOs and specialists to respond to increasingly complex,demanding challenges. They must be prepared to support the commander with a wide range of knowledge about and understanding of the communication process, the global information environment (GIE) and its potential impact on operations.

1-9. PA personnel must also thoroughly understand the fundamentals of Army operations and the strategic context within which the Army conducts operations. As the GIE compresses the strategic, operational and tactical levels of operations, PA personnel need to appreciate the linkage between public opinion, political decision-making and the national security strategy. They must understand the Army's approach to fighting, influencing events in operations other than war and deterring actions detrimental to national interests.

1-10. PA leaders must, therefore, be among the most informed people in the command. They must be thoroughly aware of all aspects of the operation. They need to know what is being reported about the operation in the global news media and how internal and external audiences are reacting to that information.

1-11. The Public Affairs Officer. The PAO's primary mission is to assess the PA situation, advise the commander on PA issues, assist him in making the best possible decisions, and translate his decision into effective PA operations. PAOs employ the decision-making process to plan, coordinate and supervise the implementation of a PA strategy that helps the commander meet his obligation to communicate with the American public, soldiers, home station communities and the Department of the Army community. PAOs analyze the situation, anticipate issues, assess implications, and develop comprehensive operations, which meet the news and information needs of internal and external audiences and facilitate media operations.

1-12. The Public Affairs Non-Commissioned Officer. The strength of the Public Affairs functional area is its non-commissioned officer corps. PA NCOs are experts on the global information environment, media operations, information strategies, and PA training. They are integral to all facets of the PA planning and decision-making process and provide the essential functional area expertise and continuity required for successful PA operations.

1-13. PA NCOs work closely with the PAO, and in many situations, a PA NCO is the commander's senior PA advisor. Therefore, PA NCOs are fully prepared to assess the PA situation, develop, synchronize and coordinate a PA strategy, implement and monitor PA operations, and measure and evaluate the success of the PA effort.

1-14. The Public Affairs Specialist. In addition to learning traditional soldier skills, PA specialists are trained to support the entire spectrum of PA operations conducted in the global information environment. They are trained on news media operations, news media facilitation, information strategies and information provision. They register media representatives, gather information, develop information products, support news media briefings, respond to news media inquiries and requests for assistance, and track and monitor news media activities throughout the area of operations. They work with news media representatives to gather accurate information and provide timely, balanced coverage of the operation.

1-15. The Department of the Army Public Affairs Civilian. Civilian PA practitioners assigned to Tables of Distribution and Allowances (TDAs) have the same skills as military PA personnel. They provide critical support during war and non-combat operations by providing a vital link between deployed forces and the home station community, and in many situations, may be called upon to deploy with the units they support, or as individual augmentees.

PUBLIC AFFAIRS ELEMENTS

1-16. Battlefield commanders have two sources for tactical PA support. The first is the PA section organic to a warfighting headquarters. The second is the PA unit, which is attached to a headquarters to augment the command's PA capability.

1-17. Conducting PA planning, facilitating news media operations on the battlefield, providing news and information, and executing PA training and support operations is manpower intensive. The austerely staffed PA sections organic to warfighting headquarters will nearly always be overwhelmed trying to meet PA requirements in war and other operations. PA staff sections, therefore, rely on early augmentation by PA units, or individual augmentation if appropriate, to accomplish the battlefield PA mission.

1-18. **Organic Public Affairs Sections.** Organic PA sections are found in warfighting headquarters at various levels including brigades, divisions and corps throughout the Army. Army PA personnel are also assigned to the organic PA sections of joint and combined headquarters.

1-19. In headquarters without organic PA sections, the commander is responsible for PA and must plan and execute PA operations or assign responsibility for PA operations as a special or additional duty to an officer or senior NCO in the command.

1-20. Regardless of the echelon, the PA staff section's primary responsibility is to assist the commander in accomplishing his mission. The staff:

- provides PA information expertise and advice
- conducts PA assessments
- provides analysis of the information environment
- conducts PA planning
- develops information strategies and guidance
- implements PA operations
- measures the effectiveness of the PA effort
- conducts PA training

1-21. The PA staff element controls augmenting PA units. It determines requirements, defines priorities and assigns missions to the augmenting unit. In conjunction with the augmenting unit commander, the staff element task organizes the unit, allocating personnel and equipment to accomplish objectives. If the PA staff element is a Public Affairs Operations Center or a Task Force Headquarters, it will coordinate Armed Forces Radio and Television Service (AFRTS) support activities for the command.

1-22. **Public Affairs Units.** PA units are fully deployable TOE organizations designed to augment the PA staff sections of warfighting units, although they can operate independently in certain limited situations. When a PA unit is deployed to augment a PA staff section, the personnel in the unit cannot be reassigned as replacements or employed as individual fillers for other public affairs elements.

1-23. PA units depend upon the unit they augment for personnel administration, finance, legal and health services, communications, food service, unit maintenance, and supplemental transportation support. PA units operating at corps and below must have the capability to transport all of their TOE equipment in a single lift using authorized organic vehicles.

1-24. There are currently four types of PA units:

- Public Affairs Detachment (PAD)
- Mobile Public Affairs Detachment (MPAD)
- Public Affairs Operations Center (PAOC)
- Broadcast Operations Detachment (BOD)

1-25. **Public Affairs Detachment (SRC 45500AA00).** The smallest of the PA units, the Public Affairs Detachment (PAD) (Figure 1-1) is commanded by a captain and includes seven PA soldiers..

Public Affairs Detachment

CDR	CPT	46A
PA NCO	SSG	46Q
JOURN	SGT	46Q
JOURN(x2)	SPC	46Q
BR JOURN	SPC	46R
JOURN	PFC	46Q
BR JOURN	PFC	46R

Figure 1-1. PAD

1-26. The PAD normally augments a division, separate brigade and armored cavalry regiments and deploys in support of combined, unified or joint operations.

1-27. The PAD commander assumes responsibilities as the PAO or deputy PAO, and the PAD PA personnel are integrated into the supported command's PA section based on operational requirements.

Historical Perspective

The 10th Mountain Division, Fort Drum, N.Y., arrived in Somalia with hundreds of reporters already there, and absolutely no public affairs personnel accompanying them. The first public affairs support arrived at the 2nd Brigade, 10th Mountain Division, 10 days later and only because the JIB dispatched one of its own PADs.

(PA After Action Report, Operation Restore Hope, 10th Mountain Division, Dec 1992)

1-28. Because of the size of the unit, the PAD provides limited:

- Personnel and equipment for digital imagery and audio acquisition.
- Personnel for media escort within the supported unit's area of operations
- Coordination of an information product distribution system that can relay news and information products to members of the supported unit, higher echelons and home station.
- Planning, developing and implementing strategy to support civilian news media and facilitate news gathering efforts throughout the supported unit's area of operations
- Personnel and equipment to coordinate, assist or conduct press conferences and briefings

- Personnel to train, advise and assist leaders and soldiers interacting with or supporting civilian news media within the supported unit's area of operations.

1-29. **Mobile Public Affairs Detachment (SRC 45413A000).** The workhorse of PA units, the Mobile Public Affairs Detachment (MPAD) is a modular, task organizable unit, which provides the full range of PA services. (Figure 1-2). The MPAD is commanded by a major and includes 20 soldiers.

Mobile Public Affairs Det

Command Cell

CDR	MAJ	46A
PA SPVR	MSG	46Z
CLERK	SPC	71L

MEDIA TEAM*

TM OIC	CPT	46A
NCOIC	SFC	46Q
JOURN	SGT	46Q
BR JOURN	SGT	46R
JOURN	PFC	46Q
BR JOURN	PFC	46R

BROADCAST TEAM*

TM OIC	CPT	46A
NCOIC	SFC	46R
BR JOURN	SSG	46R
BR JOURN	SPC	46R
BR JOURN	SPC	46R
BR JOURN	PFC	46R

PRINT TEAM*

TM OIC	CPT	46A
NCOIC	SSG	46Q
JOURN(x2)	SPC	46Q
JOURN	PFC	46Q

*This is one possible configuration. Paragraph 2 members can be configured as one, two or teams of various strengths, based on mission.

Figure 1-2. MPAD

1-30. The MPAD normally augments a Corps PA section or a Public Affairs Operations Center (PAOC). In support of a PAOC, it provides manpower and equipment to establish and operate a media center at Theater Army, TAACOM and Corps.

1-31. It may also be deployed to directly support a joint service task force or non-DoD governmental agency conducting disaster relief, humanitarian assistance, counter drug, peacekeeping, or other contingency operations.

1-32. MPADs in direct support of a gaining command PAO provide acquisition capability for print, audio and video.

1-33. Additional MPADs in direct support can expand the media escort capability of the supported PAO, augment divisions and other elements in theater and provide the PAO with staff augmentation. MPADs may be deployed forward to establish media centers or a sub-JIB in a joint environment.

1-34. MPADs have the capability to:

- Monitor and assess the perceptions of external audiences through access to civilian commercial news sources.
- Conduct assessments of the information environment, to include development of a PA estimate of the situation, as the initial part of operational planning.
- Assist the PAO in operational planning and policy and ground rules for media, coordination for logistical support to PA, and coordination of PA operations with higher and subordinate headquarters.
- Plan and develop information products, which will be produced through contracted services and/or the use of organic equipment and facilities.
- Acquire, produce and transmit information products throughout the theater, between the theater and home station, and between the theater and HQDA.
- Create and disseminate print, photographic, audio and video products for external release directly to civilian media who do not have representatives within the theater of operations. Conduct media facilitation and develop information strategies.
- Prepare commanders, staff personnel and other command members for interviews, press conferences, and similar media interaction.

1-35. **Public Affairs Operations Center (SRC 45423A000).** The Public Affairs Operations Center (PAOC) consists of command, media facilitation and post-production sections (Figure 1-3). It is commanded by a lieutenant colonel and includes 32 soldiers.

1-36. The PAOC normally augments the PA staff section at echelons above division to establish and operate a media center in support of civilian and military media representatives working in the theater.

1-37. The PAOC commander serves as the media center commander but works under the control of the PAO of the supported unit or task force.

1-38. In joint or combined operations, a PAOC serves as the Army element of the joint media operations center.

1-39. When the PAOC functions as the Army element of a joint or combined media center, the PAOC commander works for the joint or combined media center commander.

1-40. For major operations in which there is a significant media interest, the PAOC is augmented by up to three MPADs. The MPADs are either integrated into the main media center operation or tasked to operate subordinate media centers at outlying locations. A PAOC can support up to 100 news media representatives. When augmented by three MPADs, the PAOC can support up to 300 news media representatives.

Public Affairs Operations Center

Command Cell

CDR	LTC	46A
XO	MAJ	46A
PAO	MAJ	46A
CH PA NCO	SGM	46Z
PA NCO	SFC	46Q
SPLY SGT	SGT	92Y
MECHANIC	SPC	63B
ADM SPC	SPC	71L
PERSONNEL	SPC	75B
ADM CLERK	PFC	71L

ESCORT TEAM

CHIEF	CPT	46A
PRESS OFF (x2)	CPT	46A
JOURN (x2)	SGT	46Q
JOURN (x4)	SPC	46Q
JOURN (x4)	PFC	46Q

BRIEFING SECTION

CHIEF	MAJ	46A
BRIEF OFF (x2)	CPT	46A
BR JOURN	SGT	46R
BR JOURN	SPC	46R
BR JOURN (x2)	PFC	46R

A/V PRODUCTION

BR NCO	SSG	46R
PROD SPC	SPC	25V
BR JOURN	SPC	46Q

Figure 1-3. PAOC

1-41. PAOCs are modular, task organizable units having the capability to:

- Implement the theater or corps strategy to support civilian news media and facilitate news gathering efforts in theater.
- Coordinate and provide services to registered civilian news media sponsored by the command.

- Provide the personnel and equipment to coordinate and conduct media support within the theater of operations.
- Provide personnel and equipment to plan and conduct daily news media briefings.
- Provide personnel to train, advise and assist leaders and soldiers interacting with or supporting civilian news media representatives.

1-42. **Broadcast Operations Detachment (SRC 45607A00).** The BOD consists of a command element, two broadcast teams and a maintenance team. (Figure 1-4). It is commanded by a major and includes 26 soldiers.

1-43. The BOD augments a fixed or field expedient AFRTS facility under the control of a senior AFRTS facility commander, or it can establish and operate a separate radio and/or television broadcast facility to support theater level operations.

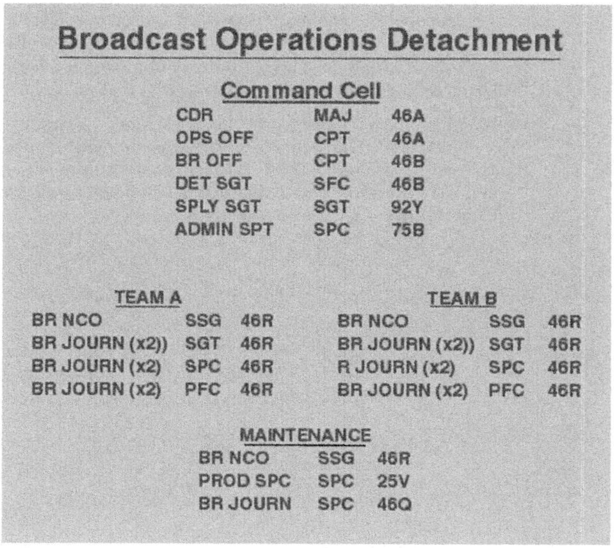

Figure 1-4. BOD

1-44. The BODs must be authorized and equipped by AFRTS to perform this mission. When deployed to perform this mission, the BODs are assigned to the PAOC supporting the command.

1-45. BODs have the capability to:

- Provide on-air broadcasters recorded materials and satellite down links to operate a 24-hour a day radio outlet.

- Provide on-air broadcasters, recorded materials and satellite down links to operate a television station.
- Originate audio and video news, feature and entertainment programming from within theater.
- Provide post production services for audio and video news and feature material supplied in unedited format.
- Provide limited audio and video materials to other public affairs operations for dissemination outside the theater.
- Acquire audio and video electronic newsgathering coverage of operations in the theater for use in internal and external information programs.
- Perform field maintenance and repair above operator level to broadcast equipment organic to the unit.
- Provide the commander with an alternate means of communications when tactical communications are not adequate or not available.

1-46. Public Affairs organizations are built around a force of soldiers who are selected and trained to articulate the goals and missions of the Army.

1-47. More than 65 percent of the total public affairs force and 85 percent of the deployable TOE unit structure is positioned in the U.S. Army Reserve and Army National Guard. These reserve units and personnel must be seamlessly integrated with the active component and focused on supporting the overall Army goals and objectives.

Chapter 2

Public Affairs Functions and Responsibilities

THE CHALLENGE

2-1. The global information environment and continually evolving information communication technologies make it imperative that information and messages be consistent at all levels. The personal comments made by a deployed soldier in a remote area of operations and the official statements released by DoD at the Pentagon must be mutually supporting. The information targeted to internal audiences must parallel the information released through the news media to the American public and other external audiences. The Army's need for security, and the soldier's and family member's right to privacy must be balanced with the Army's obligation to provide timely, accurate, complete information to internal and external audiences. The commander's information strategy must ensure that the information available in the public domain, regardless of the source, does not conflict, contradict or otherwise undermine the credibility of the command or the operation.

Historical Perspective

Civilian news coverage contributed greatly to maintaining soldier morale during Desert Storm. The coverage was generally positive; the American people were behind the operation and soldiers felt this impact. Problems arose when the coverage created rumors, and command information was not consistent with what the soldiers were seeing or hearing in the world media. Family members and non-deployed soldiers were greatly affected by news coverage, often creating problems for rear commanders and detracting from their credibility.

(After Action Report, Desert Storm 1990, Center for Army Lessons Learned)

2-2. Accomplishing this presents unique command and control challenges for commanders, PA practitioners at all levels and others involved in using information to help accomplish the mission in the most effective, efficient manner. It requires careful coordination between staff elements and necessitates continual liaison between levels of command from the tactical through the operational to the strategic.

2-3. Further complicating PA command and control challenges are PA force structure realities. The small size of the PA staff sections organic to war-fighting headquarters necessitates augmentation, especially for operations with a high level of visibility. The availability of augmenting PA units, the majority of which are located in the reserve components, and the difficulties inherent in deploying PA civilians result in heavy dependence on augmentation by individuals. This leads to the creation of ad hoc, unequipped PA elements, which have not trained together or developed relationships with other staff sections or commands, and do not have established internal or external operating procedures.

2-4. For PA personnel therefore, the critical challenge is to rapidly define command and control channels, establish lines of communication and develop operating procedures. The responsibility for doing this usually lies with the Corps PAO who normally leads the commander's PA effort. He identifies requirements, assesses resources and plans, organizes, directs, coordinates and controls the PA operation.

PUBLIC AFFAIRS RESPONSIBILITIES

2-5. Effective PA command and control establishes a public affairs organization based on analysis of mission, enemy, terrain, troops, time available and civilians (METT-TC), tailored to the situation, which reflects the commander's concept of the operation. It ensures that there are sufficient, experienced, PA personnel at each echelon to provide the commander and his force with the most effective and efficient support possible.

2-6. PA command and control begins at the DoD level. The Office of the Assistant to the Secretary of Defense, Public Affairs (OASD(PA)), retains primary responsibility for the development and consistent implementation of DoD information policies and determines who should serve as the initial source of information about operations. Although (OASD(PA)) delegates PA release authority to the combatant commander as soon as practical, it retains responsibility for approving Public Affairs Guidance (PAG), establishes public affairs policy, and coordinates and approves PA strategies and plans.

2-7. The Office of the Chief of Public Affairs (OCPA) for the Army is responsible for Army PA resources. OCPA develops PA doctrine, designs PA organizations, determines training and leader development requirements, identifies materiel needs, and manages PA personnel to ensure that sufficient assets are available, qualified and ready to conduct successful PA operations in support of any assigned mission.

2-8. Commanders supported by their PA staff personnel, plan PA operations for their assigned missions based on the situation, published in DoD directives, instructions, doctrine and guidance, and in coordination with OASD(PA). The CINCs prescribe the chain of command, organize and employ forces, give authoritative direction, assign tasks and designate objectives through component commanders, subordinate unified commanders, commanders of joint task forces and other subordinate commanders. The commander establishes responsive PA structures and ensures that they are provided with the personnel, facilities, equipment, transportation and communications assets necessary to provide adequate PA support. A failure to establish these structures results in a duplication of effort and a waste of resources. The commander is responsible for the full range of PA activities -- PA planning, media facilitation, information strategies and PA training and at sustaining base, community relations. He is also responsible for establishing, resourcing and guiding the operations of Joint Media Operation

Centers and planning all AFRTS radio and television support operations in the area of operations.

Historical Perspective

The establishment of the Joint Information Center under the auspices of the DOT Presidential Task Force on September 1, 1992 was vital to a coordinated and successful Joint Public Affairs effort. The JIC was an "umbrella" organization that served as the clearing house for dissemination of hurricane relief information to the news media. More than 10 federal agencies involved in relief operations had public affairs representatives at the JIC. Daily meetings and consistent interaction among the agencies involved resulted in a coordinated federal information effort.

(Public Affairs Lessons Learned Library, Joint Information Center, Hurricane Andrew, 1992)

2-9. Within the Army, the Corps is usually the hub for PA operations. The Corps commander, supported by his PA staff element, is responsible for the development and coordination of PA strategies, the implementation of information campaigns and the execution of PA operations based on METT-TC, the information environment, and guidance and policy received from the combatant CINC.

2-10. The Corps PAO is the principal PA advisor to the Corps commander and deploys with the lead element of the Corps headquarters. When fully deployed, the Corps PAO section operates from the Corps main command post, with a liaison officer/NCO located at the Corps plans cell. It also is responsible for establishing coordination with the PA elements of higher, lower and adjacent commands.

2-11. The Corps, through the PAO, controls the employment of augmenting Army PA units deployed in support of the operation. Up to one PAOC and six MPADs are normally allocated to augment the Corps PA section, although the size and scope of the mission will determine the actual augmentation required for each operation. The Corps PAO and his staff task organize the personnel and organizations available and allocate the equipment, communications support and facilities. When augmented, the Corps PAO operates a media operations center and establishes satellite centers as required.

2-12. Below the Corps level, the PA staff section organic to a war fighting headquarters is extremely austere. The mission of the PA section below Corps is to advise the commander by providing immediate planning expertise and guidance on issues with critical PA implications. The PA section deploys with the lead elements of the headquarters, and operates from the command's main CP.

2-13. PA units deployed to augment the staff sections organic to a headquarters are normally placed under the control of the supported PAO, who assigns the augmenting PA unit missions and tasks. He will do so in conjunction with the augmenting PA unit commander, who will retain command of his unit and ensure that his unit's personnel are not employed as individual fillers. Whenever possible, augmenting PA units should be linked with the supported command headquarters at that command's home station prior to deployment to facilitate establishment of command and control relationships.

PUBLIC AFFAIRS COMMUNICATIONS

2-14. Reliable, survivable, flexible communications are essential for effective PA command and control. In today's global information environment, information must flow to and from users, up and down the chain of command, and horizontally across the battlefield. Technology has compressed time and space and forward-deployed PA sections can be in direct communication with officials at DoD working PA strategies. The challenge is to ensure coordination and interoperability so that all elements have the communications capability necessary to effectively carry out their assigned mission, especially in today's joint, combined or interagency environment.

2-15. Deliberate, detailed planning can prevent communications shortfalls. PA planners assess their information transmission and reception needs and requirements. They then identify the communications capabilities they need access to, and determine the communications support they will need from command signal organizations. Through close coordination with the staff signal section, the identified PA communications requirements are integrated into the overall communication architecture.

INFORMATION OPERATIONS

2-16. The three central defining characteristics of the global information environment -- the facility of information acquisition and transmission, the speed of information communication and the breadth of information saturation -- combine to increase information availability. The American public, internal audiences, allies and adversaries have ready access to information. Information security is transitory and it is critical that information operations at every echelon are mutually supporting and directed at a clearly defined, decisive and attainable objective.

2-17. Credibility is essential for successful information operations. If an information source is not perceived as believable, then the desired effect of that communication cannot be achieved. Regardless of the source, target or objective of an information effort, in the GIE, credibility is founded in truth and enhanced by validation, corroboration, and consistency.

2-18. Commanders require integrated, coordinated, synchronized information operations. PA operations, which occur at, and impact on, the strategic, operational and tactical levels -- often simultaneously -- are a critical element of these operations. News media coverage of conflicting messages and information communicated by different elements of the command compromises credibility.

2-19. Integrating, coordinating and synchronizing every element of the commander's information operation -- Public Affairs, Psychological Operations, Civil Affairs, Combat Camera, Operations Security and others -- results in a synergistic information strategy. It minimizes the possibility of conflicting messages, which undermine credibility, jeopardize operations and endanger mission accomplishment.

LOGISTICS SUPPORT

2-20. Logistics is critical at all levels of command for Public Affairs mission success, during any phase of combat or garrison operations. Commanders must ensure their Joint Table of Allowances (JTA), Modified Table of Equipment (MTOE), Table of Distribution and Allowances (TDA) or Common Table of Allowances (CTA) reflects appropriate equipment levels to maintain a PA staff and media support under field and garrison conditions. Maintenance also plays an important role in Public Affairs operations. A Public Affairs element that has all its equipment cannot function properly if its equipment is inoperative, broken or deadlined. Each Public Affairs element must develop its own internal SOP in regard to logistics. (See Appendix K.)

2-21. Public Affairs staff members must be trained in the areas of supply, budget, property book, ordering, class A procurement, etc. Public Affairs must be an integral player in all mission and operational planning sessions to ensure logistical requirements are identified and resourced.

2-22. Responsibilities: The Public Affairs staff has the responsibility to identify to its resource manager, property book manager and ordering officer all fiscal and logistical requirements for field operations and home station support. Requisitions for equipment, supplies, services and allowances will be ordered and processed in accordance with appropriate Army Regulations, AR 710-2 *Unit Supply Update* and budgetary guidelines.

2-23. Requesting supplies: Commanders must ensure that equipment and components authorized by JTAs, CTAs, MTOEs, or TDAs are on hand or requested. The organization's supply operation is responsible for identifying, acquiring, accounting, controlling, storing and properly disposing of materiel authorized to conduct the mission of the unit and maintain the soldier. The organization is the foundation of the supply system. Exceptions and procedures are outlined in AR 710-2.

2-24. The Direct Support and General Support Activities provide class 1, 2, 3 (packaged and bulk), 4, 5, 7, 8 and 9 supplies directly to the using units on a customer support basis. These supplies are routinely procured through the unit supply rooms. In the event a Public Affairs element is operationally attached outside its assigned organization, it should coordinate before deployment for logistical support when possible. If prior coordination is not possible, contact for support should be made upon arrival into the theater of operations through the C-4, J-4, G-4, or S-4.

2-25. Accountability: All property acquired by the Army, regardless of source or whether paid for or not, must be accounted for, in accordance with applicable Army regulations and AR 710-2.

- Nonexpendable property is personal property that is consumed in use and that retains its original identity during the period of use. It requires formal property book accountability throughout the life of the item. It will be accounted for at the using unit level using property book procedures. Examples are desks, computers, file cabinets, chairs.
- Expendable items are property which is consumed in use or that loses its identity in use, and all items not consumed in use with a unit price

of less than $100 and not otherwise classified as nonexpendable or durable. It requires no formal accounting after issue to the user. The following classes or types of property will be classified as expendable.

- Supplies consumed in the maintenance and upkeep of the public service. Examples are oil, paint, fuel and cleaning and preserving materials.
- Supplies that lose their identity when used to repair or complete other items. Examples are assemblies, repair parts, and accessories.
- Office supplies and equipment (such as paper, staplers and hole punchers) with a unit price of less than $100.
- Durable property is personal property that is not consumed in use, does not require property book accountability, but because of its unique characteristics requires hand receipt control when issued to the user. Examples are hammers, lawnmowers, audiovisual production material and books.

2-26. Conservation of resources and property accountability is ultimately a supervisory responsibility. Property responsibility must be assigned and acknowledged in writing using hand receipts and property books as outlined in AR 735-5, *Policies and Procedures for Property Accountability* and AR 710-2.

2-27. Property book: Effective supply support at the using element or property book level requires timely and accurate processing of supply requests and receipts, accurate accounting records and adequate property control. Turn in, transfer, substitutions, hand receipt, etc., are accomplished in accordance with Army Regulations, AR 710-2 and logistical SOPs.

2-28. Budget: Budgets must be programmed in advanced. Organizations plan their budget in the previous fiscal year They must be established and managed with the appropriate command resource manager/budget analyst. When allocating funds consideration must be given to equipment replacement and upgrades, recurring supply needs TDYs, maintenance, contracts, etc. Normal operating funds are allocated/dispersed by a public affairs element operational headquarters; however, during deployments for exercises/operations funds may be available from the tasking headquarters up front or on a recuperative basis.

2-29. Maintenance: Public Affairs elements must maintain their equipment in a deployment ready state. Preventive Maintenance Checks and Services (PMCS) is an important part of the maintenance program, and is the user's responsibility. The Public Affairs element's operational headquarters provides maintenance support. For example, an embedded Public Affairs section assigned to the Headquarters Company of a separate brigade would seek maintenance support from the Headquarters Company, then the brigade maintenance section. The company/brigade's maintenance SOP would be followed for 1st, 2nd, 3rd, etc., echelons of maintenance. Maintenance for communications and data processing (computers) equipment is coordinated through the G6 (DOIM).

2-30. Local Purchase: Local purchases may be an option for procurement provided the action is in the best interest of the government in terms of timeliness, quality and cost. Local purchase requests must be made in accordance with AR 710-2. Approval for local purchase of nondevelopmental items starts at the first level of command authority and is accomplished in accordance with local policies and Army regulations. Nondevelopmental items is a generic term that covers materiel available from a variety of sources with little or no development effort from the Army. Sources include commercial items which fully meet an approved need, items being used by other U.S. services or agencies or items used by military or other agencies of foreign government. Most of these purchases are covered under the IMPAC Card program. Local guidance covers implementation.

2-31. Class A Procurement/Credit Card: Class A agents and government credit card holders must be identified and trained prior to their ability to accomplish those functions. Training is routinely accomplished at the installation level. Purchases for other than national-stock-numbered items are routinely accomplished using Class A agents and ordering officers, and the U.S. Government Credit Card.

2-32. SSSC: Self Service Supply Centers are managed at the installation or theater level. Users are required to have a valid SSSC account. Accounts are identified by DODAAC or UIC. Subaccounts are authorized IAW AR 710-2. Field resupply centers are often established at divisional-level logistics bases.

2-33. The best means of ensuring supply discipline is to be proactive and not reactive in supply operations. Enforcing compliance with regulations requires constant emphasis.

Chapter 3

Public Affairs Planning

During Operation Just Cause, PA planning and integration were inadequate. Commanders at all levels failed to involve public affairs officers in planning from fear of OPSEC leaks. The results were insufficient PA guidance provided to soldiers, family members and commanders; sometimes miscommunication to and confusion within Army family elements; and misuse of PA assets. A matter of urgent concern was the failure to plan for and use Reserve Component PA assets to relieve the pressure on an already small active PA force so that it could better handle both internal and external communication.

- Public Affairs After Action Report, TRADOC LLC, Phase II, Sept. 90

COMMUNICATING INFORMATION

3-1. Public Affairs operations assist the commander in communicating information and messages about his force and the operation to internal and external audiences. Like other operations, PA operations are conducted to bring about clearly specified, meaningful objectives, which support the commander's intent and contribute to mission success. Those objectives are defined in terms of the effect the PA operations are intended to have on target audiences -- the impact on target audience behavior that is desired -- and are measurable.

3-2. Once PA objectives are defined, PA operations are planned and executed to achieve those objectives. PA operations focus on the communication process -- an on going, dynamic, ever-changing process. The communication process is composed of elements involved in receiving, collecting, analyzing and interpreting data, identifying and analyzing audiences and formulating and transferring messages. This process is used to bring about a specified objective, while measuring and analyzing the outcome and effectiveness of the effort.

3-3. To support the commander's effort to communicate, PA professionals concentrate on five basic functions or core processes — planning, media facilitation, information provision, force training and community relations.

3-4. This chapter focuses on the systematic process for Public Affairs planning and decision-making. It addresses the information environment and the impact of information at the strategic, operational and tactical levels across the range of operations that requires public affairs considerations be totally integrated into the planning and decision-making process. Doing so enables PA personnel to prepare for potential situations, to synchronize efforts with other agencies that manage information communication, and to more successfully influence the coverage, interpretation and understanding of events. It limits the need for reactive, defensive attempts to buy time or control damage.

3-5. PA planning prepared in support of the CINC's theater campaign plan requires a series of decisions related to policy at the national level and the techniques at the tactical level. From policy to techniques, however, basic planning considerations are the same: What should the PA objective accomplish? With what audience? When? How? PA planning must not only be done at all echelons and within national policy but also within the limits of operational plans and capabilities.

TYPES OF PLANS

3-6. The amount of time available significantly influences the planning process. Two different methods of planning are described in the JCS-published Joint Operations Planning and Execution System (JOPES).

3-7. **Deliberate or Peacetime Planning** is the process used when time permits the total participation of commanders and staffs. Development of the plan, coordination among supporting commanders and agencies, reviews by staffs, planning conferences and development of proposed public affairs guidance can take many months. Deliberate or Peacetime Plans are prepared in prescribed formats--the complete operational plan (OPLAN) or the conceptual operational plan (CONPLAN).

3-8. **Time-Sensitive or Crisis Action Planning** (CAP) is conducted in response to crisis where U.S. interests are threatened and a military response is being considered. Crisis Action Planning is carried out in response to specific situations as they occur and that often develop very rapidly.

3-9. It is within the CAP process that established, working relationships between the PAO and operational planning staffs are crucial to the inclusion of PA considerations into OPLANS and OPORDS.

3-10. Both deliberate and crisis action planning are conducted within JOPES. Joint Pub 5-03.2, JOPES Volume II, describes detailed administrative and format requirements for documenting the annexes, appendixes, etc. of operational plans, and conceptual plans, the products of deliberate planning.

3-11. The purpose of JOPES is to bring both deliberate and crisis action planning into a single architecture to reduce the time required to complete deliberate or crisis action planning. This makes the refined results more readily accessible to planners, and makes it a more manageable plan during execution.

3-12. The overall procedures are the same, at all echelons, for both deliberate and crisis action planning.

- Receive and analyze the task to be accomplished
- Review the situation and begin to collect necessary intelligence
- Develop and compare alternative courses of action
- Select the best alternative
- Develop and get approval for its concept
- Prepare a plan
- Document the plan

Operation Just Cause

The basic problem---planning. No discussion concerning use of personnel can proceed without an understanding of the planning problem. Public Affairs in general was not sufficiently planned for by leaders or public affairs officers for Operation Just Cause. PAOs were not given time to plan. Only outstanding unit mission accomplishment, American public support and the hard work of public affairs personnel, prevented major PA failings in Panama. A longer duration and less popular action could have turned into a public affairs disaster.

The SOUTHCOM, XVIII Airborne Corps, and Army Special Operations Command PAOs were not informed that the operation would occur until 17 December and then they were given instructions not to discuss it with key persons on their staff. The 82d PAO did not learn of the operation until 18 December, 7th ID PAO, 19 December and OCPA **one hour** before H hour.

More critical however, was the absence of joint coordination. OASD-PA received a PA plan from SOUTHCOM public affairs in November, but the plan was never coordinated due to worries concerning possible security leaks. The XVIII Airborne Corps PAO indicated that he also knew of the operation in general terms in November but was unable to coordinate planning with the Director of Public Affairs SOUTHCOM. Because the plan was not staffed, OCPA was caught unaware. Divisional public affairs officers all indicated that they were not sufficiently drawn into planning. Some PAOs said they never saw a plan. It is obvious that sufficient public affairs planning did not occur **at any level**.

There was also a ripple effect downward caused by the lack of staffing by the Joint/OASD-PA. In addition, senior leaders at division level and above did not draw PA officers into planning to maximize the limited planning time that was available. Commanders strongly complained about poor balance of coverage in the media, inability to send command information at the same pace as civilian media reporting, and lack of sufficient public affairs guidance; yet senior leaders are reluctant to draw PAOs into the planning process to prevent these problems from occurring. Army leaders must come to grips with this dilemma.

(Public Affairs After Action Report TRADOC Lessons Learned Collection Phase II. Sept 1990)

3-13. An OPLAN is a complete, detailed plan. It includes a description of the concept of operations from the commander's perspective and presents additional annexes provided by various staff sections which identify specific functional area requirements, restrictions, limitations, or considerations. The inclusion of a public affairs annex is essential to successful integration of PA principles and guidance into the OPLAN.

3-14. A CONPLAN is an abbreviated operational plan, which requires considerable expansion or alteration to convert it into an OPLAN or OPORD. Detailed support requirements are not included. The commander determines what annexes will be included to complete the CONPLAN.

3-15. A Public Affairs Estimate is an assessment of a specific mission from a Public Affairs perspective. It is an examination of critical Public Affairs factors, their influence on the planning and execution of operations, and their potential impact on mission success. The senior PAO at each echelon is responsible for consolidating information and preparing the PA Estimate. A sample PA Estimate is included at Appendix C.

3-16. The Public Affairs Annexes to OPLANS or CONPLANs provide the details and instructions necessary to implement Public Affairs media facilitation, news and information provision, and force training operations. It is coordinated with all staff agencies, especially those that significantly impact the information environment -- Psychological Operations, Civil Affairs, Signal, and Military Intelligence -- to ensure that Public Affairs activities are synchronized with other activities.

DELIBERATE PLANNING

3-17. A commander continually faces situations involving uncertainties, questionable or incomplete data, or several possible alternatives. As the primary decision maker, the commander, with the assistance of the staff, must not only decide what to do and how to do it, the commander must also recognize if and when to act. How the commander arrives at a decision is a matter of personal determination. However, superior decisions (those, which offer the best solution, decisively, at precisely the correct time,) result from the commander's thorough, clear, and unemotional analysis of facts and supported assumptions. This is done through the "deliberate planning process."

3-18. To support the commander's decisions and command objectives, the PAO must develop a thorough, clear, comprehensive public affairs strategy. This strategy allows the PA to link public affairs considerations into planning for contingency, future and current operations. With the PA strategy, the PAO defines the public affairs perspective of the operation, and identifies how the Army public affairs involvement in this operation supports strategic goals. It provides the intent for PA operations and the Army approach to meeting the information needs of critical internal and external audiences. It is the framework for defining and developing the PA scheme of operations.

3-19. Based on the PA strategy, PA plans are developed for integration into OPLANs. A PA plan is produced by the operational commander's PAO, and it details the media facilitation, news and information provision, and force training and support procedures which will be employed in support of the operation.

3-20. The first crucial step in fulfilling the PA strategy requires the PA Plans officer/NCO to establish and maintain a routine, ongoing relationship with operational planners within the organization. The PA plan is coordinated with key staff agencies, integrated into the OPLAN as a PA Annex. Synchronization with these other activities ensures services and support required by the PAO is provided and multiplies the impact of the PA plan. This process is followed at subordinate echelons as planning guidance is communicated down the operational chain of command.

3-21. There are five phases in the deliberate planning process. Items in parenthesis identify PA actions performed within each phase:

- **Phase I** - Initiation. The task assigning directive outlines the major combat forces available for planning, gives general planning instructions, lists assumptions for planning, and specifies the product document such as an OPLAN, CONPLAN. (PA planners begin assessing the information environment, its impact on operations and the PA requirements to operate within a specific arena.)

- **Phase II** - Concept Development. (Using the supported CINC's mission statement and concept of envisioned operations, the supporting PA planners analyze the mission, formulate tentative courses of actions and develop the PA Estimate for the operational scenario and requirements.)
- **Phase III** - Plan Development. Subordinate commanders use the CINC's concept and the allocated major combat forces as the basis to determine the necessary support, including forces and sustaining supplies for the operation. (The PA planners provide the CINC with recommendations for public affairs assets required, phasing of PA forces and support into the theater of operations, and perform a transportation analysis of their movement to the destination to ensure that the PA segment of the entire plan can feasibly be executed as envisioned. For the supported CINC's PA requirements, above those organic to the tasked major combat elements, the supporting commands [force providers] of each service, as much as possible, identify real-world PA assets to take part in the plan and sustainment to meet requirements. The supporting command identifies PA requirements in OPLANs, OPORDs, and taskings, through operational channels, to major subordinate commands.)
- **Phase IV** - Plan Review. The review process is more than a single phase in deliberate planning. The Joint Staff performs or coordinates a final review of operations plans submitted by the combatant CINCs. It is a formal review of the entire operation plan. Approval of the plan is the signal to subordinate and supporting commands to develop their plans in support of the CINC's concept. (PA planners do not wait until the plan is approved before beginning to develop their supporting plans; they have been involved in doing this, while coordinating with their command's planning staff. In the meantime the CINC has been building the overall plan.)
- **Phase V** - Supporting Plans. The emphasis in the Supporting Plans Phase shifts to the subordinate and supporting commanders. (This is the phase in which PA planners begin to concentrate on how to meet tasks identified in the approved operation plan by preparing Public Affairs Annexes to supporting plans. This input outlines the actions and relationships of assigned and augmenting PA assets.)

INFORMATION ENVIRONMENT ASSESSMENT AND PA ESTIMATE

3-22. Planning fosters effective application of knowledge, logic, and judgment. Analysis of the information environment (IE) starts the process used to develop an estimate of the situation.

3-23. The IE analysis provides the basis for the development of all PA operational plans and is a channel for integration of strategic, operational and tactical planning guidance. It is a method of identifying factors within the information environment that have potential implications for the planning and execution of Army operations. PA planners study and evaluate the dynamics of the area information environment to identify specific public affairs operational considerations.

3-24. Analysis of the information environment focuses on research into the following areas, which will be put into the "Situation and Considerations" portion of the PA estimate:

- Information infrastructure
- Media presence
- Media capabilities
- Media content analysis
- Public opinion assessment
- Information needs assessment
- Impact assessment/courses of action (COA)

3-25. An analysis of the IE using this approach builds a complete picture of the conditions facing commanders and their PA forces, providing them the tools necessary to anticipate trends, actions, issues, and conflicts. The PA staff officer or NCO conducts research and assessment for the estimate then evaluates, prioritizes, and suggests courses of action that public affairs can best support, while considering the information environment. To acquire all the information necessary for an accurate picture of the operational environment, the staff officer or NCO must work closely with intelligence, civil affairs, psychological operations, military police, visual information and other staff sections involved with information gathering.

3-26. The PA assessment must include those aspects under the control of the commander, as well as those the commander cannot control. This can only be achieved with a thorough integration of PA planning at all stages and into all aspects of the planning and decision-making process. Although a variety of techniques may be used in the analysis of the IE, the PA assessment should address the following primary categories.

- **Information Channels and Infrastructure**. This element focuses on an assessment of the information infrastructure. It addresses the resources, communications facilities, organizations, and official and unofficial information channels available within the area of responsibility (AOR). It addresses the means to transmit and receive unofficial information. It addresses specific requirements for American Forces Radio Television Service (AFRTS) information services and the availability of assets to meet theater requirements. It identifies the availability of host nation telephone service for voice and data transmission, the accessibility of audio/video channels, the prevalence of private communications devices such as cellular telephones, facsimiles, computers with modems, radios and televisions, and the nature of the information available through these information channels. It addresses alternate means of voice and data communications, whether military or government contracted, for use in the absence of host nation information channels and infrastructure. Much of the information required for this category may be obtained through civil affairs or psychological operations elements assigned or attached to the command and U.S.Information Service offices supporting consulates or the embassy within the area of operations.

- **Media Presence**. This is an assessment of the media presence in the area of operations prior to the introduction of American forces and an assessment of the expected level of media presence commanders should anticipate once deployment begins. It includes a description of the type of media (print or broadcast), the visibility of the media (local, national, or international; American or foreign), and the focus of the news media present (news or entertainment) covering the operation. The assessment of the media presence should address the authority under which media representatives are operating (open or closed borders, and free press or controlled press) and the reporters' degree of access to the theater of operations.
- **Media Capabilities**. This element is an assessment of the media's information collection, production, transmission and communication capabilities in the AOR. This element analyzes the technological capabilities of the media representatives present within the AOR. It describes their level of sophistication (if they must transport products out of the area of operations for transmission to parent media or do they have self-contained interactive satellite telecommunications access). It also addresses the media's level of logistics support and its potential impact on Army commanders who are required to provide the media free and open access to the AOR. It includes information about their transportation assets, resupply channels, and equipment maintenance requirements. Additionally, the media's general ability to provide their own security should be assessed.
- **Media Content Analysis**. Media content analysis is an assessment of news coverage, the media's agendas and an analysis and prioritization of the potential strategic and operational issues confronting the command. Media content analysis assesses what is being said, by whom, and how it is being presented. It is a constant process that must begin well before planning for a specific operation begins and continues through Mobilization, Deployment, Employment, Sustainment, and Re-deployment. Content analysis reveals the meaning, tone, and accuracy of messages, how the information was presented, and the cumulative affect of the information. A media content analysis will provide an evaluation of the quantity of coverage, both in and out of theater, and the nature of that coverage. This will assist the commander to understand the strategic context, the measure of success and the definition of an end-state for the operation as viewed from outside the command and the Army itself. It will also be an essential element of friendly information (EEFI), as explained in FM 3-13 (100-6), in determining objectives and strategies for communicating the Army perspective, and for working to achieve a balanced, fair and credible flow of information.
- The specific methods for conducting a media content analysis are explained in Appendix O.
- **Public Opinion**. A public opinion assessment surveys the national and international attitude about the operation and the command, leaders and soldiers conducting it. This assessment looks at the perceptions held by major audience and coalition groups, and the relative solidity or strength of those attitudes. It addresses the perceptions held by international

audiences: those traditionally allied with the United States and those traditionally considered to be adversaries of the US. The public opinion assessment should include as a minimum, consideration of the following groups:

- American public (general)
- Civilian political leadership
- Coalition and allied forces and their general population
- Host nation citizens
- International public
- Internal command audience
- Home station community
- Specific special interest groups (if needed)
    - In determining the effects of the media on public opinion, there are three general types of evidence which explain behavior response: direct indicators; indirect indicators; and post-event sampling. Direct indicators are evidence that provide a direct link between the information received by the public and the behavioral response. These indicators include but are not limited to: personal interviews and surveys to estimate awareness and understanding of an issue; dissident group marches, meetings, advertising and other activities; monitoring internal and external law, order and discipline activity; and chain of command after action reports, staff journals and duty logs. Indirect indicators are evidence that identifies behavioral response generated by separate events or activities which appear to be the result of reception of media information. These indicators include: cause-effect estimates from information products and sources other than the military or civilian commercial media; interest level in news media products; shifts in social or economic trends; shifts in political support.
    - Post-event sampling considers the qualitative and quantitative statistical evidence that identifies the level of and nature of awareness and behavioral response to information. This includes the results of surveys, interviews, group observation, probability and non-probability samples, which will identify if and how the public was influenced by information products or messages.

- Information Needs. This is an assessment of the information needs and requirements of the previously identified key publics. It analyzes and prioritizes key external and internal audiences and assesses their news and information expectations. It identifies the types of information that should be made available to soldiers, their family members, other home station community audiences, the American public, and the host nation local populace. It will identify other audiences, such as allied or adversary leaders and publics that will be interested in available "cross-border" information.

PA ESTIMATE AND PA GUIDANCE COORDINATION

3-27. The purpose of the PA Estimate is to determine whether the mission can be accomplished and to determine which COA can best be supported by public affairs. In preparing its estimate, the Public Affairs staff:

- Reviews the overall mission and situation from the public affairs and information environment perspective.
- Examines all public affairs factors impacting on or impacted by the mission.
- Analyzes each COA from the public affairs perspective.
- Compares each COA based on the public affairs functional analysis.
- Concludes whether the mission can be supported by public affairs, and from the public affairs perspective, which COA can best be supported.

3-28. The Public Affairs Estimate summarizes the information environment, prioritizes the major issues confronting the command and predicts anticipated outcomes in detail. It measures the effectiveness of previous and current information strategies, and based on this evaluation, identifies possible courses of action to support command PA objectives. The PA Estimate also contributes to the development of Public Affairs Guidance (PAG) for specific operations or missions. PAG is a primary tool that guides commanders and PA leaders in the application of doctrine and policy during operations. PAG provides the PA force at all echelons standard operating procedures.

3-29. But to be effective, PAG must be developed with the needs of the front-line PA force in mind. PA planners must be able to "see" and "feel" the battlefield. They must have an understanding of the information environment and how it will change throughout the operational continuum. They must be aware that all the resources available at the planning headquarters may not be available or feasible in the theater of operations. Issues that need to be addressed include information release authority restrictions (national, theater or local). These restrictions often place the PA leader in a difficult situation -- one in which an overwhelming number of news media on the scene will seek answers to legitimate questions about unfolding events -- activities that the PA leader cannot discuss. The result is a loss of credibility for the Army.

3-30. DOD policy requires that proposed PAG be provided to the Assistant to the Secretary of Defense for Public Affairs (OTASD-PA) by the unified, specified and other major commands for all operations. This requirement includes major joint training exercises that could attract national and international attention. Subordinate command PA leaders should conduct continuing PA assessments as a part of mission training for theater-specific contingencies in anticipation of PAG development requirements there.

PAG DEVELOPMENT

3-31. Upon receipt of a "warning order," the commander, through the PA staff, will begin development of proposed PAG. In reality this warning order may be preceded by a telephonic "heads up" call from a PA planner at a higher headquarters that allows PA planners to begin working on proposed PAG before the hard copy of the warning order arrives. This proposed PAG should be based on the warning order or other planning guidance, the proposed command operations plan (OPLAN), and the PA Estimate. Once the proposed PAG is developed, it is staffed through command staff. Once approved by the commander, it is forwarded through major command and Unified/Specified command PA channels to DoD.

3-32. Commanders of Unified/Specified commands should ensure the proposed PAG is coordinated with appropriate elements and functional staffs within the theater of operations. This includes governmental and non-governmental organizations such as the State Department and its embassies, civil affairs, country assessment teams, host governments, allied force public affairs teams, the four U.S.military services and their subordinate commands.

3-33. PA leaders at all levels, specifically major command and above, should work to approve PAG as quickly as possible in order to provide subordinate PA leaders the opportunity to develop and implement PA strategies to support their commands. The format for PAG is included as an appendix to this manual at Appendix E.

PA ANNEX DEVELOPMENT

3-34. Once the PA estimate and proposed PAG are completed, and the other staff officers have completed their estimates, the commander selects a course of action. The commander then outlines it to the staff. The commander may select one of the proposed COAs, a combination of two or more, or a completely new one. The PA staff must then be prepared to enter the plan development phase (Phase III) which requires development of a PA annex. A format for a PA Annex is included at Appendix D.

3-35. The operation-specific approach to conducting public affairs activities is called a PA scheme of maneuver. This PA scheme summarizes the commander's PA intentions, and details the media facilitation, news and information provision, and force training and support procedures, which will be employed to support a particular operation.

3-36. The PA scheme consists of the PA estimate of the situation, higher command PA guidance, and the selected course of action. It is coordinated with key staff agencies, integrated into the operation plan through the development of a PA annex, and synchronized with the other activities to be executed as part of the basic plan.

3-37. The PA scheme, when included in the PA annex, should not only identify public affairs force requirements for the operation, but more importantly, it must provide the commander a visual picture of how public affairs will support the commander's concept of operation as outlined in the plan.

3-38. The PA activities addressed in the PA scheme of maneuver are:

- **Media Facilitation**. Media Facilitation is activities executed to support news media efforts to cover the operation, facilitate the timely, accurate, balanced provision of information which communicates the Army perspective, and minimizes the media disruption of operations or endangerment of mission accomplishment. Media facilitation is accomplished by the early establishment of a media center as the focal point for media representatives seeking to cover the operation. Normal media center operations include scheduling briefings, coordination for interviews; responding to media queries; coordinating unit visits and media escort requirements; and resolving media - military incidents. To prepare for encounters with the media, commanders must accept and understand the role of the news organizations and the journalists in the theater, and their capabilities in getting information from the battlefield or area of operations. Commanders must provide media access to the force, keeping in mind the impact their technology will have on operations security. Commanders must identify and provide support and resources to assist the media in their mission.

- **Information Strategies**. Activities executed to fill the news and information needs and expectations of internal and external audiences. Proliferation of personal computers, the World Wide Web, the Internet, on-line services, fax machines, E-mail, cable television, direct broadcast satellites, copy machines, cellular and wireless communication and many other information technologies have created an endless stream of data and information that flow into a world filled with images, symbols, words, and sounds. Public affairs specialists acquire information using a variety of sources. Because of the volume of information and the vast number of potential distribution mediums, the PA staff uses a systematic acquisition strategy. They acquire information from participants, leaders, developed sources, the media, research and development, intelligence, culture at large, and subject matter experts. Print, video, audio and electronic information products are provided to deployed soldiers, home station audiences such as family members and the home station community and news media representatives using contracted services and organic military assets. They communicate the Army perspective and contribute to timely, balanced coverage of the operation.

- **Force Training and Support**. Activities executed to assist members of the DA community in interacting with media representatives. Force training and support are conducted to educate soldiers, family members and DA civilian employees on their rights and responsibilities with respect to news media representatives attempting to provide coverage of an operation and related issues. It focuses on helping them to respond when they encounter news media representatives seeking interviews, photo opportunities, responses, reactions, interpretations or comments on an operation, policies or events. The intent of force training and support is to assist members of the community and media representatives in approaching each other with mutual respect. Training for public affairs personnel expands on soldier and unit leader training. It stresses individual as well as collective tasks with an aim of developing units fully prepared to accomplish the range of public affairs missions. It integrates public affairs into the battle staff and trains PA planners to assess the operation environment from a public affairs perspective, produce a PA Estimate, develop the PA Annex and PA Guidance.

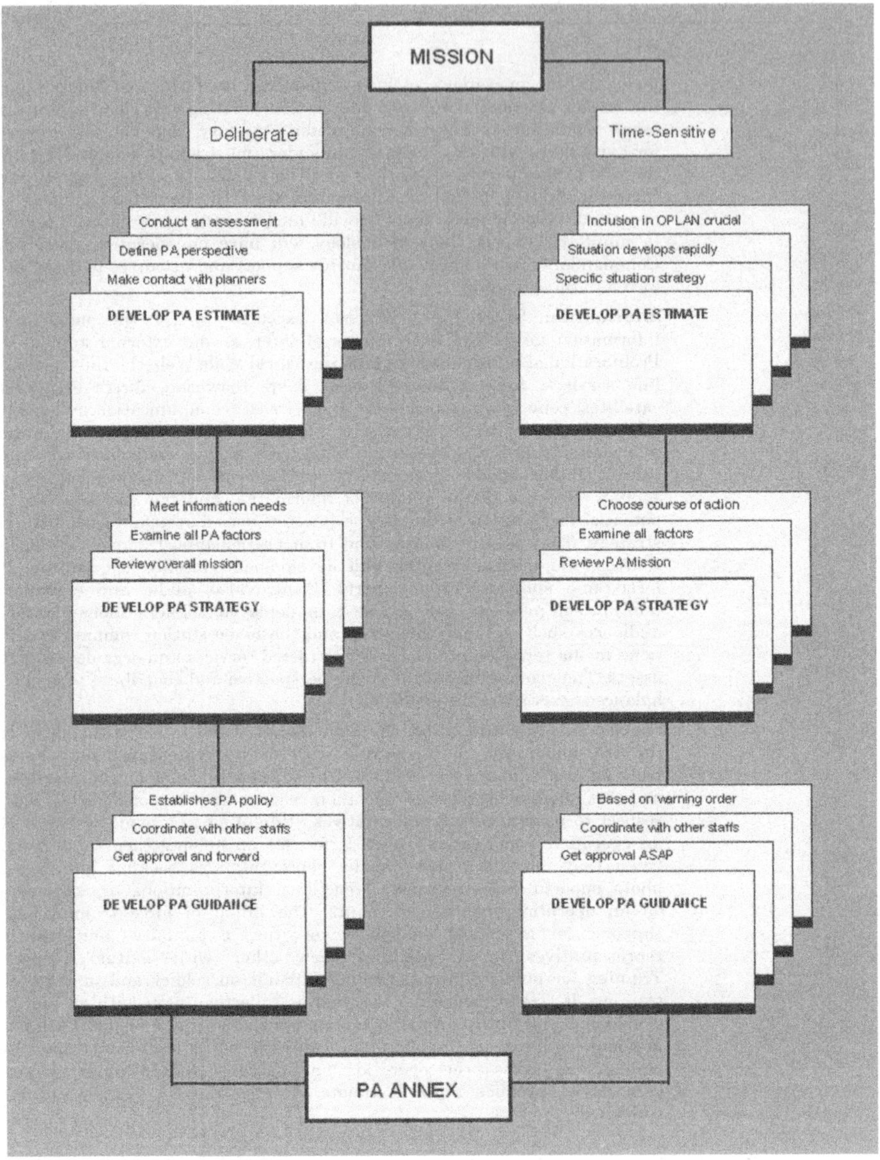

Figure 3-1. PA Planning

PA INTEGRATION INTO STAFF PLANNING

3-39. Concurrent with formulation of the PA staff estimate, PAG development, and production of the PA Annex to the OPLAN, PA planners must be an integral part of the staff planning process, especially on the following matters.

3-40. **Force Planning**. In force planning, the PA staff works with the J3/G3/S3 staff. Force planning consists of PA force requirements determination, force list development and refinements in light of PA force availability and PA force shortfall identification and resolution. In force list development, the PA assets needed to meet the mission are identified. Force availability is considered based on the strength and readiness of organic PA units, their personnel and equipment. Identification of PA force shortfalls addresses the lack of organic or mission-capable PA assets and the additional requirements and augmentations for PA units and personnel needed to accomplish the CINC's concept of operations. All taskings for unit or personnel augmentation must be validated and requested through the J3/G3/S3 operational channels. Tasking authority does not exist PA staff to PA staff or PA staff to subordinate unit.

3-41. **Support Planning**. To plan for logistical support of PA units and personnel assigned to carry out the CINC's concept of operations, the PA staff coordinates with and identifies support requirements to the J4/G4/S4. Specific logistical areas of concern include support in all classes of supply to the PA force, availability and authorized levels of support to civilian media, local purchase and contract support, property accountability, and vehicle transportation and maintenance support as tasked for through the J3/G3/S3.

3-42. **Transportation Planning**. PA forces move from their home station to a specified destination in the theater, either as part of their parent organization or a task-configured PA unit. This movement involves planning by several echelons of command, possibly stops at several intermediate locations en route, and a schedule constrained by a variety of operational requirements and priorities. Key staff for the PA planner to interact with include the command's transportation officer, movements control officer, and staff officers within the J3/G3/S3 and J4/G4/S4 that have staff supervision in this area. Key items PA planners need to track in this process are:

3-43. **Time-Phased Force and Deployment Data (TPFDD).** The TPFDD is the JOPES database portion of an operations plan. It contains time-phased force data, non-unit related cargo and personnel data, and movement data for the operation plan. The Appendix 1 to Annex A of the operation plan is the Time-Phased Force and Deployment List (TPFDL) which identifies types and/or actual units required to support the operation plan and indicates origin and port of debarkation or ocean area. It may also be generated as a computer listing from the TPFDD. PA planners must ensure that the TPFDD/TPFDL contains their unit line numbers (ULNs) for units, personnel, or cargo. Assets not listed on the TPFDD do not deploy. It is crucial to the planning process that the PA staff closely coordinate with the J3/G3/S3 and J4/G4/S4 to ensure that PA assets are reflected on the TPFDD or included as organic/attached assets to parent units with validated ULNs on the TPFDD.

3-44. **Destination (DEST)** - the geographic location where the force is to be deployed/employed.

3-45. The distances between the port of debarkation (POD) within the theater of operations to the destination (DEST), to the port of support (POS), to the marshaling area or assembly area. Where troops land at the APOD (Aerial Port of Debarkation) or SPOD (Seaport of Debarkation) they may be substantial distances from the port where the PA element's equipment arrives in theater and operations begin.

3-46. Transport of equipment must be planned for it to be available at the earliest possible date within the theater of operations. Thus, the PA planner must set a realistic, achievable required delivery date (RDD). This must be aligned with the CINC's required date (CRD). Planners begin with the RDD to establish two interim dates, the earliest arrival date (EAD) and the latest arrival date (LAD). Once these dates are established, then the ready to load date (RLD) and the available to load date (ALD) are established at home station to meet the earliest departure date (EDD).

3-47. **Communications/Automation Planning**. Key planners on the coordinating staff for communications and automation planning are the J6/G6, G3/S3 and CE officer. Specific concerns include priorities for radio/telephone communications, satellite uplinks and downlinks, number of telephone links/trunks allocated to PA requirements, E-mail access, and inclusion into the Communications Electronics Operating Instructions (CEOI). In addition, consider possible development of web pages or sites, like BosniaLink, the Task Force Eagle Homepage, the Desert Voice in Kuwait or Task Force Falcon in Kosovo.

3-48. **Information Environment**. When formulating PA plans and coordinating on the overall plan with the commander's staff, PA planners work closely with the staff element having supervising responsibility for each separate issue.

- Operational Security - G3/S3
- Psychological Operations - G3/S3
- Civil Military Operations - G5/S3
- Combat Camera Operations - G3/S3
- Armed Forces Radio and Television Operations - G3 and Armed Forces Information Service

APPLYING METT-TC

3-49. To function as part of a deployed or deployable organization the PAO and PA NCO must think and state requirements in terms that the rest of the organization can understand. Moreover, the PAO must fit the operational PA requirements into the operational planning procedure of the organization about to deploy. Operators think in terms of mission, enemy, terrain and weather, troops, time available and civilian considerations or METT-TC. METT-TC is used to envision how the operation will occur, to identify potential risks or hazards, and to define troop and equipment requirements.

3-50. **Mission** - alert, marshall, deploy, operate (internal information, media facilitation, information provision) redeploy

3-51. **Enemy** - rumors, disinformation, propaganda, OPSEC

3-52. **Terrain and weather** - theater of operation, theater of war, area of responsibility, intermediate staging base, homestation and weather condition

3-53. **Troops** - embedded assets, units (PADs, MPADs, BODs, PAOCs) AFRTS, HTNRs, ABS, NBS, JIBs, Star and Stripes, stringers, surrogate PAOs, Adjutants, and S1s. All other AG services, signal, USIS, DoS and homestation forces and audiences

3-54. **Time available** - timeline, transition, reports

3-55. Essentially, the information required for a METT-TC analysis is provided by the PA estimate of the situation, which contains the selected course of action and detailed descriptions of PA actions to be performed. These requirements are then translated into the command's planning language and format, resulting in the PA Annex to the OPLAN or OPORD.

3-56. At theater level and above, the PA annex is normally Annex F to the OPLAN. At corps and below, commanders can tailor their plans to fit specific needs or preferences, so the PA annex may fall in another location among the annexes. Regardless of where it is located, the PA Annex is used to provide information about the conduct and execution of public affairs operations in support of the basic OPLAN. The PA annex outlines the situation, identifies the specific PA mission and explains the concept of the operation. It also provides detailed information and guidance PA personnel need to conduct successful PA operations at the operator level. A sample PA Annex format is included in this manual at Appendix D.

3-57. Phase IV, Plan Review, consists of staff coordination and plan adjustment or correction.

3-58. The final phase of the planning process, Phase V, Supporting Plans, follows the same course as the first three phases, with attention aimed at the specific aspects of the overall plan. These supporting plans focus on conducting specific operations, which must be successful in order to guarantee success of the larger mission. PA support to these supporting plans is as important as PA coordination and input to the main campaign plan.

STANDING OPERATING PROCEDURES

3-59. Integral to the operational effectiveness of PA sections are their standing combat operating procedures. These routine procedures ensure that all members of the section are working in concert toward the same PA objectives and that PA activities are easily blended into the actions of the command's staff.

3-60. PA SOPs differ from PA plans and PA annexes to OPLANS in that they specifically detail and describe how PA is conducted within a certain command or unit. They are routine procedures and actions that apply to each section or unit.

3-61. The senior PA NCO prepares the staff section or unit PA SOP. PA units designated to support or augment specific commands in the execution of contingency missions should use SOPs from these supported commands.

3-62. SOPs should address:

- Preparation for combat. Stockage, prepackaging, and maintenance of vehicles, equipment, and expendable and nonexpendable supplies.

- Vehicle load plans.
- Alert and mobilization actions, routines and procedures.
- Composition of quartering and/or advance parties and rear echelons.
- Organization for combat, including detailed delineation of duties for each individual, shift compositions, and plans for reconstitution in the event of combat losses.
- Operations center and media center layouts (theater, corps and/or division main/rear CPs).
- Procedures for preparing, disseminating and disposing of records, reports, estimates and orders.
- Physical, document, and tactical security.
- Communications procedures. These steps include radio/telephone operating procedures unique to the command, message routing and preparation formats, and operation of communications and data transmission equipment.
- Movement and displacement.
- Operations under NBC conditions.
- Field Maintenance.
- Personal hygiene, rest, and morale, welfare and recreation requirements and procedures during deployment.
- Post-operations and reconstitution procedures. Maintenance, restocking and packaging composition of advance and rear parties; disposition of records, and preparation of after-action reports are included.
- A PA SOP outline is included in this manual at Appendix K.

POST-MISSION PLANNING CONSIDERATIONS

3-63. During mission planning and preparation, Public Affairs planners should consider debriefing and other post-mission activities. These activities normally include:

- Collective debriefing of the operational element on all aspects of mission execution, including lessons learned.
- Collection of maps, notebooks, logbooks, plans, annexes, duty officer/NCO logs, serious incident reports, news releases, tapes and transcripts of news briefings and conferences, and all other information products pertinent to the mission after action report.
- Maintenance and storage of unit and personal equipment.
- Individual debriefing of key personnel.
- Other reconstitution measures as required.

3-64. Upon completion of these activities, the operational element begins pre-mission sustainment training or prepares for its next mission. The planning staff begins review of lessons learned for integration into future plans. See Appendix W for information on producing PA Lessons Learned for the Center for Army Lessons Learned.

Chapter 4

Media Facilitation

"It is likely that small pools of news media will be assigned directly to operational units to cover all facets of activity. With few exceptions, there will be no security review of media copy or audiovisual products. The policy will be to maintain security at the source. It is important to support the efforts of the media and our dealings with them should not be confrontational, but professional and courteous."

- GEN Binford Peay
Commander, U.S. Central Command [1994]
FM 3-61 (46-1), Public Affairs Operations

PRINCIPLES OF INFORMATION

4-1. In the past 20 years the Army has undergone a fundamental shift in our approach to dealing with the news media. In response to the perceived treatment by the press during the Vietnam War, we have gone from adopting an exclusionary tactic for the conduct of the invasion of Grenada in 1983, to managing the controversial pool system for covering the initial stages of the Gulf War in 1991, and more recently evolving into an almost completely open access policy in Somalia (1993), Haiti (1994) and Bosnia (1995) operations.

4-2. The results of this policy evolution are the "DoD Principles of Information," which form the foundation for the PA function of media facilitation. The basic approach that DoD and the Army take to media facilitation is contained in Appendix A, The DoD Principles of Information and Appendix B, The Guidelines for Coverage of DoD Combat Operations.

4-3. Simply stated, media facilitation is providing assistance to civilian and military news media representatives covering an operation. The objective of media facilitation is to support news media efforts. This includes providing accurate, timely, balanced, credible coverage of the force and the operation, while minimizing the possibility that media activities will disrupt the operation.

4-4. Media facilitation includes assisting media entry into the area of operations, registering media representatives, orienting them on the ground rules for coverage and ensuring that they understand the security policies and constraints under which they must agree to operate if they desire Army support.

4-5. Media facilitation also involves arranging interviews and briefings, coordinating unit visits and escorts, and assisting media representatives with transportation, messing, billeting, communication support, safety and equipment. Media facilitation involves the early establishment of a media center as a focal point for media wishing to cover an operation, for Army personnel seeking assistance with media representatives in their area, and for resolution of problems or incidents resulting from media-military interaction.

4-6. A primary strategic goal of any Public Affairs staff is to support an operational commander in achieving a constant flow of complete, accurate and timely information about the mission and U.S. forces.

4-7. The PA staff accomplishes this goal by making information fully and readily available within the constraints of national security and OPSEC, and by facilitating inclusion of civilian and military news media representatives in military units whenever possible.

PA STAFF AND MEDIA CENTER RESPONSIBILITIES

4-8. **PA Staff Sections.** The prime focus of the PA staff is staff coordination. The staff will be the element tasked with executing the media facilitation strategy.

4-9. The staff provides PA planning and operational guidance to the PAO. They ensure leaders within the command understand the commander's media relations policies, and serves as the command ombudsman in the settlement of conflicts between the media and the military.

4-10. As an active participant in the command's information planning element, the staff coordinates with G2, G3, G5, PSYOP, U.S. Information Service, and other staff elements in developing the commander's information strategy to ensure synergy and to reduce the probability of conflicting messages.

4-11. **Media Operations Centers.** Currently, joint, combined and Army media centers fulfill the requirement for a focal point for the news media during military operations. In essence, the media operations center (MOC) is a command post for media support efforts. It serves as both the primary information source, and as a logistical support and coordination base for commercial news organizations covering the operation.

4-12. Media centers are organized when large numbers of news media representatives are anticipated to cover military activities. Media centers may be formed for all types of operations or for any stage within an operation.

4-13. When operated by unified/specified commands, these media centers may be called a Joint Information Bureau (JIB). At the combined commands, they are called an Allied Press Information Center (APIC), Coalition Press Information Centre or Combined Information Bureau (CIB). At theater level and below, they are simply referred to as Media Operations Centers.

4-14. MOCs support the commander and are subordinate to the command's PAO. They provide the commander a professional, immediately available, fully trained organization designed to respond to national and international civilian media interest in American military operations.

4-15. In addition, the media operations center provides the following functions:

- A single point of contact and information source for media within the theater
- Briefings and enforcement of media guidelines and ground rules
- Primary information release authority for the senior PAO
- Coordination of news media coverage with corps, divisions, brigades, etc.
- Coordination with all service branches for each service, agency or country
- Identification and communication of host-nation sensitivities to all personnel in theater
- Preparation for and conducting press briefings and news conferences
- Registration of news media personnel
- Media Operations Center Staffing and Organization

4-16. Organization and personnel staffing of media operations centers are determined by the responsible command in coordination with the PAO and his staff. APIC staffs should be a proportionate representation of the forces, with representation from all services involved in the operation. This will be determined by CPA at the unified command. Regardless of the echelon establishing a media operations center, the organizational model is functionally designed and remains relatively the same.

4-17. MOCs normally consist of two major elements: a Headquarters Group and a Media Operations Center Group.

4-18. The headquarters is made up of the command group and support staff. The command group contains the commander, deputy commander and/or executive officer, and the sergeant major. The support staff is normally task organized to support tailored forward deployed MOC teams or sub-MOCs when the APIC operates as other than a single element. The support sections provide administrative support, conduct lease and purchase contracting, setup, operate and maintain the unit's equipment, and conduct the day-to-day operation of the MOC. The support sections are responsible for the execution of MOC communications, supply operations, administration support, vehicle maintenance, security and other support functions as required.

4-19. The Media Operations Center consists of a Plans Section, a Media Support Section and an Information Operations Section.

4-20. The plans section is responsible for all MOC media planning. It establishes MOC requirements and determines operating procedures and policies. It maintains channels of communication with OASD(PA) and the JPAO (or senior command PAO). It is responsible for recommending and assisting in the development and dissemination of PA Guidance. It monitors available major U.S., international and local television and radio broadcasts and print publications providing coverage of the operations, conducts news media analysis and evaluates the effectiveness of MOC operations.

4-21. The Media Support Section (MSS) is the primary point of contact for news media representatives (NMRs) in an area of operation seeking information or assistance in covering the force and the operation. The MSS receives and registers NMRs, briefs NMRs on the media ground rules and security procedures or concerns, and orients them on the force, the operation and other pertinent issues (special safety or host nation considerations).

4-22. The MSS orchestrates the command's news briefings and coordinate for subject matter experts to explain and discuss operations and capabilities. The MSS is also responsible for coordinating for appropriate, knowledgeable escorts, unit visits, and service member interviews. It assists the Joint Force or other senior PAO in preparing service members for interaction with the news media. Finally, it provides support to Joint Force elements and service component PA elements seeking assistance with NMRs.

4-23. The Information Operations Section is responsible for monitoring plans and operations from within the command's operation center and assessing the PA implications of events occurring throughout the area. It ensures that the MOC has current situation information, is aware of issues of potential media interest, and can obtain any operational information necessary for the development of responses to media inquiries in a timely manner.

4-24. The IO section ensures that PA operations are synchronized with other combat functions and promote early coordination of PA, CA and PSYOP functions.

4-25. **MOC Staffing**. Currently, media operations center staffing requires augmentation, either by PA-trained individual fillers or by Army PA units. As fully independent units, the Public Affairs Operations Center (PAOC) (SRC 45423A000) and Mobile PA Detachment (MPAD) (SRC 45413A000) are currently organized, trained, and prepared to fill this role.

4-26. In fact, these Army PA detachments are specifically designed to function as an Army media operation center in theater, corps, or division-controlled operations. MPADs can be combined to form media sub-centers in forward battle areas.

4-27. PA personnel from non-deployed commands and installation PA sections may be called upon to augment news media centers however, requests for individual augmentation should be coordinated through operational channels. Reserve and Guard unit personnel can be used to augment on a voluntary basis.

4-28. An example of a media operations center is included at Appendix L.

MOC OPERATIONS

4-29. In major operations -- actions conducted by unified commands -- a Joint Information Bureau will usually be the first to deploy for this purpose. A JIB will be staffed by public affairs personnel from the services represented in the joint force; participating services may establish their own media centers subordinate to the JIB to disseminate information about their particular missions.

4-30. As the operation unfolds, the Army plans for and contributes to a replacement PA organization for the JIB which consists of individual PA personnel from each of the services and Army PA detachments.

4-31. Media Center operations will be based on five primary assumptions:

- Accurate information is available in a timely manner and adheres to the DoD Principles of Information in Appendix A.
- Current trends in communications technologies within the information environment will continue to reduce the news media's reliance on military support and assistance when covering operations and will continue to increase the availability of information to a worldwide audience.
- Media representatives will be in an area of operations at the start of, and in most cases, before an operation begins.
- Media interest and coverage in non combat operations may be higher at the outset, and barring a significant event which renews national or international attention or interest, will taper off over time. During a high-intensity conflict, media interest could remain high.
- Military PA elements require access to complete information, state-of-the-art communication equipment, and must possess sophisticated coordination channels in order to pre-empt speculative, inaccurate or biased reporting.

4-32. Media centers will support and be responsible to the senior commander of the operation on a 24-hour basis. Media centers are usually established by unified command CINCs to support the news media in an area of operation.

4-33. During the first 24 hours after arrival in a new theater of operation, a media center can provide limited media support services.

4-34. Within this first operational day, the MOC must:

- Establish a "hasty media center" as the initial focal point for the news media until additional media support forces arrive.
- Establish communication with OASD (PA), each service's PA chain of command, and with units operating within the theater.
- Request operational information release authority within the theater.
- Establish command structure/lines of authority.
- Coordinate with appropriate authority for leasing and purchasing contracts.
- Begin to register news media personnel in the area
- Provide basic media support (coordination of media access to subordinate units and media escort as resources permit).
- Assist or conduct command news briefings and conferences.
- Coordinate Subject Matter Expert (SME) interviews.
- Be capable of assisting in the transmission of media products.

4-35. Media Support -- Initially, the media operations center will need to provide varying degrees of support to news media personnel including specialized equipment (flak vest, NBC gear, helmets), transmission of media products, etc.

4-36. This support may include but not be limited to:

- Coordinate media contact with units or individuals to include SME interviews
- Provide a single point of contact for information on operational issues
- Provide news releases, fact sheets, copies of transcripts for news briefings/conferences and copies of archival file products
- When other means are not available, the media center may provide coordination for transportation (to and from interview sources), transmission of media products and food and billeting
- Provide limited media escort within the area
- (SOPs for MOCs should be pre-established for each theater of operation and used for media operations within that theater.)

REGISTRATION OF MEDIA

4-37. Principle to supporting the commander's information strategy is the inclusion of news media representatives (NMR) within Army units from the earliest pre-deployment stages of all operations. The personal safety of media representatives, as acknowledged by the media themselves, is not a reason for excluding them from operations.

4-38. However, all media requesting support or access to units to cover Army operations must be registered. This includes freelance journalist, military media representatives, such as those who are assigned to Armed Forces Radio and Television Service, Stars & Stripes newspaper and other Armed Forces Information Service (AFIS) personnel who are not supporting units on the battlefield.

4-39. **Registration versus Accreditation**. Accreditation is the verification and validation that a person represents a legitimate commercial news organization. This means that accrediting governments or military organizations will physically verify the affiliation of an applicant with a specific news organization.

4-40. This is difficult to perform amidst an ongoing operation, especially when deployed far away from CONUS. It is generally accepted that, when overseas, the decision to accredit news media is made by the host nation's government in coordination with the combined or unified commander.

4-41. When accreditation isn't required by the host nation, responsibility for this determination is held by the combined or unified commander. Accreditation is normally performed at Corps level or higher.

4-42. Accreditation is a major problem for many commands because they are forced to determine the legitimacy of smaller, lessor-known news organizations and freelance journalists without news organization affiliation.

4-43. Unless it is absolutely required by host nations, the American military will attempt to avoid accreditation.

4-44. Registration, however, is merely an accounting tool, which provides PAOs the ability to know what media are represented in the theater, where they are located, and their movement around the theater. This information is helpful in planning and conducting media logistical support and transportation, and in preparing subordinate commands for media encounters. It is also helpful to commanders who might want to provide newsworthy events to the media.

4-45. Registration also identifies which news media have asked for military assistance and access, and have agreed to the command's media ground rules.

4-46. **Registration Requirements.** The registration process is conducted in five basic steps:

- Verify the identity of the media representative (including checking for valid passport/visa, professional media organization membership card, media ID card, other military press credentials, etc.).
- Have them sign an agreement to abide by the established media ground rules for the operation in exchange for granting support, access to units, information and other privileges. If required, revoke credentials for those who violate the ground rules. (Enforcement of this requirement is essential.)
- Have NMR agree to and sign a liability waiver that frees the military of responsibility if the NMR is killed or injured as a result of covering the operation. (An example of a waiver of liability is at Annex I).
- Give NMRs proof of registration (memorandum, press badge or other identification).
- Maintain a roster of registered NMRs and monitor their movements during the time they are receiving military support.

4-47. NMRs who refuse to agree to the military ground rules and who are not registered will receive only the support and information assistance as provided to the general public.

4-48. NMRs should be informed that registration and acceptance of media ground rules will entitle them to better access to units and subject matter experts, and provision of military ground and air transportation when possible.

MEDIA GROUND RULES

4-49. Media ground rules will assist inprotecting the security and the safety of the troops involved while allowing you the greatest permissible freedom and access in covering the story. All interviews with news media representatives will be on the record.

4-50. Security at the source will be the policy. (An example of media ground rules is in Appendix X.)

4-51. The following categories of information are releasable:

- Arrival of major U.S. units when officially announced by a U.S. spokesperson. Mode of travel (sea or air) and date of departure from home station

- Approximate friendly force strength figures, after review by host nation government
- Approximate enemy casualty and POW figures for each action operation
- Non-sensitive, unclassified information regarding U.S. air, ground and sea operations (past and present).
- Friendly force size in an action or operation will be announced using general terms such as multi-battalion or naval task force
- Specific force/unit identification/designation may be released when it has become public knowledge and no longer warrants security protection
- Identification and location of military targets and objectives previously under attack
- Generic origin of air operations such as land or carrier based.
- Date/time/location of previous conventional military missions and actions as well as mission results
- Types of ordnance expended will be released in general terms rather than specific amounts
- Weather and climate conditions
- Allied participation by type of operation (ships, aircraft, ground units, etc)

4-52. Information Not Releasable

- Information about future military plans, activities or operations
- Vulnerabilities or weaknesses on command, control, personnel or the operation
- Friendly unit and command strengths, on-hand equipment or supplies; the presence, activities and methods of operation of specifically designated units or equipment
- Information on friendly force security and deception measures and countermeasures, and intelligence collection activities
- Specific information on friendly force current operations and movements, deployments and dispositions
- Information on in-progress operations against hostile targets
- Information on nuclear, biological or chemical weapons, equipment or training

MEDIA POOLS

4-53. Journalists, as a group, are strongly opposed to media pools in any form. The media pool is seen as a restriction placed on the media representatives and their ability to provide coverage of the news. They are grudgingly tolerated, and should be only used as a last resort when space onboard military transportation is limited, access to an area must be controlled, and after all other possibilities have been explored and eliminated. Even under conditions of open coverage, pools may be appropriate for specific events. Both the Army and the news media are in agreement, however, that limited access is better than no access at all.

4-54. When a pool system is required, the military PAO will identify the maximum size of the pool that can be supported. The news media representatives on the scene will select media pool members. A roster of media personnel registered with the Army PAO will be used to identify the media representatives eligible to participate. The pool should consist of, but not be limited to, a minimum of one video crew (camera operator, sound technician and reporter), one still photographer (wire service, newspaper, or magazine), one radio reporter, and one newspaper or wire service reporter. Special consideration must be given to international reporters as well. While this is a fair and representative pool structure, it is the media themselves who must determine the make-up of the pool. Some news events and situations may lend themselves more to print, or conversely television reporting, and the media representatives may choose to select an unbalanced pool.

4-55. All pool members must be willing and able to meet deadlines and supply information products (video, audio, still media, and text) in a timely manner to all media representatives who are entitled to material generated by the pool. The military media center will also have access to this information and will make it available to all other requesting news media organizations.

4-56. Consistent with its capabilities, the military will supply PAOs with facilities to enable timely, secure, compatible transmission of pool material and will make these facilities available whenever possible for filing independent coverage. In cases when government facilities are unavailable, journalists will, as always, file by any other means available. The military will not ban communications systems operated by news organizations, but electromagnetic operational security in battlefield situations may require restrictions on the use of such systems.

4-57. Once a media pool has been selected, the media pool will select a team leader. It is the responsibility of this team leader to ensure that members of the media pool meet their obligation to share information. The Army PAO will not involve himself in settling internal disputes of the media pool.

4-58. Finally, the pool is an option of last resort. It should be disbanded as soon as free and open access to the operational area can be allowed, normally within the first 24 hours of an operation.

THE DOD NATIONAL MEDIA POOL

4-59. The DoD National Media Pool was established to prevent recurrence of problems encountered with media coverage during Operation Urgent Fury in Grenada in 1983. During the first 24 hours of Urgent Fury, more than 600 reporters attempted to gain access to the operation. The large numbers overwhelmed the limited Public Affairs elements available to assist them.

4-60. In 1985, the Secretary of Defense established the DoD National Media Pool, a civilian news element of approximately 16 media representatives from various national news organizations, with the mission of covering an operation from its initial stages until open coverage could be allowed.

4-61. The pool members remain on call in Washington, D.C., and are available for immediate worldwide deployment. Their products are shared by the open news media until the pool is disbanded and access is granted to the entire news community.

4-62. Supported commanders are responsible for providing operational support to the DoD National Media Pool. At a minimum, the pool members will require:

- Daily, comprehensive and unclassified operational news briefings.
- Access to ongoing combat operations. The media are aware of the personal risks involved in covering combat operations. They will not be denied access to them based on risk to their personal safety.
- Reasonable access to key personnel. All information gathered from these personnel is unclassified and on the record.
- An escort -- usually a lieutenant colonel or colonel -- to coordinate pool support and access requirements.
- Transportation and itinerary planning and coordination that will allow media to gain access to the theater of operations and to disperse pool members throughout the operational area.
- In today's global information environment, when news media can report live from almost anywhere in the world in almost any environment, the technological capabilities of most news organizations decreases the importance of the DoD National Media Pool once word of an operation has spread.
- When the DoD media pool is operational, PAOs will attempt to provide the same information support concerning theater operations to all other media in the operational area.
- However, their primary responsibility is to the DoD Media Pool. After the DoD media pool is dissolved, all media in theater will be dealt with in an equitable manner with respect to information and support provided.
- As soon as open access to the operational area can be allowed (normally within the first 24 hours of an operation), the DoD National Media should be disbanded.

NEWS BRIEFINGS

4-63. There are several reasons for holding news briefings, in addition to the daily operational news briefing required at the unified command level.

- Credibility: The physical presence of a briefer and his willingness to meet the issue head on leads to a much more credible presentation
- Uniformity: All media get the same information at the same time.
- Expression of concern: A briefer represents the face of the command, which shows more concern than an impersonal news release, especially in situations where there is loss of life or extensive damage.
- Complexity of material: Where material is technical or complicated, the news briefing makes the subject matter more easily understandable. The question and answer session that accompanies a news briefing saves time in call-backs by news reporters needing clarification.

4-64. News briefings should be done daily during an operation and when important events dictate. They should:

- Get out a specific message
- Explain complex or technical matters
- Reach a large number of media interested in the same subject matter area.

4-65. PA specialists should think about media deadlines and set the time to help the media meet those deadlines. Be sure to invite all media within the area in a timely manner.

4-66. A knowledgeable and articulate spokesperson should be chosen to present the material. This should be the subject matter expert (SME), but may be the PAO or the commander. At the very minimum, a person of prominence within the command should be selected.

4-67. Other SMEs may be in attendance at the briefing to field technical questions. The SME interviews should be at the request of an individual media representative and the time should be set to facilitate the media to meet those deadlines.

4-68. Consideration should be given to the appearance of the presentation, the message, space, lighting, electrical needs, suitable setting, chairs, tables and press packets. All handouts should be reviewed.

4-69. Appendix H provides briefing and press conference formats.

Chapter 5

Information Strategies

The defining trend of the 1990s, from corporate boardrooms to private living rooms, is connecting everyone and everything to everyone and everything else. The American public is faced with many choices: interactive television, cellular phones, modems, faxes, personal digital assistants, an assortment of daily newspapers, online computer information services and other easily accessible information services -- if it can be connected, it seems to be in demand. The computer has invaded the American home. In this decade it has become routine that news and information can now be tailored to fit the individual needs of the consumer using on-line services, distributed electronically and received instantaneously at his or her computer. With the sophistication, power and miniaturization of these technologies improving each year, and the cost dropping at an equally rapid pace, the public will expect access to these devices and services and the information they carry. These emerging technologies have contributed to the refocusing of the Public Affairs mission. It has resulted in shifting the information provision function from an emphasis on producing specific products (such as post and field newspapers and radio/television news programs) to focusing on the processing of our themes and messages and their intended effects -- the function of information communication, rather than the form. This chapter explains the objectives of information strategies, identifies and explains the elements of information strategies, and describes the relative advantages and disadvantages of present day and emerging information communication channels available to PA organizations. Most importantly, it explains how best to use these information channels to satisfy the information needs of the various target audiences as we enter the information age.

INFORMATION STRATEGIES

5-1. Information Strategies is the sum of all actions and activities, which contribute to informing the American public and the Army. The responsibility for this activity is assigned to an element within each PA section, which focuses entirely on accomplishing the information strategy mission. This section is usually called the Public Affairs Information

Services Section. At all echelons, it employs numerous techniques to provide news and information to internal and external audiences. The Army provides an expedited flow of complete, accurate and timely information, which not only communicates the Army perspective, but also attempts to educate audiences and engender support for the force.

5-2. Using a combination of contracted services, organic military assets, and government and commercial communications networks, Public Affairs organizations provide information to news media representatives, deployed soldiers, home station audiences and the American public. The Information Services Section within a Public Affairs organization coordinates information efforts and develops informational products (such as digital text, graphics, and photos, printed publications, audio/video news releases and graphic imagery) into consolidated campaigns designed specifically to present the Army's perspective. This means that Army Public Affairs communicates information to create an informed American public and Army force, assist them in gaining a clear understanding of the strategic, operational and tactical situation.

5-3. To accomplish this, the Information Services Section must develop information objectives or goals during the planning process prior to an operation. These information goals are similar to the PA standards of service and support, which appear in Appendix Q, in that they establish a basis for determining successful information communication operations.

5-4. These information objectives should include:

- Ensuring an understanding of the role of America's Armed Forces in American society.
- Ensuring an accurate perception of the particular military situation or mission.
- Ensuring an understanding of individual and unit roles in mission accomplishment.
- Establishing confidence in America's Army to accomplish the assigned mission in accordance with our national values.
- Establishing confidence in and support for American soldiers.

5-5. By establishing a comprehensive information strategy program, Public Affairs can assist in mission accomplishment by increasing audience understanding of the situation and establish confidence in and support for the force. This contributes to unit cohesion and provides commanders with increased range of action, free of distractions and limitations.

5-6. This is best accomplished by three basic types of information campaigns:

- **MISSION**. Both external and internal publics need to know what the mission is, what they're being asked to do and why. They need to know not only the organization's mission, but also how it fits into the big picture -- the political/strategic-level situation, and why it is important.
- **ROLE.** All military members and civilian employees need to have an understanding of their job and how it relates to mission accomplishment. The general public needs to have an accurate understanding of the military's role and its ability to accomplish

the mission. This understanding results in confidence in the force and demonstrates American unity and resolve.

- **MORALE.** Military members need to have access to news and information about current events and the activities available to them while deployed. They also need to have access to information from civilian commercial news sources. This is important because, in addition to being more credible, it allows the deployed force to see how the operation and their participation in it are being portrayed for the American public. In order to better understand the mission, their role in it, and give it his or her full effort, they have to know what effect the operation is likely to have at the local, regional, national and international levels. The opportunity to involve themselves in educational and other activities is necessary to quality of life and morale. A well-informed service member is more effective.

5-7. The general public is interested in soldiers, their lifestyle, how they are being treated and their ability to accomplish a given mission. Information about these topics provide reassurance, confirming that soldiers maintain professional and ethical values and are being cared for adequately.

Historical Perspective

During Operation Desert Shield/Storm, many commanders developed innovative methods of sharing Command Information products produced in theater and in the rear. The products greatly enhanced morale at both ends. Some commanders and PAOs used the products as issue management tools to dispel rumors in theater and at home station. The products included field newspapers, newsletters, videotapes, audiotapes, etc. They also let soldiers returning from the area meet with family support groups to answer any questions.

(After Action Report 1991)

RESPONSIBILITIES

5-8. Commanders at each echelon are responsible for Public Affairs operations and support. Public Affairs officers and noncommissioned officers at various levels assist commanders in the discharge of these responsibilities. Public Affairs staffs are responsible for accomplishing the Public Affairs information communication mission. This responsibility includes Public Affairs operations in all subordinate, assigned or attached commands.

5-9. PA is only one of many information channels available to a commander. PA information provision cannot substitute for a commander's personal involvement in his "Command Information" program.

5-10. All public affairs practitioners have access to all information that is not classified or violates Operational Security or the Essential Elements of Friendly Information (EEFI) for use in preparing information products. Commanders must ensure the EFFI is up-to-date based on current situations and operational guidance. Public Affairs personnel at all levels must produce and release accurate information packages based on DoD directives and Army policies.

5-11. **Strategic Level Commands** are responsible for providing public affairs guidance to subordinate units. They develop central themes and messages and provide umbrella guidance to subordinate PA staffs. They must also provide subordinate commands with information useful in preparing information products for internal and external release. They are additionally responsible for marketing public affairs information products to subordinate commands, home stations, the Army as a whole, as well as the general public. Strategic level Public Affairs staffs are the primary coordination point for the Armed Forces Radio and Television Service and the geographical manager for radio and television services including personnel, down links, facilities and equipment.

5-12. **Operational Level Commands** are responsible for communicating public affairs guidance to subordinate units. They expand on information campaign themes and messages, and provide additional information products to subordinate command PA staffs. Additionally, they provide subordinate commands with information useful in preparing information products for internal and external release. They are responsible for gathering and producing public affairs information products for release. In the event the Operational Command is the senior command in an area or theater, it assumes the responsibilities of the strategic level command.

5-13. **Tactical Level Commands** are responsible for gathering information products for release through their next higher headquarters to home stations, the Army as a whole, as well as the general public. These commands are also responsible for coordinating the dissemination of information and information products received from senior commands down to subordinate commands. In the event the tactical command is the senior Army command in an area or theater, it assumes the responsibilities of the operational-level command and will be augmented to accomplish these additional functions.

INFORMATION STRATEGY PROCESS

5-14. The Information Services Section uses all available means to gather complete, factual, unbiased information for use in information campaign development. The information is developed, converted into the most appropriate product form based on the information needs/target audience assessment and information communication channel availability, and then transmitted to the intended audiences. This is called the Information Provision Process.

5-15. Although the information strategy process follows a deliberate cycle, it is a continual process. Information campaigns are also conducted simultaneously, with personnel examining external and internal information needs, carrying messages from concept through execution and program evaluation, to accomplish specific PA objectives. The cycle has four phases -- acquire, process, protect and distribute. (See Figure 5-1.) Evaulation is a key component in the cycle. It must be conducted throughout the four phases. This ensures the campaigns are meeting their objectives, and are altered if they do not.

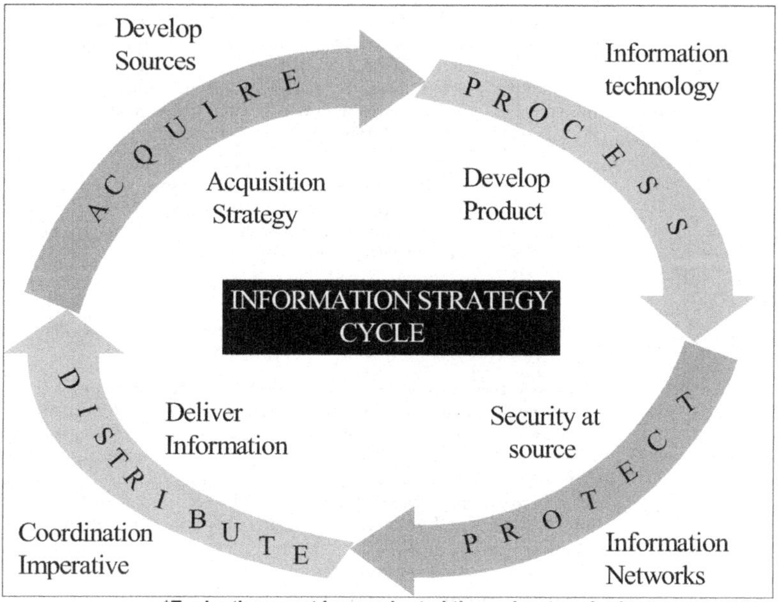

***Evaluation must be conducted throughout each phase**

Figure 5-1

ACQUIRE

5-16. During this phase, a ISS identifies and assesses several factors: the situation, the environmental factors, the mission, and the target audiences. They determine the information needs of the various target audiences. They then begin to gather information based on the information requirements of these audiences.

- ***Information Needs Assessment and Audience Analysis.*** Identification of the target audiences and target audience populations and densities must be developed during the planning stages of operations with target audience assessments. Upon assessment of target audiences and consideration of which type of information product will best serve each audience, commanders must ensure adequate public affairs personnel, equipment (to include communications), resources and funds are available and included in OPLANS/OPORDS to achieve mission success. An assessment of soldier information needs is crucial to the information package development and the selection of appropriate products.
- ***Information Gathering.*** Information gathering is the first step in producing packages for release to internal and external audiences. Public affairs personnel gather information through

operational, administrative, logistical, battle, staff, command and support channels. They also acquire information through the media, research, leaders and the culture at large. Information may be obtained via electronic and telecommunication systems in addition to written documents and oral communications. While information often comes from superior and subordinate commands, it may also be obtained laterally.

- Sources of information must be valid and diverse enough to provide a broad overview. Public Affairs staffs must ensure the source does not speculate, nor speak out of his area of expertise. However, soldiers' experiences and personal opinions may lend credibility and provide a "grass-roots" view.

PROCESS

5-17. During this phase, the ISS begins to develop products, prepare them for release and determine methods for distrubition.

- **Product Development.** Development of information products is performed to some degree by PA personnel at all Public Affairs levels and is an on-going process. The term "product," as used in this process, means the message for intended communication, regardless of the format or communication channel proposed. Initial production development may be command directed or initiated by the Public Affairs staff or provided by other Army agencies. The command's resources and the target audience's requirements will determine the product type.
- **Prepare products**. Information packages should be prepared in accordance with Public Affairs Guidance, Soldiers' Manuals, Field Manuals, SOPs and Army Regulations. Ultimately, the Public Affairs staff is responsible for content.
- **Media Forms/Methods.** Soldiers use a variety of technologies to gather information and produce information products. During the gathering process, PA soldiers conduct interviews, attend briefings, withdraw data from government and commercial computer databases, bulletin boards, and e-mail systems. They acquire text, graphics, photography and motion video from government and commercial Internet systems. The nature, distribution, usability and flexibility of public affairs systems are crucial in the processing of information.
- Professional quality systems should be used whenever possible. For printed products, preferred systems include computers, desktop publishing, word processing, laser printing, etc. Reproduction may be Army-contracted, Army-funded or reproduced using the command's assets. Video and audio products intended for release to news media should attempt to be broadcast-industry standard.
- Electronic newsgathering and editing systems should be used when available. Visual products should be generated by modern methods including digital imagery and computer graphics.
- **Print.** Articles released to home station for military publications and family support group publications; for marketing to civilian publications; field publications, e.g. newsletters, with and without photography.

- **Video.** Raw electronic news gathering video and printed news scripts for release to military and civilian outlets.
- **Audio.** Radio interviews; features; internal command information scripts; radio news; news reports for release to military and civilian outlets.
- **Visual.** Digital imagery; photographs; slides, view graphs; graphics for release to military and civilian print and broadcast media.
- **Digital.** Each of the categories described above may be developed and distributed electronically, either through commercial information services directly on the Internet, or by using tactical Army communications systems (SINCGARS and ATCCS). Modern technology in use on the battlefield has made digital transmission the preferred method for all types of products.

PROTECT

5-18. **Security at the source.** No information strategy is complete without a clear cut understanding of how to protect the information. Both sides can benefit from information and use information simultaneously against each other. Pieces of the right information can have a dramatic impact on the outcome of an operation. Public Affairs professionals will continue to protect vital information by practicing security at the source and following established operational security guidelines.

5-19. In addition to protecting the raw and completed information products, public affairs personnel must also take the necessary steps to protect information networks.

DISTRIBUTE

5-20. **Dissemination.** Public Affairs information packages should be released in the format most easily used by the recipient. While this is not always possible, a product stands less chance of being used if it is incompatible with the recipients' equipment. For example, a video product released on Hi-8 to a TV station that works exclusively with Beta SP has less a chance of being aired than a video story in a compatible format.

5-21. Public Affairs information packages must be expedited to the users by the most technologically advanced reliable method. Great consideration must be given to the speed and reliability of the mode of dissemination. This must be included in target audience assessments and conducted during planning stages of operations. Articles and photos may be sent from deployed locations to home stations via computer systems and telephone lines. Yet based on the quality of the telecommunications system, it may be more reliable to use the mail or a courier. As technology improves PA capabilities, Public Affairs will incorporate those improvements into the information gathering and dissemination system to increase its potential to reach an ever-growing, information-hungry public. For example, the emergence of smaller, more powerful satellite link ups can provide PA elements the ability to reach targeted audiences sooner and from more locations.

5-22. Internal information packages must be available to soldiers at all levels of command. Public Affairs must develop and coordinate a distribution scheme with the commander, the general staff, and with signal as the proponent for physical distribution of certain commercial news and information products. An efficient distribution system will also ensure prompt delivery of public affairs products. The public affairs staff must conduct periodic quality control checks and update the distribution scheme as necessary, based on changing population densities or information products. Electronic means are the preferred mode of distribution, however additional methods include, but are not limited to contract delivery, AG distribution and the military postal system.

5-23. **Products for distribution to deployed soldiers.** Publications produced by other military agencies intended for deployed soldiers must be given the same distribution considerations as commercial information packages. For example Stars & Stripes, Soldiers Magazine, Army Trainer and home station post newspapers contracted for delivery to deployed soldiers must be given the same distribution considerations as other publications. However, a separate distribution scheme may be required.

INFORMATION PROGRAM EVALUATION

5-24. The final step in the Information Provision function is the evaluation of our communication efforts. Evaluating communications programs is research, which boils down to a series of questions:

- Did we achieve our objectives?
- Were our policies and programs effectively communicated?
- Was the operation affected positively by our efforts?
- Did the American public support our soldiers? Was unit cohesion enhanced?
- What audiences received our messages? What was the impact of our communication programs on these audiences?
- Research is the foundation of the Information Program Evaluation. Corrections and changes in courses of action should be based on solid factual information. Methods for conducting Information Program research are discussed in greater detail in Appendix P.

NON-HIERARCHICAL AND INTERNET INFRASTRUCTURE AS A RESOURCE

5-25. Modern technology has provided us with an advanced form of communications structure called non-hierarchical structure. The advantage of non-hierarchical communication structuring is that every "node" in a communications web shares information with every other "node". Each node on a network can identify itself and "find" others in the network in order to communicate specific information. Public Affairs elements act as information nodes, gathering, developing and sharing information vertically and horizontally on the battlefield.

5-26. While this technology was originally intended for command and control, it is essential for other functions on the battlefield as well. Well-coordinated public affairs operations will leverage this capability to move information -- sending messages around the battlefield, to and from home station, and up and down the chain of command.

5-27. The end state of this technology effort is that both organizations and individual soldiers on the battlefield will possess this capability.

TELECOMMUNICATION SYSTEMS

5-28. Information technology provides the means to collect, process, display, develop and disseminate information in an unparalleled manner. This technology has begun to revolutionize our approach to information provision.

5-29. The PA specialist now has the ability to access desired information on a certain issue and tailor and develop this information into a message for dissemination -- all from a personal laptop computer.

5-30. This "from anywhere to anywhere" capability allows the PA specialist at all echelons to accomplish his mission of presenting the Army's perspective -- framing issues and informing targeted audiences.

5-31. There are two telecommunication systems available to Army PA specialists: DoD's internal secure communication network, known as the Defense Data Network (DDN) and the worldwide commercial information network -- Internet. PA specialists must be familiar with both.

DEFENSE DATA NETWORK

5-32. Defense link is an entry point to Internet sites operated by the Office of the Secretary of Defense, the armed services and related defense agencies. It provides a means to search and download Department of Defense directives, obtain transcripts of important speeches and connect to other sites dealing with specific military operations.

5-33. The Department of Defense organizations are extensively represented on the Defense Data Network. Each of the military services has a centralized directory of its Internet sites. Army Link, as its name suggests is a catalog or hotlist of more than 280 Army home pages. The sites are indexed alphabetically and by subject area.

INTERNET

5-34. The Internet is a network of networks. Originally, the Internet was developed by DoD with the goal of building a automation system linking government agencies together which would continue to function in case part of the network ceased to exist. From this inception, the Internet grew, as other agencies and activities noticed the benefit of linked telecommunications. The system now includes organizations from all segments of society: government, defense, business, education, health services, etc. There are more than 30,000 networks participating on the Internet and more than 150 million individual users. And more are connecting everyday.

5-35. Army Public Affairs personnel have access to the Internet through their local Directorate of Information Management (DOIM).

5-36. On the Internet, there are six main activities: "Hypertext" documents with multimedia links on the World Wide Web can be viewed, files can be downloaded with a file transfer program or "FTP," data can be located, communication via e-mail, reading and posting messages to news groups and bulletin boards, and logging in to remote computers using Telnet.

5-37. You can do most of the same things in the Internet that you can do on any of the commercial information services, but you do them in the context of a much larger network that isn't centrally organized or controlled.

5-38. **World Wide Web.** The World Wide Web, also known as WWW, W3, or simply the Web, is one of the most popular activities on the Internet. The Web allows you to view documents that feature graphics, "hypertext" (text which contains highlighted keywords which are linked to other documents or information sources available on Web) and multimedia links. The hypertext and multimedia links are tied to other documents or information forms that might be on the same computer, one across the country, or a machine on the other side of the world.

5-39. The WWW contains thousands of main menus, called "home pages" which identify the various sub-categories at each web site. Web users can search using key words or locations through the lists of home pages for specific web sites.

5-40. **Gophers.** A gopher is a menu driven system that offers text only. Users of the web can access gopher sites and retrieve information. It is a predecessor of the Web.

5-41. **News groups.** They are discussion groups built around a particular topic. Some are managed, others are not. News Groups take two forms: Mailing List: A list of E-mail addresses to which messages are sent. You can subscribe to a mailing lists typically by sending an email to the contact address with the following in the body of the message: the word subscribe, the name of the list, and your email address. Discussion groups: A particular section within the USENET system typically, though not always, dedicated to a particular subject of interest. Also known as a newsgroup. USENET is a collection of the thousands of bulletin boards residing on the Internet. Each bulletin board contains discussion groups, or newsgroups, dedicated to a myriad of topics. Messages are posted and responded to by readers either as public or private E-mails.

5-42. **E-mail.** E-mail is a means of interpersonal communication that falls somewhere between the immediacy of a phone conversation and the more thoughtful but slower exchange of ideas previously done by writing letters and memos. The specifics of using it vary greatly according to the mail software being used.

5-43. **Commercial On-line Information Services.** Users cannot directly dial up to the Internet using a modem, but must gain access through an internet service provider. Many offices, universities and large businesses provide access for their employees. For access at home, a user can either subscribe to one of the commercial on-line services, which provide Internet access in addition to their other features or to one of the dedicated Internet server companies located in most cities.

Chapter 6

Training

"You must remember that whether you wear one stripe or six, one bar or silver eagles you automatically become an "Army spokesman" when you are approached by the press. Within 24 hours the words of that Army spokesman can be flashed world wide, particularly if they can be construed as criticism....... Everything you say should have the ultimate aim of furthering that effort. Your approach to the questions of the press should emphasize the positive aspects of your activities and avoid gratuitous criticism. Emphasize the feeling of achievement, the hopes for the future, instances of outstanding individual or unit performance and optimism in general. But don't destroy your personal credibility by gilding the lily. As songwriter Johnny Mercer put it, "You've got to accentuate the positive and eliminate the negative."

- U.S. Continental Army Command memo to advisers in Vietnam.
The Military and the Media,
William M. Hammond, 1988

TRAINING OBJECTIVES

6-1. The cost of learning in combat is too high. Therefore, commanders and staff section chiefs must conduct cost- and time-effective staff training during peacetime. In order to conduct effective training, every consideration regarding PA activities should be evaluated.

6-2. Evaluation should include, but is not limited to, analysis, planning and execution of the operation. The PA staff, in conjunction with other members of the organization, must train in the initial planning of the operation. In order to train effectively, the following considerations should be evaluated:

6-3. **Analysis.** Specified and implied tasks are identified at home station for deployment, mission and redeployment requirements.

- Feasibility of all tasks are determined and assessed.
- Consider any PA guidance issued by higher headquarters and determine specialized training requirements.
- Consider all phases of host nation operations and the effect on your operation and organization.

- Determine the availability of required assets and equipment for all phases of the operation.
- Consider the availability and use of internal and external assets -- joint, RC and combined -- to augment known shortfalls.
- Determine any transportation requirements for your organization or attached personnel.

6-4. **Planning**. Planning includes PA annexes, command directives, DOD guidance, regulations and directives stipulated by the commander's intent and mission requirements.

- PA annexes should contain a brief, general description of the situation and intent to conduct operations.
- Planning should also consider enemy forces and host nation personnel and their relationship to the operation.
- PA planning should be considered for all attachments and detachments currently known or listed under the task organization. This includes planning for assets and requirements from home station to redeployment.
- Training should be conducted emphasizing procedures for handling the transmission of all information. Theater-unique requirements may call for special training scenarios.
- All logistical and administrative requirements need to be addressed in the planning and training process.

6-5. **Execution.** The staff, soldiers and media understand the commander's intent in terms of his command position and operational mission. The operational area ground rules are included in the PAG and are widely disseminated throughout the organizations.

- Encourage local media to accompany deployed units.
- Adequate vehicles have been identified and assigned to transport PA personnel and media throughout operational areas. This includes vehicles, aircraft and available transportation assets for transporting all media personnel and products. Consideration of transportation should include all support activities in the forward area of operation.
- Public Affairs personnel in the media center will prepare releases of specific events to disseminate in the area of operations. News media inquiries are anticipated, received, researched and answered as quickly as possible.
- Local security review policies are clear and will not delay the transmission of information.
- Executions of the theater policy for registration policies and local media pool operations have been completed. Non credentialed media are identified and escorted to the rear for registration. Once registered, media personnel will be allowed to cover stories and interact with organizations in the area of operations.

6-6. The staff section must be able to:

- Cope with the unexpected. For example, media arriving at the unit level without PA escort or media not registered with the joint media operations center.

- Separate fact from fiction. For example, media having misinformation which they believe as fact and trying to confirm it.
- Coordinate well amid confusion. For example, be able to access the PA staff at the MOC.
- Think clearly while under great stress. For example, have a unit PA representative (UPAR) who will handle escorted/nonescorted media for the unit.
- PA elements develop training plans from assessments of their performance proficiency on their METL.
- Detailed training plans for Public Affairs Detachments are contained in ARTEP 45-500-10-MTP, for the Mobile Public Affairs Detachment in ARTEP 45-413-30-MTP and for the PAOC, ARTEP 45-326-50-MTP.
- The training plans in those documents can be readily adapted to train other types of PA elements, such as a division PA staff.

PRINCIPLES

6-7. All PA training must be based on the training principles in the FM 25-series:

- **Train as combined arms and services team**. Do not train in a vacuum. Train with the unit you support.
- **Train as you fight**. Conduct realistic training. PA leaders must move soldiers out of the office and into the field to conduct training, including joint and combined operations. Set up a media operations center and conduct operations for a day.
- **Use appropriate doctrine**. All PA leaders should be familiar with FM 3-0 (100-5), *Operations*; FM 3-13 (100-6), *Information Operations*; and the PA manual, FM 3-61 (46-1), *PA Operations.*
- **Use performance-oriented training**. Performing tasks under field conditions with appropriate evaluation enables your soldiers to perform better under actual conditions.
- **Train to sustain proficiency**. PA units must take advantage of all training opportunities to sustain proficiency, participating in all CPXs, FTXs and deployments that their supported units are involved in.
- **Train to challenge**. Training for PA tasks must create the kind of pressure PA soldiers will face in actual situations. Ensure it is challenging but realistic.
- **Train using multi-echelon techniques.** The entire PA chain-of-command must participate in training for it to be realistic and effective.
- **Train to maintain**. Soldiers and leaders must keep equipment in a high state of readiness in support of training and for deployment.
- **Make commanders the primary trainers**. The unit commander is the primary training manager for the unit. The commander assigns primary responsibility to officers for collective training and to NCOs for soldier training.

TRAINING OPPORTUNITIES

6-8. Regardless of type, size or configuration, PA elements/units must train with the units they support. PA leaders must evaluate the ability of their units to support their real-world missions successfully and determine areas of weakness.

6-9. PA leaders must determine what training opportunities they can participate in realistically and coordinate for support from higher headquarters. If support is not provided, then unit leaders must create their own unit training program that simulates the tasks they will be asked to do when deployed.

6-10. For example, PADs should deploy with brigades to NTC or JRTC to practice their unit mission and capabilities and be evaluated accordingly. If the PAD is tasked with garrison duties, these duties must be retasked to allow the unit to train and maintain.

6-11. PA elements and units should take advantage of the excellent training opportunities offered by the JRTC and NTC. Both centers conduct media-on-the-battlefield training for visiting maneuver units. Public affairs elements should accompany the maneuver brigades and establish field media centers to take maximum advantage of these resources. It is imperative that PA elements be able to function effectively in austere environments. They cannot perform to their maximum capability unless they train accordingly. The key is repeated, tough, realistic training.

Historical Perspective

An initial PA team was sent to Somalia to set up a media operations center in December 1992. Mogadishu at the time was a war zone with many fronts. Initially, U.S. forces set up operations at the airport. Under the constant roar of C-5s, the PA team tried to organize a MOC. They had a very difficult time dealing with the lack of power and water. Eventually they were able to set up in a schoolyard. A "JIB-in-a-box" arrived with computer equipment, but it was all 220-volt — power in Somalia is 110-volt.

For the duration of the operation, personnel at the MOC learned to live and work under austere conditions, where food, water, power, and communications were constant problems. They eventually were able to set up a briefing room with benches, an old couch and a bulletin board for posting media opportunities, and conducted media operations under austere conditions.

(After Action Report 1992)

6-12. Home station CPXs and FTXs are fundamental training events that provide PA elements relatively low-cost opportunities to practice the full spectrum of wartime PA tasks on a reduced scale.

6-13. A visit by even a single reporter can be used to exercise a broad range of media support tasks including establishing a field media center; arranging interviews; preparing subject matter experts, individual soldiers and commanders to meet the media; preparing fact sheets; responding to media queries; registering media; arranging escorts and transportation, etc.

PA TRAINING FOR NON-PA AND PA PERSONNEL

6-14. Public affairs training is not limited to training for PA soldiers. PA elements/soldiers also have a responsibility to train non-PA soldiers and family members in military/media relations. Organic PA elements must develop training programs for their supported units. These can take the form of classes as well as pre-deployment briefings for units and family members.

6-15. Training for commanders and subject matter experts. Unit commanders and individual subject matter experts must be prepared to be interviewed. It is a PA responsibility to ensure they are familiar with the ground rules and know to restrict their statements and comments to their area of expertise. Details on how to do this are contained in the Soldier's Manual task, "Prepare a Spokesman to Address the Media," #224-176-4013.

6-16. Training for unit soldiers. Individual soldiers must be advised of the inevitability of media presence during military operations. Classes for units can be part of sergeant's time training and consist of teaching soldiers what they should and should not talk about when meeting the press. All soldiers should be encouraged to represent themselves as soldiers and encouraged to speak about the jobs they perform for the Army.

6-17. Training for family members. PA training for family members consists of educating them on their rights and responsibilities when interacting with the media. Family members often know more about specific operations than should be revealed to the media. They must be advised not to discuss information, which may be used by the enemy against their spouse's unit, such as details about troop movements, destinations, missions etc. They must also be advised that they have the right to refuse to talk to the media. Family member briefings should be a standard element of pre-deployment family support group activities.

6-18. Basic soldier skills. Public affairs soldiers are soldiers first and public affairs practitioners second. It is imperative they are well trained in basic soldiering skills. Public affairs elements and units must make time to train to standards on common soldier tasks that allow them to effectively shoot, move, communicate and survive on the battlefield.

6-19. Readiness. All PA elements and units must have, and exercise, detailed load plans. Soldiers should be aware of the importance of maintaining all equipment and vehicles in a state of readiness for deployment.

6-20. PA NCOs must ensure soldiers participate in preparations for overseas movement (POM) so they are administratively and medically prepared for worldwide deployment. Soldiers should be aware of the importance of wills, shots, powers-of-attorney, personal data and allotments.

6-21. Staff Section and Unit Training. A state of operational readiness to conduct combat or non-combat operations must be attained and maintained. This level of readiness is accomplished by preparing individuals, shifts and staff sections to perform assigned tasks and other duties at the desired standard of proficiency in advance of assigned operations.

6-22. Standards of performance must be set so the section can evaluate its performance. Examples of these standards include:

- Understanding DOD's policy statement -- Commanders will ensure maximum unrestricted disclosures of unclassified information to news media representatives consistent with operational security, guidance from higher headquarters and the privacy of individuals concerned.
- Escorted and unescorted media -- If the media has a PA escort, you may agree to an interview after the escort explains some basic ground rules. If not escorted, ask media personnel to accompany you to the command post, NCOIC or OIC and contact higher headquarters/public affairs officer.
- Conducting an interview -- Military personnel have the right to deny media interviews. If a soldier elects to provide the media with an interview, he should only discuss those things which he has direct responsibility or personal knowledge, and ensure an escort (PA or non-PA) is present. See Appendix N for more details.

THE INTERVIEW PROCESS

6-23. The Interview Process:

- Prepare for the interview. Consider the type of questions the media will ask, and think what your answer will be. When possible, ask for PA assistance (UPAR or PA personnel).
- Relax and be yourself. Imagine that the reporter is someone you know and talk with him in a relaxed manner.
- If cameras are present, ignore them and talk directly to the reporter. Be brief and concise. Remember, a TV news story will use only 10- to 15-second answers.
- If you need time to respond, ask the reporter to restate the question. A simple pause before answering the question is sufficient.
- Use simple language and avoid military jargon (i.e., military abbreviations or acronyms). If you must use military terms, explain what they mean.
- Use appropriate posture and gestures. Relax and be yourself.
- Answer only one question at a time. If asked multiple questions, answer the most important one first, or answer the one you're most comfortable with.
- Always try to end your comments on a positive point. This is your opportunity to tell your unit's story.

6-24. Things not to do:

- Do not allow media to videotape recognizable landmarks nearby, sensitive equipment, interior of tactical operations centers or other sensitive areas. Use OPSEC as guide on this matter.
- Do not answer speculative questions or give opinions concerning real or hypothetical ("what if") situations.
- Do not use the expression, "No comment." A more appropriate comment would be, "We don't comment on future operations." or

"I'm not qualified to respond to your question." or "That information is classified, so I can't discuss it."

- ***Never*** lie to the media.
- Consider everything you say to the media as "on the record." ***Never*** make "off the record" comments.
- Don't lose your temper when media representatives ask questions you consider inappropriate or foolish. Such questions are usually rooted in ignorance rather than in malice.
- Do not discuss operational capabilities, exact numbers or troop strengths, numbers/types of casualties, type of weapons systems or future plans. Use general terms like approximate, light, moderate or heavy.
- Don't repeat a negative phrase in response to a media representative's negative question (e.g., Q: Since your unit is poorly trained, can you really deploy? A: We're well trained and ready to go.).
- Staff section and unit training should be integrated into headquarters FTXs and CPXs in order to perform adequately within the unit's operational staff during real-world missions.

TRAINING EXERCISES

6-25. Training exercises will vary from major FTXs to CPXs and Tactical Exercises Without Troops (TEWTs). Each training opportunity provides the staff element realism, the opportunity to experiment and the ability to face situations. These challenging training exercises enable soldiers and units to tell the Army's success stories.

6-26. Training exercises also enable the commander and staff to:

- Emphasize the tactical SOP.
- Execute plans in a dynamic, hostile environment.
- Exercise bold solutions.
- Exercise contingency plans.
- Experience possible defeat without the penalty of combat loss.
- Work toward goal-oriented performance standards by team building while providing PA support.
- Training exercises also allow PA personnel to become operationally, strategically and culturally aware, and puts them in a unique position to interface with the news media personalities which shape the perceptions of our national and international audiences.

TRAINING FOR MEDIA FACILITATION

6-27. Facilitating media in the military environment includes three phases -- planning, preparation and execution.

6-28. **Planning.** The Media Operations Center receives requests for registered media to visit units in their areas of operation. The MOC will contact a unit and coordinate the time and location for the visit, as well as notifying a PA or nonPA escort for a mission. The escort will need to know the unit's location and whether the unit 's UPAR (if applicable) will travel with the group or meet them at the unit. The unit will need to know the number of media visiting, the duration of the visit and the status of the unit for the last 24 hours (i.e., has the unit been in contact with the enemy and has there been any significant actions).

6-29. The escort will plan the route to the unit. The number of media going will determine the number of escorts and vehicles needed for the trip.

6-30. **Preparation**. The JIB or escort will notify the media and tell them the time and location of departure and estimated time of return. This allows the media to plan for their supplies and equipment for the trip. The escort should meet with the media and update them on the unit's status, find out if the media has any specific requests before departing, cover safety points and OPSEC requirements, and ensure they have everything they need. This will also provide the escort with a direction for the types of questions or issues the media may address. The information concerning the unit's situation will allow the escort to develop a list of Q&As for the commander and individuals whom the media will interview.

6-31. The escort should meet with the drivers and review the route to the unit, cover contingency plans and determine an inspection time before movement. The escort should also contact the unit to reconfirm the visit and their location, as well as providing a list of the names and agencies of media visiting them. The escort should meet the media at the vehicles to insure they have what they need for the trip.

6-32. **Execution.** The escort should leave as scheduled and travel as the tactical or non-tactical situation dictates.

6-33. When arriving at the unit, the escort should have the driver park the vehicle as directed by the unit and have media personnel wait at the vehicles (if possible, with military supervision). The escort needs to request that the media not take photographs of the unit's position for operational reasons. Let the media know you are going to get the commander and will return as soon as possible. If the situation is hostile, get the media inside the perimeter and secure them away from equipment and the command post.

6-34. The escort should meet with the UPAR and let him/her know who is in the media group. Ensure that the unit will be able to provide the interviews the media requested.

6-35. The escort should brief the commander and let him know the ground rules, covering possible Q&As. Answer any questions or concerns he may have about the interview. Let him know that you will be present during the interview to assist him. When possible and if mission requirements permit, the commander should be available for the media. Review the unit's OPSEC requirements to ensure you understand what cannot be photographed.

6-36. The escort should determine where the commander wants to meet the media. If the area is not appropriate for conducting interviews (may be based on PA experience or media request), provide an alternative recommendation to the commander.

6-37. The escort should move the media to the interview point and introduce the commander to the group. After the interview, the escort should facilitate the rest of the coverage with the help of the UPAR. Do not allow the media to linger in the area of operations and become mission detractors. Wrap up the visit, return to your vehicles and depart. Upon returning to the JIB, the escort should be available to back brief the PAO.

6-38. The requirement to maintain proficiency in the full range of public affairs collective and individual capabilities and skills places a high priority on tactical training for both AC and RC PA soldiers. Tactical training participation enhances knowledge of battlefield requirements, increases unit cohesion, and forms the basis of experience needed for operational planning, mobilization, deployment and mission success.

6-39. PA will continue to face expanded missions in the joint arena. Quality PA training provides sufficient numbers of trained PA personnel to conduct joint and combined PA missions. Training must emphasize the joint perspective, enhance interoperability, and contribute to each combatant CINC's PA Mission. Exercise participation is critical to the training of PA personnel.

6-40. Trainers must fully incorporate a broad array of PA activities into all types of exercises and war gaming. These activities can be injected into computerized battle simulation as the training exercise driver. They should also be used in BCTP command post exercises, and JRTC, CTC, and NTC rotations. Seminars, area assessments and TEWTs are all forms of training that also provide relevant, realistic training.

Chapter 7

Community Relations

Public opinion about the Army is greatly influenced by the actions of each command. What the command does for its local community or fails to do affects the perceptions and attitudes of the American people, upon whom the Army depends for its support and existence. This applies not only to official acts but also to unofficial acts, which by their commission or omission affects public opinion. This principle also applies to individual members of the Army, their dependents and Army civilian employees in their personal contacts with the civilian community. Conducting community relations is a vital element to successful public affairs operations. Commanders and public affairs officers (PAO) must seize on key opportunities to gain and maintain links to internal and external publics.

EFFECTIVE COMMUNITY RELATIONS

7-1. PA personnel act in concert with veteran's groups, civic leaders and local populations to increase understanding and build support for Army activities. Army support to and participation in public events is based on the fact that the Army belongs to the American people. Common ownership requires that Army resources be used to support events and activities of common interest and benefit.

7-2. Effective community relations requires:

- Command supervision at all levels.
- Appreciation of public opinion and attitudes toward the Army
- Planning definite actions and positive policies.
- Implementing programs in a competent, professional and responsible manner.
- Constant evaluation of continuing programs to measure their effect upon the public and the command.
- Sharing the results of the program.

7-3. Commanders must maintain continual liaison with persons and organizations in the local community to help resolve common problems and develop cooperation and understanding between the installation and the local community. Community relations develop an effective two-way channel of communication between the Army and the community. PA does this by capitalizing on opportunities for better relations and resolving potential and actual areas of conflict.

7-4. Community relations projects or programs may be supported by use of exhibits, equipment and facilities. Exhibits consist of displays such as mission exhibits models, devices and other information and orientation materials at conventions, conferences, seminars demonstrations, exhibits, fairs or similar events. Also included are exhibits displayed on military installations during open house programs.

COMMUNITY RELATIONS ACTIVITIES

7-5. The goals of community relations is to develop an open, mutually satisfactory, cooperative relationship between the installation and the community. A successful community relations program improves the community's perception of the Army and its appreciation for the installation and the soldiers, family members and civilians who are part of the installation. It is based on openness and honesty. Community relations objectives are community assistance and social improvements for the community in which the military must work and live.

Historical Perspective

Fort Eustis, Va. started a pilot program in 1992 called Operation Self-Enhancement to give high-risk middle school students the opportunity to visit the post and focus on careers, teamwork and self-esteem. The program was so successful that it has become an annual event. Students receive light military training through an array of testable tasks and obstacles presented by members of a cadre team. This training helps students build their self-esteem and self-confidence and affords them the opportunity to interact with positive role models. This also gives the students a better idea of what the Army is about.

7-6. Community Relations activities include:

- **Speakers Bureaus.** Speakers are an effective means of developing understanding of the Army, stimulating patriotic spirit and informing the public about the activities of the installation, its units and its soldiers. Commanders should establish an installation speakers bureau and encourage soldiers of all ranks to participate in the installation program.
- **Community Liaison.** Maintaining liaison through informal community relations councils can establish and maintain open communications with community officials and organizations. Councils can be charged with a variety of responsibilities, such as developing and promoting new ways for members of the command to actively participate in local community activities, capitalizing on opportunities for better relations and resolving potential and actual areas of conflict. Community liaison can also involve recognition of private citizens, local community leaders and citizen groups and organizations for their support of the Army by public service awards. Commands can further community liaison through membership in civic, business and professional organizations when

the goals and objectives of those organizations are beneficial to the Army and their programs and projects are consistent with Army interest.

- **Ceremonial Units.** The band, color guard and other ceremonial units participating in public events are excellent ways to accomplish community relations objectives. These representatives of the Army serve as ambassadors to the civilian community and promote patriotism, interest in the Army, and awareness of the professionalism of our forces.
- **Exhibits.** Exhibits and displays of Army equipment, historical materials, models, devices and other information are other community relations activities that can enhance understanding of the Army and the installation. They provide an excellent opportunity for interaction between our soldiers and members of the local community and can communicate the professionalism, readiness and standards of our forces.
- **Open House.** Open houses may be scheduled to coincide with Armed Forces Day, the Army Birthday, service branch birthdays or anniversaries which mark the history of the installation, a unit or community events or in support of media day. An Open House gives the local community an idea of who we are and what we do. They also have the opportunity to visit us on the installation-- at our job site.
- **Physical Improvements.** Community service physical improvements focus on ensuring that the physical infrastructure is as safe as possible and provides the fullest possible range of support to the population. These activities encompass a wide range of programs that do not compete with the services provided by contractors and businesses in the local civilian community.

7-7. Some examples of physical improvements are:

- Construction projects that enhance the recreational, educational, environmental or cultural facilities of the community, such as building community picnic areas and hiking and biking trails.
- Demolition projects that enhance the safety and appearance of the community, such as the removal of unstable playground equipment.
- Projects that create or enhance a safe, clean environment, such as the removal of debris from a community wildlife area or painting a community recreation center.

TOWN HALL MEETINGS

7-8. Town Hall meetings provide installation commanders with an unfiltered means of communicating ideas to internal and external communities. This tool for conveying important information and ideas about the command cannot be underestimated in its effect and should not be planned haphazardly.

7-9. Commanders, PAOs and staff directorates must work together to produce an effective community relations product.

7-10. Prospective town hall meeting planners must understand, and properly apply, the correct type of town hall meeting. With a focus on the type of meeting and probable audience, the planner can begin the process of planning and conducting the event.

7-11. As part of the plan, the planner must determine the likely audience for the meeting, including attendees from internal and external audiences. He must also evaluate possible attendance by key publics. The planner should develop a standing operating procedure (SOP) to ensure each mechanism of the process is in place for the scheduled event.

7-12. Finally, post-event analysis is imperative to accurately assess its effectiveness.

7-13. The PAO must develop systems to quickly assess the feedback data and activate a follow-up plan that will maintain confidence from the community that town hall meetings are meaningful events.

TYPES OF TOWN HALL MEETINGS

7-14. Installation commanders can stage various types of town hall meetings. The commander must determine which meeting type, or hybrid, is appropriate for disseminating information and gaining useful feedback from internal and external publics.

7-15. The following meeting program structures have inherent strengths and weakness; knowing the potentials for message delivery will assist the commander in making his decision:

- **Commander - Expert Format**: This meeting (Figure 7-1), is characterized by the commander attending with key staff members facing a live audience. Typically the commander and his staff will give presentations and then field questions from the audience. Usually, attendance is open to the public.
- The primary advantages to this format center on the open nature of the meeting.
- This meeting provides the commander with an opportunity to provide detailed presentations with time being a minimal constraint. He also receives instant feedback from the types of questions from the audience and the passion with which questions are asked.
- Meetings in this format are likely to be seen as the most easily accessed by the internal and external publics.
- Among the disadvantages of the format is the ability to reach large audiences and control the conduct of the meeting.
- Unless the meeting is taped for later airing on the commander's cable access channel, the audience is often narrowly focused.
- Although the possibility exists that large audiences will attend, it is also possible small or narrowly focused audience will limit the general effect the commander seeks.
- An open meeting can also become the forum for unruly or disgruntled audience members to incite others or attempt to draw the commander into an open confrontation.
- This factor can be mitigated through the use of question time-limits and use of a moderator (other than the commander), but cannot be

totally eradicated. (As with any other public event, security must be a consideration in the planning process).

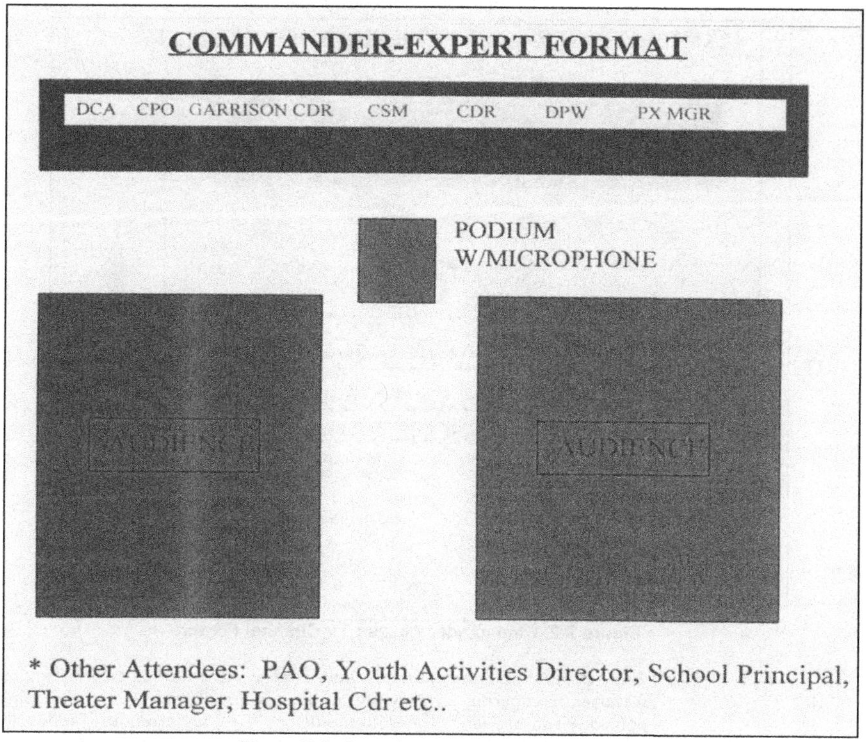

Figure 7-1. Commander Expert Format

7-16. Commander Access TV Channel Format: This meeting format (Figure 7-2) uses the commander's cable access channel to air the event. Normally, the commander, CSM and Garrison Commander (and other staff members as necessary) give a presentation. No live audience is in attendance. The commander provides a set of phone numbers, allowing questions to be called in. The staff operates the phone bank, accepts the questions, directs the questions to the appropriate staff agency, and delivers the answers by 3x5 card to the on-air panel. The panel members read the questions/answers to the viewing audience.

7-17. This format offers advantages focused on control and distribution of the product. The venue and the setting are completely controlled by the commander and his staff. The staff screens questions and, thus, no surprises will occur. Indeed, if questions/answers are given to the commander, he may choose to either not answer or return the card for more information.

7-18. Other advantages include an ability to re-run the meeting as often as desired and provide copies to local cable providers, many of which are interested in using the product for airing.

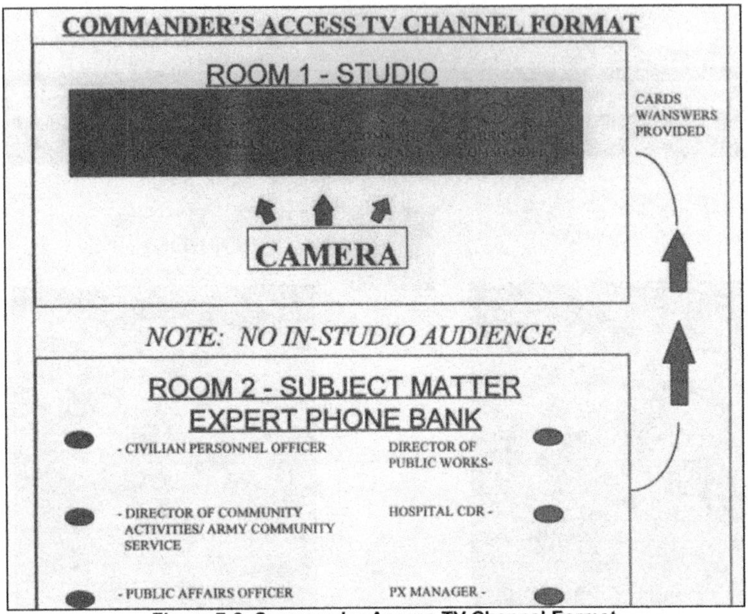

Figure 7-2. Commander Access TV Channel Format

7-19. Disadvantages to this format center on a difficulty in gauging feedback, competing for viewing audiences on television, and providing personal contact with the commander. With no attending audience, collecting feedback data is problematic. Follow-up questionnaires, using statistical probability methods, provide the only means of gaining reliable feedback. The likelihood of viewers to "channel-surf" is high due to the specific nature of questions from callers. Holding the interest of viewers, all of whom have multiple other viewing options available, is difficult. Finally, by appearing on television, in uniform, with a phalanx of staff members, the commander risks appearing to be speaking from "the mount." Audiences may view the commander as speaking down to them and being out-of-touch to their concerns.

7-20. Key members from the internal and external publics form a roundtable discussion (Figure 7-3). The topics are set by published agenda, with some time left for open discussion. The meeting results are published for general distribution. Media are usually invited as participants.

7-21. Roundtable discussions with key publics offer advantages to the commander by providing information to the individuals who represent overlapping and wide constituencies. By setting an agenda, the commander can deliver focused messages with a high likelihood of the messages later reaching targeted audiences. It allows the commander and potentially key staff officers to deliver presentations and provide follow-up information to those in attendance.

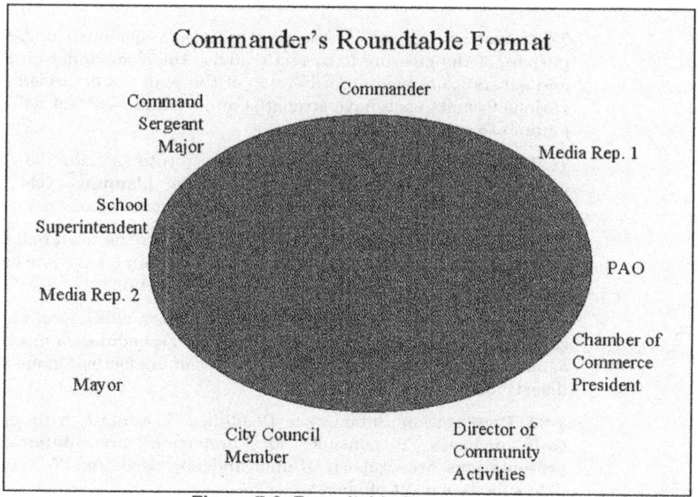

Figure 7-3. Roundtable Format

7-22. The principal disadvantage is the distance between the commander and the publics. The publics will receive information about the meeting second-hand and in a potentially filtered state. Gaining feedback may also prove challenging (but not necessarily impossible). Other disadvantages may be in the limits of the audience makeup; key publics, which are not invited, may resent the slight.

7-23. Commander-in-the-round format. A room is set-up that will allow the commander to be in the middle of all the attendees. Normally, he will stand and walk around in the circle formed by attendees. The meeting is usually open to the public.

7-24. The primary advantage to this format is the close contact the commander shares with his audience. Audience members may feel that barriers are lowered because the commander is close in proximity and no staff members are buffering their access. The commander also can realize feedback very quickly and can gain some appreciation for the resonance of his ideas with the assembled audience.

7-25. Such closeness with the audience can also be a significant disadvantage. Limiting surprises and controlling potentially unruly audiences is extremely difficult in this format. Further, because of the commander's reliance on his personal notes and memory, his ability to provide detailed information and multi-media presentations may be restricted.

7-26. Mitigation against such limits is dependent on the site design. Other disadvantages include problems associated with other "open" formats, including a possibly non-representative makeup of the audience.

7-27. Before deciding on which format is appropriate for the information, the commander must have a plan for what must be achieved.

7-28. Characteristics of an effective plan:

7-29. Answers the "why?" questions. The planner must understand the purpose of the meeting to correctly advise the commander on the format and substance to begin coordination of the plan. As described above, the various formats each have strengths and weaknesses that will assist the planner to shape the meeting.

7-30. Routinely Scheduled. Normally, meetings can be scheduled quarterly or monthly. Use the installation planning cycle to ensure proper coordination and notification of public meetings.

7-31. Site Plan. Checking and securing a site for the town hall meeting is dependent upon the type of meeting selected. Plan for the site early in the process and establish the layout of the site in detail.

7-32. Calendar Check. Before scheduling events, check local and regional calendars for possible conflicts. For example, scheduling a meeting on the same night as the local high school homecoming football game may prove disastrous.

7-33. Presentation Submission Deadline. Coordinate with appropriate staff agencies responsible for preparing presentations. Ensure presentations are properly staffed and approved. As PAO, establish a firm submission cut-off date.

7-34. Focused Presentation. Inform the staff of the commander's intent for the meeting. If the commander wants a particular theme addressed, ensure staff agencies adhere to the parameters of the intent.

7-35. Media Invitation List. Invite local and regional media including print and electronic outlets. Develop relationships with individual reporters and provide background material as necessary. (Local newspapers, radio stations and television stations can often assist in publicizing meetings as well).

7-36. Publicity Plan. Ensure all available avenues are used to publicize meetings. Included in this process are the post newspaper and radio station (where available), normal distribution, staff meetings, E-mail delivery and chain of command communications. Take special care to invite key publics by individual invitation and phone call follow-up.

EVALUATING FEEDBACK

7-37. Assessing the effectiveness of town hall meetings is essential to developing community-related policies and courses of action for the command. PAOs can use standard statistical measurements using survey techniques to gauge the level and intensity of views of the various publics. Other analyses can be derived from follow-up media content analysis, letters to the commander and post newspaper, and reactions at the meetings. None of the methods described here will render a perfect picture. PAOs must exercise good judgment and personal insight when advising the commander of analytical results.

7-38. **Response Follow-up**. Investigating and responding to issues raised at town hall meetings are critical to public perceptions of the level of care the commander applies to community operations. Just as a maneuver commander sees battle damage assessment (BDA) as crucial to determining the efficacy of fires, commanders and PAOs must determine the true productiveness of community relations programs. Commanders will often feel compelled to promise action, such as investigation or immediate problem resolution, during a meeting. Staff agencies are normally the conduit for actions (only occasionally will the commander personally provide the requisite service).

7-39. PAOs should provide the staff oversight of the response mechanism. PAOs can devise a recurring memorandum that provides the commander (usually through the chief of staff), details about the status of actions. All staff agencies should receive updates routinely.

7-40. Along with the response mechanism, PAOs can use command information and media relations channels to inform the publics of problem resolutions. For example, if an issue raised at a town hall meeting indicated that the local recreation center was routinely opening two hours late each day, describe the measures taken to alleviate the problem. The PAO can use the post newspaper or radio station to provide lists of problems and resolutions.

CONCLUSION

7-41. Community relations and activities are vital to instilling and maintaining the confidence of internal and external publics in our great Army. Commanders and PAOs cannot leave the prospect of successful relations to chance. Too much is at stake. Careful selection of the type of town hall meetings to be used must be taken. The event must be planned with the attention to detail required for all military operations. Each part of the plan has unique importance and cannot be overlooked. To ensure the effectiveness of the operation, PAOs are compelled to build a clear mechanism for evaluating outcomes. Finally, the command must provide conspicuous follow-up responses to issues raised to complete the process. Town hall meetings provide an excellent opportunity for commanders and the various publics to interact and improve community institutions.

COMMUNITY ASSISTANCE

7-42. Community assistance applies the skills, capabilities and resources of the needs and interests of the local community. Providing support for and participating in events and activities which are beneficial to both the Army and the community, builds on a long tradition of "America's Army" helping American communities. Identifying opportunities, which advance the interests of both the Army and the community, is an important objective for every commander.

7-43. Community assistance projects and operations must impact positively on the unit or individual soldier, enhance unit or individual readiness and contribute toward the common good of the community. Army commanders must ensure that their initiatives are not competitive with local resources or services and do not benefit any particular interest group and will not result in monetary or service remuneration in any form for unit members or the unit as a whole.

7-44. Increasing public awareness and understanding of the Army, inspiring patriotism and enhancing the Army's reputation as a good neighbor is a goal of community assistance. Community assistance activities can help build unit morale and esprit de corps. These activities also provide an excellent opportunity for soldiers to serve as role models, which not only enhances recruiting efforts, but also serves to motivate soldiers by promoting their self-esteem and furthering their sense of service to the nation.

READINESS ENHANCEMENT

7-45. Certain community assistance activities enable a commander to train soldiers, enhance individual and unit readiness, maximize use of assets and foster a positive training environment where soldiers can become involved in realistic, "hands-on" training opportunities. Projects should be selected which exercise individual soldier skills, encourage teamwork, challenge leader planning and coordination skills, and result in measurable, positive accomplishment. Finally they should enable a unit to exercise use of its equipment, resulting in training opportunities that can generate greater operator efficiency for future missions.

7-46. Examples are:

- Medical Readiness Program. The Medical Readiness Program is an activity in which Army medical unit personnel, together with state medical emergency officials, plan and provide support in the form of diagnosis, treatment, and preventive medical, dental, and veterinary care to citizens in remote areas of the U.S. or its territories. The program is designed to enhance the unit's medical readiness, provide unit training opportunities and serve the public in locations where medical care is not otherwise available. The program may not compete with local private medical care that may be available.
- Air Ambulance Participation. The Military Assistance to Safety and Traffic Program. (MAST) is a proven example of Army support to civil authorities. This program permits the utilization of Army aviation assets to conduct emergency air evacuation and recovery efforts.
- These projects contribute to the health and welfare of the community, making the Army an integral partner in community progress and development. They can enhance the ability of the local community to support itself and its people, to provide the best possible services to the citizens and to promote a positive, healthy safe environment.
- Community service activities are those which focus on improving the community, its infrastructure and its ability to serve the local population.

7-47. **Social Improvements.** Community service social improvements, which focus on making the social environment as healthy as possible, provide the widest range of support to the population. They encompass many projects including:

7-48. Support to youth programs, such as scouting and programs that provide assistance to special need audiences such as the Special Olympics.

7-49. Examples are:

- The Civilian Youth Opportunities Program (Challenge). This is a youth program directed at attaining a high school diploma, providing job training and placement, improving personal and social skills and providing health and hygiene education and physical training. Soldiers work with civilian leaders to provide a comprehensive support package, ranging from choosing appropriate clothing to attending residential training facilities.
- Youth Physical Fitness Clinic Program. The National Guard encourages fitness and combines academic and athletic achievement by helping schools conduct competitions in selected athletic events. This program also establishes a separate scholar-athlete category for those students with a 3.5 or higher grade point average.

7-50. Involvement in ventures and projects that enhance the educational or cultural climate of the community, such as adult literacy, reading or community theater programs.

7-51. Examples are:

- Civilian Community Corps. This program provides managerial, organizational and technical skills for disadvantaged Americans seeking the skills for success. Through this program, the Army helps participants become productive citizens. In exchange, participants perform a wide range of community service activities that improve the foundation of American society. This program encourages intra-governmental cooperation on the federal level. It also encourages partnerships with industry, education, state, federal and local governments.
- Science and Technology Academies Reinforcing Basic Aviation and Space Exploration Program. (STARBASE) This program is an innovative partnership of professional educators, military personnel and corporate sponsors. It promotes science, mathematics and technology basics for primary through secondary schools. Using National Guard resources to spark student interest, the program develops strong self-esteem, provides excellent role models, promotes positive attitudes and develops goal-setting skills.
- The guiding principle behind community relations efforts is that the installation and the community have a common interest in providing the best possible support for each other. A cooperative relationship exists, because soldiers stationed at the installation receive life support from the community while many of the civilians who make up the community receive life support from the installation. The interdependence of the military installation and the civilian community can involve economics, education, health care, basic services, quality-of life issues and many others.
- The community relations goal of local commanders is to develop an open, mutually satisfactory, cooperative relationship between the installation and the community. These efforts improve the community's perception of the installation and the soldiers, family members and civilians who are part of the installation.
- Participation in community relations activities is an effective method for projecting a positive Army image, making the best use

of assets, providing alternative training opportunities and enhancing the relationship between the Army and the American public. Activities vary widely, ranging from individual soldier involvement to full Army participation. They are characterized by detailed coordination between the military command and community authorities. They fulfill community needs that would not otherwise be met, enhance soldier and unit morale, skills and readiness and improve the mutual support between the military and civilian communities.

Chapter 8
PA Organizations

THE BRIGADE

8-1. The Brigade PAO is the lowest level to which the Army has assigned organic Public Affairs Assets.

8-2. Working as both a special staff officer and as a member of the Brigade's planning team, the BDE PAO acts as the spokesperson for the unit, advisor to the Commander, and provides Public Affairs guidance and planning to commanders at all levels.

8-3. To support and conduct Public Affairs Operations within the Brigade, the BDE PAO has the following functions, organization and equipment:

8-4. Specific functions of the Brigade PAO, as outlined within the core competencies are:

- **Public Affairs Planning**
  - Advise the commander and staff on PA implications of plans and actions.
  - Write Brigade PA annex and matrices using operational tools, terms, graphics, and concepts
  - Execute the plan
  - Plan for future operations.
  - Command and control attached Public Affairs assets within the brigade AO.
  - Tactically communicate with PA units and supported combat units in the Brigade's Battlespace.
- **PA Training**
  - Train and supervise stringers to assist their commanders in conducting their internal information programs.
  - Train soldiers, family members, DACs and everyone habitually associated with the unit to comfortably and successfully communicate with the media.
  - Utilize Train the Trainer.
  - Evaluate public affairs training programs of subordinate commands.
  - Train subordinate commands to facilitate the media.

  - Train leaders and soldiers to protect information products and information systems from compromise and intrusion by practicing security at the source.
- **Media Facilitation**
  - Assist media to gain access to units, soldiers and commanders.
  - Evaluate subordinate command's media facilitation plan.
  - Assist subordinate commanders to coordinate transportation.
  - Conduct situation briefs as needed.
  - Monitor media operations within the Area of Operations. (Maintain accountability of the media to preclude fratricide. Accomplished through reports from escorts or through electronic tagging and monitoring.)
  - Respond to and mediate potential media problems; react quickly to coordinate and facilitate information issues in Brigade's Battlespace.
- Information Strategies
  - Monitor and analyze the local Military Information Environment.
  - Provide public affairs support to the G5 / S5 for the development and implementation of civil affairs programs.
  - Support higher echelon Public Affairs requirements for information.
  - Gather Open Source Information to help build relevant information.
  - Assist Commanders to conduct internal and external information to include Hometown News Release Program.
  - Monitor local news media products (Visual, print and audio) and analyze for PA implications.
  - Monitor and Analyze the local Military Information Environment.
  - Protect digital images, information products and PA and non-PA information systems from compromise and intrusion.

8-5. Organizational Structure:

- Captain , 46A.
- Sergeant, 46Q

8-6. Equipment:

- Computer with compatible software and communications hardware
- Access to the Tactical Internet and tactical battlefield radio communications
- Access to FM, Satellite, and video and cellular communications
- Access to Army Battle Command System, (MCS\P or CSS\CS)

8-7. Transportation:

- HMMWV

8-8. Addititional Support Requirements. PAO requires linguistic support from the Civil Affairs soldiers or contract civilians attached to the Brigade.

SAMPLE PA EXECUTION MATRIX BRIGADE PUBLIC AFFAIRS OFFICER

	PHASE I	PHASE II CROSS LD	PHASE III DEFEAT	PHASE IV SEIZE OBJ. LION	PHASE V RECONSIT.
BRIGADE	PAD CIVILIAN MEDIA				PAD AND CIVILIAN MEDIA
1-66 AR	INTERNAL	EMBEDDED MEDIA-- PRIORITY OF EFFORT			EXTRACT MEDIA
1-4 AV	AVIATION WEEKLY	SAME	SAME	SAME	AVIATION WEEKLY DEPARTS
1-5 IN (L)	OWN THE NIGHT MAG.				
SCOUTS		MEDIA REPS LINK PRIOR TO DEPLOY			
BSA	FOOD PREP QTRLY	SAME	SAME	SAME	SAME
MEDICAL	SAFETY STORIES				MEDIA - INJURED SOLDIERS

THE DIVISION

8-9. The Division PAO is the next level to which the Army has assigned organic Public Affairs assets. The division is largely self-sustaining and capable of independent operations. The division is a unit of maneuver, organized with varying numbers and types of combat, combat support (CS) and combat service support (CSS) units.

8-10. The division may be armored, mechanized, medium, light infantry, airborne or air assault; each can conduct operations over a wide range of environments.

8-11. Working as both a special staff officer and as a member of the division planning team, the DIV PAO acts as the spokesperson for the division, advisor to the Commander, and provides Public Affairs guidance and planning.

8-12. To support and conduct Public Affairs Operations within the division, the DIV PAO has the following functions, organization and equipment:

8-13. Functions: Assumption #1: DMAIN is not in country

- Gather, analyze and disseminate open source information, focusing on global media, to increase the situational awareness throuhout the command.
- Gather and disseminate multimedia products throughout the command, and to external and home station audiences.
- Conduct Public Affairs research and write the PA Estimate of the situation.
- Write the PA plan for OPORDERS, OPLANS, and TACSOPS.
- Monitor and analyze American and foreign public sentiment of current operations from available media sources for PA implications and advice.
- Monitor and analyze battlespace communications (visual, audio, FM, satellite) for PA implications.
- Monitor news media products (Visual, print and audio) and analyze for PA implications.
- Facilitate Media Operations.
- Subfunctions:
  - Assist media to gain access to units, soldiers and commanders.
  - Coordinate air and ground transportation on a non-interference basis.
  - Conduct daily situation briefs as needed.
  - Respond to media queries.
  - Validate media's credibility, expertise, knowledge, purpose and legitimacy.
  - Train and supervise unit-level PA representatives (Command Information NCOs) to assist their commanders command information programs.
  - Train soldiers, family members, DACs and everyone habitually associated with the unit to comfortably and successfully communicate with the media.
  - Survey soldiers, family members, DACs and other members of the internal audiences to measure effectiveness of the command information program.

8-14. Assumption #2: DMAIN is in country

- Advise the commander and staff on PA implications of plans and actions
- Gather, analyze, and disseminate Open Source Information, focusing on global media to increase the commander's situational awareness.
- Gather and disseminate multimedia products throughout the command, and to external and home station audiences.

- Conduct Public Affairs research and write the PA Estimate of the situation.
- Write the PA plan for OPORDERS, OPLANS and TACSOPS
- Monitor and analyze American and foreign public sentiment of current operations from available media sources for PA implications and advise
- Monitor and analyze battlespace communications (visual, audio, FM, satellite) for PA implications
- Monitor news media products (Visual, print and audio) and analyze for PA implications
- Coordinate and integrate all information-related functions (PSYOPS, CA, VI, Joint, Combined and Interagency PA) into the PA plan
- Conduct Information Operations
- Act as a conduit for CI products from the field, sanctuary and commercial sources for input into the commander's information program. Provide command information to soldiers, family members and Department of the Army Civilians.
- Facilitate Media Operations
- Subfunctions:
  - Coordinate air and ground transportation on a non-interference basis
  - Assist filing stories, video and photographs on a non-interference, reimbursable basis
  - Conduct daily situation briefs as needed
  - Respond to media queries
  - Validate media's credibility, expertise, knowledge, purpose and legitimacy
  - Conduct Primary and Secondary accreditation
  - Primary--Full accreditation of non-accredited media
  - Secondary--Process media previously accredited at Corps and higher
  - Survey soldiers, DACs and other members of the internal audiences to measure effectiveness of the command information program

8-15. Organizational Structure:

- At a minimum an embedded PA division section has:
  - Major 46A and two Captains 46A
  - Master Sergeant 46Z
  - Specialist 46Q
  - Specialist, 46R
  - PFC 46Q

8-16. Equipment:

- Computer with compatible software and communications hardware and tactical fax machine.

- Access to the Tactical Internet and tactical battlefield radio communications (SINGARS, MSRT etc.)
- Access to FM, Satellite, video and cellular communications
- Access to Army Battle Command System, (MCS\P)

8-17. Transportation:

- Light Division--HMMWV
- Armor and Mech Division-- HMMWV and Trailer

8-18. TAC 1 AND 2 Missions and Functions

TACTICAL COMMAND POST PA SECTION

8-19. The TAC CP, Public Affairs Section is task-organized based on METT-TC. It is the center of gravity for immediate internal and external communication, resolution of Public Affairs issues and violations of ground rules by media representatives. It acts as a conduit to the Information Operations Cell in the DMAIN adding immediacy to the Open Source Information process.

8-20. Public Affairs personnel deploy with the lead elements during any contingency and remain with the TAC CP to assist the commander, provide internal command information to deployed soldiers and limited external information to home station. They conduct media facilitation to expedite the flow of information to the America public while freeing the commander and his soldiers to conduct their mission.

8-21. The TAC CP, Public Affairs Section maintains connectivity with commanders and global information sources. It contributes to and monitors the common relevant picture, and synchronizes collection and dissemination efforts of soldiers far forward to internal and external audiences.

8-22. Particularly during split-based operations, the PAO task organizes his section to best serve the commander, his soldiers and the American public's need for information.

8-23. Assumption #1: DMAIN is not in country

- Act as the division spokesman.
- Advise the commander and staff on PA implications of plans and actions.
- Execute the plan.
- Assist DMAIN to gather Open Source Information to build the common relevant picture.
- Conduct Information Operations.
- Gather and disseminate multimedia products throughout the command and to external and home station audiences. Act as a conduit for CI products from the field, sanctuary and commercial sources for input into the commander's information program.
- Monitor and analyze battlespace communications (visual, audio, FM and satellite) for PA implications.
- Monitor local news media products (Visual, print and audio) and analyze for PA implications.

- Coordinate, integrate, and synchronize all information-related functions (PSYOPS, CA, VI, Joint, Combined, and Interagency PA).
- Synchronize Public Affairs assets (internal and external) within AO.
- Facilitate Media Operations:
  - Assist media to gain access to units, soldiers and commanders.
  - Coordinate transportation on a non-interference basis.
  - Assist filing stories, video and photographs on a non-interference, reimbursable basis.
  - Conduct daily situation briefs as needed.
  - Respond to media queries.
  - Monitor media operations within the AO.
  - Respond to and mediate potential media problems; react quickly to coordinate and facilitate information issues throughout the AO.
  - Validate media's credibility, expertise, knowledge, purpose and legitimacy.
  - Provide seamless connectivity for media accountability throughout the AO.
  - Conduct Primary and Secondary accreditation.
  - Primary--Full accreditation of non-accredited media.
  - Secondary--Process media previously accredited at Corps and higher.

8-24. Assumption #2: DMAIN is in country

- Act as the division spokesman.
- Advise the commander and staff on PA implications of plans and actions.
- Execute the plan.
- Assist DMAIN to gather Open Source Information to build the common relevant picture.
- Monitor battlespace communications (visual, audio, FM and satellite) for PA implications.
- Facilitate Media Operations:
  - Assist media to gain access to units, soldiers and commanders.
  - Conduct daily situation briefs as needed.
  - Monitor media operations within the AO.
  - Respond to and mediate potential media problems; react quickly to coordinate and facilitate information issues throughout the AO.
  - Provide seamless connectivity for media accountability throughout the AO.
  - Conduct Secondary accreditation.
  - Secondary--Process media previously accredited at Corps and higher.

8-25. The division public affairs section provides public affairs support to the division commander and to divisional units deployed in support of combined or joint operations. The division PAO has operational and tactical control over all PA TOE organizations assigned or attached to the division and coordinates closely with embedded PA sections within brigades or divisions to carry out PA operations.

8-26. The division PA staff, when deployed, is augmented by one PAD and one MPAD per three combat brigades. The division public affairs section, if augmented only by a Public Affairs Detachment, operates the division media operations center.

8-27. Traditionally, divisions have operated as part of a corps. In corps operations, divisions normally comprise 9 to 12 maneuver battalions, organic artillery battalions and supporting CS and CSS units. Divisions perform a wide range of tactical missions and for limited periods are self-sustaining. Corps augment divisions as the mission requires.

Chapter 9

Information Operations

INTEGRATING PUBLIC AFFAIRS

9-1. Information Operations involve a variety of disciplines and activities that range from electronic warfare and physical destruction through cyberwar and information campaigns. Public affairs is a related activity of IO, and contributes to overall operational success, both real and perceived.

9-2. Successful integrated IO requires coordination of themes, messages and activities in order to leverage the massing of information effects. When synchronized with other military operations, IO is a combat multiplier.

9-3. Information campaign objectives cannot be neatly divided by discipline, such as PA, CA and PSYOP. The responsible organization cannot be easily determined solely by looking at the medium, the message or the audience. For example, information about weapons turn-in policy and collection sites may be disseminated through a variety of means. This could include direct contact by Civil Affairs personnel with local populations and their leaders; PSYOP print and broadcast products; news releases, press conferences and other media facilitation coordinated by PA.

9-4. In accordance with joint doctrine (Joint Pub 3-61, Doctrine for Public Affairs in Joint Operations), public affairs are an operational function designed to contribute to the overall success of joint operations. For Public Affairs, the audience may be internal or external, but the objective is constant: Soldiers, participants and the public must understand objectives, motives and the nature, scope and duration of friendly actions. The relevant audiences important to the commander are not limited to soldiers and the American public, but are also international as well as local to the operation.

9-5. Synchronized information operations contribute to military campaigns in a variety of ways. These contributions may:

- Gain or sustain support for the U.S. or coalition position
- Reduce the need for combat forces
- Influence events with non-lethal means
- Counter propaganda and disinformatio
- Discourage adversary offensive operations
- Deter hostile actions

- Undermine illegitimate regimes
- Support the maintenance of coalitions

9-6. Information Operations during peacekeeping operations:

- Promote peaceful cooperation
- Lower friendly force requirements
- Counter propaganda and disinformation
- Reduce friction leading to hostilities
- Gain and maintain the initiative
- Shape opponent plans and operations

9-7. These goals may not be achieved solely by tactical level information operations, but rather, may be theater and national-level issues that are reinforced by tactical-level message dissemination. This requires horizontal and vertical integration of themes and messages to achieve a massing of information effects.

IO STAFF ORGANIZATION

9-8. Composition of the Information Operations battle staff/coordination council or other such element is flexible and tailored to the operation and desires of the commander. (See figure 9-1). Notional IO staff structures are included in FM 3-13 (100-6), Information Operations.

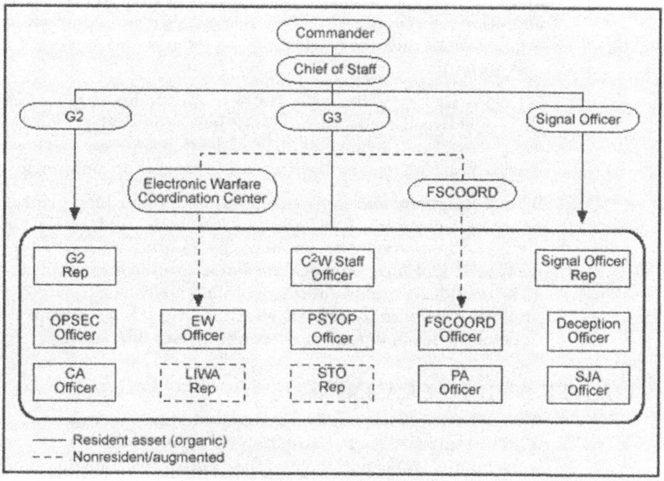

Fig 9-1 Notional IO Battlestaff

9-9. The IO cell is often headed by the G-3 or his designated representative, and includes representatives of a variety of organizations. The staff may include, but not necessarily be limited to the G2, G3, G5, G6, PAO, PMO, JAG, PSYOP, Electronic Warfare Officer, Political Advisor, Joint Military Commission representative, Fire Support Coordinator and Targeting Board representative.

9-10. The successful accomplishment of a specific mission may require the close coordination and synchronization of the range of information activities as well as maneuver elements. While the IO cell is lead by the G3 or his designate, IO coordination is the responsibility of the Land Information Warfare Activity field support team. While providing a public affairs representative to the IO Cell, PAOs must maintain a clear and direct link with the commander.

LIWA

9-11. The Land Information Warfare Activity, an INSCOM element, provides commanders with field support teams (FST) that serve as IO advisors in addition to effecting the synchronization and coordination of the range of activities that support IO. LIWA field support teams do not serve as functional area specialists, but rather, coordinate the activities of those elements. For example, the LIWA FST members may be from the military intelligence and PSYOP branches, but do not serve as the commander's intelligence analyst or PSYOP planner, or for that matter, Public Affairs advisor. They do, however, coordinate the actions and products of these and other activities in support of the IO plan.

PA SUPPORT TO INFORMATION OPERATIONS

9-12. PA participation in IO involves no completely new tasks but does require a broadened scope of operations. PA support to IO requires analysis of the Global Information Environment (GIE) and the operational environment, as well as synchronization of efforts with other organizations and agencies to ensure themes and messages are consistent and deconflicted.

9-13. PA in IO requires PA staffs to be fundamentally proactive rather than reactive. Often, actions may be taken and products developed to assist command achievement of a desired end state. This is more than merely reacting to events with a press release or conference

9-14. PA actions and events that support IO include print and electronic products, news releases, press conferences and media facilitation. PA advises the commander on how the operation is being perceived and portrayed and also provides guidance to unit commanders and soldiers. This includes regular talking points and themes for commanders and preparing soldiers to interact with the press. It's a means of emphasizing selected issues and positions--speaking with one voice.

PA ESTIMATE AND IO PLANNING

9-15. The starting point for PA contributions to Information Operations is the Public Affairs Estimate. (See Appendix C). The PA estimate consolidates information on the audiences, media presence, public opinion, personnel available and PA guidance.

9-16. This is not a static document created at the beginning of an operation, but must be continually updated to reflect changes in the operational situation and environment. Issues to consider include:

9-17. **Audience analysis.** Who are the audiences, both internal and external? What are their information needs? How do they get their information: television, radio, newspapers or word of mouth? Is the media state-run or independent? Does the audience population have telephones, cell phones, fax machines or Internet connections? These devices are frequently found even in developing countries and must be considered during the analysis of information channels.

9-18. **Media presence.** What media representatives and organizations are in the area of operation? Are they radio, television or print? Are they state-run or independent? What is their political slant? Are they pro- or anti-coalition? Are they receptive to coalition information products such as news releases or other print or electronic products? Is the local media interested in live interviews with U.S. commanders and soldiers?

9-19. **Public Opinion.** What are the opinions/beliefs of the local populations; of the international community; of the U.S. national population?

9-20. **Personnel available.** What is the available Public Affairs force structure (PADs, MPADs, BODs, CPIC/JIB staff, unit organic PA staff and individual additions), translators, Combat Camera and administrative staff.

9-21. **PA guidance.** What guidance has been received from higher levels? Official positions on theater issues are naturally not developed at the tactical level. What is the theater strategic/national command authority position? This is often coordinated and deconflicted at all levels via conference calls and other communication means.

IO CAMPAIGN CYCLE

9-22. There are four stages to an IO campaign cycle: capability development, assessment, planning and execution. The execution stage is accompanied by evaluation-- during and after the mission -- in order to adjust operations as needed and after the operation to gather lessons learned.

9-23. Capability development:

- Identify local resources and available external support. Theme and message delivery can take many forms, including radio/television, handbills, leaflets, loudspeakers, soldiers, displays, the Internet, internal information products, USIA, Voice of America, print and electronic news releases, press conferences, direct contact with parties, leaders, officials and citizens. Direct contact may include military liaisons, Civil Affairs personnel, diplomatic contact, or any form of personal interaction.
- Establish processes and procedures
- Collect, organize and store relevant information

9-24. Assessment

- Perform mission analysis
- Obtain commander's guidance
- Define IO goals and objectives
- Conduct risk assessment
- The assessment phase includes a mission analysis, clarification of the commander's guidance, initial identification of IO goals and objectives and a risk assessment. Goals and objectives include a

determination of the desired end state and what must be done to achieve it. This may mean inducing others to take or not take certain actions, or have the information to make certain decisions that will support the goals of the operation. Public Affairs is not in the business of shaping beliefs and attitudes of populations, but can provide factual information that enables people to make informed decisions.

- Risk Assessment
  - Consequences for command if information operation fails?
  - Potential unintended effects?
  - Operational success too reliant on IO?
  - Can IO campaign be used against U.S. or coalition?
  - Force protection issues?
  - Compromise or loss of impartiality?

9-25. Planning

- Develop and coordinate themes
- Determine the best implementation means
- Delineate tasks and responsibilities
- Identify feedback and measures of effectiveness channels
- Prepare implementation order
- During the planning phase, specific themes are developed and coordinated with all members of the IO cell. Message/theme delivery methods are determined and specific tasks and responsibilities are assigned. The IO cell may use a synchronization matrix to effectively manage IO events. For example, this matrix will indicate specific actions, events or products each member organization of the IO cell will execute or produce to support the plan. For example, a specific event may require PSYOP leaflets and broadcasts, PA press releases, news conferences and interviews with soldiers and Civil Affairs meetings with local officials and community leaders. These activities are coordinated on the matrix, ensuring deconfliction of resources, messages and products.
- Measures of effectiveness and feedback indicators vary widely and should be identified in the planning process. They may include questions raised by the media, editorials and commentaries, statements by public officials, postings to internet newsgroups and forums, demonstrations and protests, statements during meetings, responses given to public opinion surveys, behaviors during specific events, as well as other SIGINT and HUMINT collection and analysis.
- The product of the planning phase is a synchronization matrix and execution schedule. The matrix is then coordinated with the overall synchronization matrix, ensuring that IO is coordinated across the BOSs.

9-26. Execution

- Conduct the mission.

- IO monitoring must be conducted throughout the execution of the event and during follow-up review, feedback and evaluation.
- If necessary or possible, alter mission if evaluation determines it is not successful or unexpected responses occur.

9-27. **Evaluate**

- Assess the effectiveness of the operation
- Determine preventive methods, document lessons learned and apply to next operation

Appendix A

DoD PRINCIPLES OF INFORMATION

The DoD Principles of Information are contained in DoD Directive 5122.5, Change 1. They chart the course for all DoD Public Affairs activities, and apply to the full continuum of day-to-day activities and operations. It is the commander's responsibility to ensure that all planning for military activities and operations efficiently and effectively achieve the goals set by these principles.

DOD PRINCIPLES OF INFORMATION

A. Timely and accurate information will be made available so that the public, Congress, and the news media may assess and understand the facts about national security, defense strategy, and on-going joint and unilateral operations.

B. Requests for information from organizations and private citizens will be answered in a timely manner. In carrying out this policy, the following principles of information apply:

(1) Information will be made fully available, consistent with statutory requirements, unless its release is precluded by current and valid security classification. The provisions of the Freedom of Information Act and the Privacy Act will be complied with in both letter and spirit.
(2) A free flow of general and military information will be made available, without censorship or propaganda, to the men and women of the Armed Forces and their family members.
(3) Information will not be classified or otherwise withheld to protect the government from criticism or embarrassment.
(4) Information will be withheld only when disclosure would adversely affect national and operations security or threaten the safety or privacy of the men and women of the Armed Forces.
(5) The Department's obligation to provide the public with information on its major programs and operations may require detailed public affairs planning and coordination within the Department and with other government agencies. The sole purpose of such activity is to expedite the flow of information to the public; propaganda or publicity designed to sway or direct public opinion will not be included in Department of Defense public affairs programs.

Appendix B

THE GUIDELINES FOR COVERAGE OF DOD COMBAT OPERATIONS

DOD MEDIA GUIDELINES

The DoD Media Guidelines, issued as Change 3 to DoD Directive 5122.5, provide the following guidelines for coverage of DoD combat operations:

A. Open and independent reporting will be the principal means of coverage of U.S. military operations.
B. Pools are not to serve as the standard means of covering U.S. military operations. But pools may sometimes provide the only feasible means of early access to a military operation. Pools should be as large as possible and disbanded at the earliest opportunity--within 24 to 36 hours when possible. The arrival of early access pools will not cancel the principle of independent coverage for journalists already in the area.
C. Even under conditions of open coverage, pools may be appropriate for specific events, such as those at extremely remote locations or where space is limited.
D. Journalists in a combat zone will be credentialed by the U.S. military and will be required to abide by a clear set of military security ground rules that protect U.S. forces and their operations. Violation of the ground rules can result in suspension of credentials and expulsion from the combat zone of the journalists involved. News organizations will make their best efforts to assign experienced journalists to combat operations and then make them familiar with U.S. military operations.
E. Journalists will be provided access to all major military units. Special operations restriction may limit access in some cases.
F. Military public affairs officers should act as liaisons but should not interfere with the reporting process.
G. Under conditions of open coverage, field commanders will permit journalists to ride on military vehicles and aircraft whenever feasible. The military will be responsible for the transportation of pools.
H. Consistent with its capabilities, the military will supply PAOs with facilities to enable timely, secure compatible transmission of pool material and will make these facilities available whenever possible for filing independent coverage. In cases when government facilities are unavailable, journalists will, as always, file by any other means available. The military will not ban communications systems operated by news media organizations, but electromagnetic operational security in battlefield situations may require limited restrictions on the use of such systems.
I. These principles will apply as well to the operations of the standing DoD National Media Pool system.

Appendix C

PUBLIC AFFAIRS ESTIMATE

SAMPLE PA ESTIMATE

FM 6-99 (101-5)

____

Classification

Headquarters
Place of issue
Date, time, and time zone
Message Ref. no.

PUBLIC AFFAIRS OPERATIONS ESTIMATE NO. __________

References: Maps, charts, or other documents.

Time zone used throughout the estimate:

1. MISSION

This paragraph lists the command's restated mission from a public affairs perspective.

2. THE SITUATION AND CONSIDERATIONS

The paragraph describes the strategic and operational media environment in which the operation is being conducted and identifies the critical factors that might impact on the command's mission -- the "action and reaction" within global media channels. It identifies the media environment across the operational continuum, describing it from "austere" for low media interest and capability in a limited AOR communication infrastructure to "dynamic" for high media interest and capability in a high-tech AOR infrastructure. At a minimum, this paragraph must include:

a. Information environment. This paragraph describes the characteristics of the operation and the information environment in the area of operations. It identifies any activities or issues affecting the over-all mission and the command's public affairs objectives.

b. Media presence. An assessment of the news media presence in the theater of operations prior to deployment and the likely presence of additional news media during the conduct of operations. This assessment should address the authority under which media representatives are operating and the degree of control that can be imposed on their efforts.

c. Media capabilities. An assessment of the media's information collection and communication technology, specifically identifying their level of visual information acquisition and satellite communication capabilities. It includes an analysis of the logistics support, transportation assets, and host-nation communications infrastructure available to them.

d. Media content. An assessment of the global media's presentation of information and their agendas, and an analysis and prioritization of the potential strategic and operational issues confronting the command in the news media. This media content analysis will provide an evaluation of the quantity of coverage and the nature of that coverage.

e. Public opinion. Assessment of national and international attitude about the operation and command, leaders, and soldiers conducting it. This paragraph should include both the perceptions held by major audience groups and the relative solidity or strength of those attitudes. A public opinion analysis should include as a minimum an analysis of the following groups:

- American public
- Civilian political leaders
- Coalition and allied forces and their publics
- International audience
- Internal command audience
- Home station public

f. Information channel availability. An assessment of the information channels available for the communication of information in and out of the AOR. It identifies the means available to the commander for receipt, transmission, and dissemination of voice, data, text, graphics, and digital visual imaging. It describes command, coalition, and local national facilities and equipment available, to include an analysis of available telephone lines for voice and data transmission, the accessibility of audio and video channels, the prevalence of private communications devices such as soldier-owned cellular telephones, facsimile machines, computers, portable radios and televisions, still and video cameras, and the nature and flow of the information possible through these channels.

g. Information needs. This is an assessment of the information needs of the previously identified key publics. It analyzes key internal and external audiences and assesses theirs news and information expectations. It identifies the types of information made available to these key audiences.

h. Personnel situation. Describes the present dispositions of public affairs personnel and units that affect the PA mission, and the assets needed and available. State known personnel problems, if any, that may affect the PA situation. Consult the personnel estimate for details. (Examples of personnel include shortages of PA NCOs and skilled operators for equipment.)

i. Public Affairs situation. This summarizes current PA objectives and identifies specific courses of action for each objective. At high levels of command, detail information in a summary with a reference to an annex to the estimate. Subparagraphs will include all current (PAG) from OSD (PA).

j. Logistical situation. State known logistic problems, if any that may affect the PA situation. See logistic estimate for details. (Examples of logistic problems include the lack of transportation and adequate facilities.)

k. Assumptions. Until specific planning guidance from the commander becomes available, you may need assumptions for initiating planning or preparing the estimate. Modify these assumptions as factual data or planning guidance becomes available.

3. ANALYSIS OF COURSES OF ACTION

Analyze each course of action based on the public affairs objectives in paragraph 2i. Indicate problems and deficiencies. At a minimum, subparagraphs should include media facilitation and support, news and information provision, and force training and support.

Analyze each COA from a PA point of view to determine its advantages and disadvantages for conducting PA. The detail in which the analysis is made is determined by the level of command, scope of operations, and of urgency of need.

4. COMPARISON OF COURSES OF ACTION

Compare each course of action. List advantages and disadvantages of each course of action under consideration. Include methods of overcoming deficiencies or modification required for each course of action.

5. CONCLUSIONS

a. Indicate whether you can support the command mission (restated in paragraph from the public affairs viewpoint.

b. Indicate which COAs you can best support.

c. List major public affairs deficiencies, which commander must consider. Include specific recommendations about methods of eliminating or reducing their effects.

____

(Public Affairs Officer)
ANNEXES: (As required)

DISTRIBUTION:

(NOTE: The headings listed in this assessment are for example only. Use headings appropriate to your command's operations.)

____

Classification

PA Operations Estimate

Appendix D

PUBLIC AFFAIRS ANNEX

SAMPLE PA ANNEX

FM 6-99 (101-5)

____

(Classification)

(Change from oral orders, if any)

Copy ___ of ___ copies
Issuing headquarters
Place of issue (may be in code)
Date-time group of signature
Message reference no. _____

ANNEX__ (PUBLIC AFFAIRS) TO OPERATIONAL ORDER ______

References: Maps, charts, and other relevant documents

Time zone used throughout order:

1. SITUATION

A brief general description of the situation, information affecting public affairs support, which paragraph 1 of the OPORD does not cover, and intended purpose of this annex.

a. Friendly forces. Outline the higher headquarters' plan (and PA annex) and adjacent unit PA plans. Provide information on friendly coalition forces, which may impact the PA mission. Note PA resources supporting the unit. (Who, where, when), (higher, allied and adjacent headquarters).

b. Attachments and detachments. Identify all augmenting PA units supporting this command and all attached/assigned subordinate units. Include effective dates, if applicable.

c. Enemy forces. List information not included in the OPLAN/OPORD, which may impact the PA mission. (Who, where, when, disinformation, rumors, propaganda and OPSEC).

d. Media. Identify media in the area. (who, where, pools, US. international, local-host country).

e. Assumptions. List any additional assumptions or information not included in the general situation, which will impact the PA mission.

2. MISSION

Clearly, concisely state the public affairs mission. (Internal information for deployed and non-deployed forces, media facilitation and staff operations).

3. EXECUTION

a. Concept of operation. Briefly summarize the public affairs operation plan. Include PA priorities. (Intent --access, information, welfare, morale, will to win) (Concept--who, where, what, why, when) (Specifics--task to a subordinate; who is to do what, where, when, covers non-PAs too, actions with media: credential, train, transport)

b. Outline of PA tasks. Identify and assign supporting PA tasks to each element of subordinate and supporting units. Assign specific tasks to elements of the command charged with public affairs tasks, such as requirements for PA augmentation.

c. Coordinating instructions. Give details on coordination, task organization and groupings. List instructions, which apply to two or more subordinate elements or units. Refer to supporting appendixes (PA assessment) not referenced elsewhere. (Public Affairs Guidance, media in country, media enroute with US forces, media contact report, handover checklist, and task organization).

4. SERVICE SUPPORT

a. A statement of the administrative arrangements applicable to this operation. If they are lengthy or are not ready for inclusion in the OPORD, these arrangements may be issued separately and referenced here.

b. A statement of the logistical arrangements applicable to this operation. Specific coordination should be included if possible, but arrangements may be issued separately and referenced here, if they are too lengthy. (Class I-IX and water), (Services: billets, medical, laundry and mortuary), (Transport: ground, air, TOE, tasked rented/leased, contracted).

5. COMMAND AND SIGNAL

List signal, visual imaging and satellite communications policies, headquarters and media center locations or movements, code words, code names, and liaison elements. (PAO location, media center, JIBs, sub-JIBs, phones, faxes, e-mail and web page).

ACKNOWLEDGE:

NAME (Commander's last name)
RANK (Commander's rank)

OFFICIAL:
APPENDIXES: (List PA assessment appendix)
DISTRIBUTION:

Classification

Appendix E

PUBLIC AFFAIRS GUIDANCE

SAMPLE PA GUIDANCE

Public Affairs Guidance (PAG) is the operational tool that guides commanders and their public affairs officers in the application of doctrine and policy during major military operations, exercises, and contingencies. The information below is tended to assist local commanders in preparing and obtaining approved guidance.

DoD policy requires that PPAG be provided to the Assistant Secretary of Defense-Public Affairs (ASD-PA) by the Unified and Specified commands and others, as required for all major operations.

This requirement includes major training exercises that could attract national and/or international attention. PPAG may not be used without ASD-PA approval.

Upon receipt of the warning order, the commander, through his PAO, should request PAG from high headquarters. PAG may be included in alert notification or operational orders (see Part Three: Operational Planning). Commanders of major units/commands will direct their PAOs to prepare PPAG to forward the proposal through MACOM and Unified/Specified command PA channels to ASD-PA.

Commanders of Unified/Specified commands should ensure that the PPAG has been coordinated with appropriate organizations within the theater of operations whenever possible (e.g., embassies, country teams, host governments, subordinate commands).

Upon receipt of the PPAG, the ASD-PA coordinates and staffs the PPAG within the DoD and Department of State.

The ASD-PA then issues a message either approving, modifying, or disapproving the PPAG. PPAG is broken down into subject, references and then eight paragraphs (Information/explanation, Purpose and coordination for PPAG, PA Approach, Public Statement, Q&As, Contingency Statement, Miscellaneous Information, Point of Contact). The format for PPAG follows:

SUBJECT

The subject line of the PPAG should state "PROPOSED PUBLIC AFFAIRS GUIDANCE - followed by the exercise and/or event name (U)." For coordination, it is best if the subject is unclassified.

If an exercise or event is so sensitive that the actual name cannot be used, an unclassified short title should be used; e.g., "PROPOSED PUBLIC AFFAIRS GUIDANCE - CC-(U)."

REFERENCES

Pertinent messages or other documents shall be cited in the reference section. If the PPAG is based on PA policy in the Significant Military

Exercise Brief, then the SMEB message Date-Time-Group (DTG) shall be listed.

EXERCISE INFORMATION

The **first paragraph** of the PPAG shall explain the references, the exercise, and any significant existing or anticipated problems associated with the exercise. The information in this paragraph is not for release so may remain classified after the PAG is approved for release. This paragraph may restate some PA information from the SMEB.

COORDINATION INFORMATION

The **second paragraph** shall explain the purpose of the message; identify it as being fully coordinated and theater-approved; request ASD-PA approval and specify the date it is required for use. If the PAG is transmitted to the ASD-PA before it is fully coordinated, it is the responsibility of the submitting command to ensure that the ASD-PA is promptly informed of the results of the remaining coordination. The submitting command should always follow-up a PPAG message with a phone call to ensure that the primary addressee(s) is aware that the message is en route. When the submitting command is a supporting CINC from outside the supported CINC's AOR, the supported CINC is responsible for theater coordination.

PA APPROACH

The **third paragraph** shall discuss the public affairs approach for the exercise; i.e., active or passive. This may be a restatement of the PA policy indicated in the SMEB.

- Active Approach
  - For this discussion, an "active approach" involves efforts made to stimulate public or press interest such as distributing press releases and inviting the press to observe the exercise. If an exercise or event is to be publicly announced, this paragraph shall state who will make the announcement, the method of announcement, and preferred time, and date for the announcement. If unusual circumstances prevail, the rationale for the recommendation should also be included. Part I and II exercises shall normally be announced by the ASD(PA) by issuing a news release (blue top). Other lesser exercises or training deployments, if announced by the ASD(PA), normally shall be made by release of a memorandum for correspondents (MFC). The preferred release time and/or date of exercise announcements is 1200 Eastern Standard Time (E.S.T.) on either Tuesdays or Thursdays in conjunction with the normally scheduled DoD press briefing. If a combined announcement is desired with a host country, complete details of the methods, time, and procedure shall be included in this paragraph. The active approach is recommended whenever possible to ensure appropriate media coverage of specific commands and/or units.
- Passive Approach
  - A "passive approach" is where no action is taken to generate media and/or public interest in an issue or activity beyond answering specific inquiries. If a passive approach is desired,

the PPAG shall so indicate and specify that the PAG is for response to query (RTQ) only. It shall also specify who is authorized to respond; e.g., "Only OASD(PA) may RTQ," or "All of the following addressees may use this PAG for RTQ only." To de-emphasize an event, it is best to authorize release or RTQ at the lowest possible level.

PUBLIC STATEMENT

The **fourth paragraph** shall contain a statement that explains the exercise and/or event. The statement shall be for public release in an active PA approach or for RTQ in a passive PA approach. For ease of coordination, each paragraph of the statement shall be identified as a sub-paragraph of the message; for example: The following statement is for initial public release: (TEXT FOLLOWS): QUOTE.

EXERCISE (NAME).....

THE EXERCISE WILL.....

PREVIOUS MILITARY.....

FOR ADDITIONAL INFORMATION,

CONTACT.....(UNQUOTE).

As indicated above, the last paragraph of the statement shall identify points of contact where additional information may be obtained.

QUESTIONS AND ANSWERS (Q&AS)

The **fifth paragraph** shall contain a list of proposed Q&As to enable the user to respond to the majority of anticipated questions. They should all be contained in one paragraph and should be numbered sequentially; e.g., Q1, A1; Q2, A2; Q3, A3, etc. Q&As are for use in both active and passive PA approaches, but are strictly for RTQ only and shall not be given to media as handouts.

CONTINGENCY STATEMENT

The **sixth paragraph** of the PPAG shall contain a contingency statement to be used efore release of the final PAG. Usually, the contingency statement should be that we don't discuss exercises before they have been formally announced. However, this approach can be modified, as appropriate, depending on the circumstances of the exercise. If a contingency statement is not required, so state in Paragraph 6 of the PAG.

MISCELLANEOUS INFORMATION

The **seventh paragraph** shall contain other pertinent information to include the following items (when a certain sub-paragraph is not applicable, so state):Media Information Centers (e.g., Joint Information Bureaus (JIBs), Press Information Centers (PICs), exercise PA elements, etc.) Discuss whether centers are joint or combined; delineate who is responsible for the establishment; give generic description of its composition (e.g., U.S. Army desk (O-4 and E-6/E-7), U.S. Navy/Marine desk (USN O-4/O-5, and USMC E-5/ E-6), etc.); establish the center's functions (coordination of all exercise media and/or PA activities, clearance of U.S. military-generated news material before release, production of news material for release, escort of accredited news media representatives); etc.

- Command Relationships
  - Designation of sole approving authority for all exercise-related news materials; procedures for the release and/or clearance of information (to include list of addressees for notification in case of accident and/or incident); request for participating commands and/or units to ensure that the media center is action and/or information addressee on all messages with potential PA impact (to include incident and/or accident reports); hometown news release requirements and/or instructions (passive PA approach may make hometown releases inappropriate); etc.
- Media Coverage
  - State whether media coverage is encouraged or solicited, giving rationale; news media transportation instructions; point of contact (POC) and procedures for handling such requests; requirements for news media representatives (valid passport, working media visa, local accreditation requirements, funds for food, lodging, return travel (if military air is not available), etc.); instructions regarding assistance to continental United States (CONUS)-based units for handling request from news media for accompanying travel before and following public announcement of the exercise; etc.
- DoD National Media Pool
  - Each exercise is a potential opportunity for activation and deployment of the DoD National Media Pool to cover exercise activities. As a minimum, planning should include arrangements for local ground and/or air transportation, special clothing or equipment to be provided, messing, billeting, protection of media equipment and gear, local escort requirements, and communications support for filing of pool products. Sponsoring commands shall indicate whether the exercise should be considered for a pool deployment. Identify the primary POC should the pool be activated.
- Internal Media and Audiovisual Coverage
  - Provide instructions on assistance that will be provided to this effort; degree of freedom of movement (to include whether escorts are necessary); screening of visual information (VI) materials upon completion of exercise; sponsoring command POC for handling internal information matters; etc. Also include guidelines for Armed Forces audiovisual teams documenting the exercise.
- Media Opportunities
  - If known well enough in advance, provide chronology of potential exercise events that would be of interest to media.
- Miscellaneous PA Considerations

- Indicate any other proposed PA activities or considerations; if there are none, then so state.

POINTS OF CONTACT

The **eighth paragraph** shall state the originating POC's name and phone number.

DECLASSIFICATION INSTRUCTIONS

Declassification instructions shall be the last part of the message and in accordance with subsection 4-207 of DoD 5200.1-R (reference (d)).

Appendix F

PUBLIC AFFAIRS OPERATIONS BRIEFINGS

BRIEFINGS

The purpose of a briefing and the desired response or result determines the briefing technique. Basically, there are four briefing types: the information briefing, the decision briefing, the mission briefing, and the staff briefing.

The Information Briefing. The information briefing informs the listener and deals primarily with facts and background information. The Information Briefing contains an introduction to the subject and the scope of the subject area. It then presents the high-priority information requiring immediate attention and complex information involving complicated plans, systems, statistics, or charts. It may also explain controversial situations or information, which require elaboration.

The Decision Briefing. The Decision Briefing includes many of the elements of the information, but goes further by seeking a decision from the decision-maker. At the beginning, the briefer clearly states that he is seeking a decision; at the end, he requests the decision.

The Mission Briefing. The Mission Briefing gives special instructions, amplifies the mission, elaborates on new orders, or assigns taskings to subordinate elements. This briefing usually follows the five-paragraph operations order format. But the briefer may also choose the information briefing format.

The Staff Briefing. The Staff Briefing informs the commander and staff of the current operational situation. Its purpose is to generate a coordinated or unified effort and in a tactical environment. It serves to keep the entire staff aware of each section's activities, thus aiding coordinated action. While there is no specific briefing format, commanders usually tailor this type of briefing to fit their information needs. PAOs address the major PA activities and the PA implications of the operational situation and other staff sections activities.

In a commander's staff briefing, the PAO is responsible for providing a summary of the "global information environment (GIE)." The PAO should present this information at the beginning of the briefing, following the intelligence summary. The PAO's GIE summary, combined with the G2/S2's intelligence preparation of the battlefield (IPB) briefing, helps to complete the picture of the operational situation, which the other elements of the command must react.

Preparation. After conducting the initial analysis of the situation, the briefer gathers and organizes information, prepares visual aids, and then rehearses, briefing only what is essential. Before developing the briefing, the briefing officer or NCO must know and understand:

- the higher headquarters mission
- the higher commander's intent
- the commander's personal guidance and intent
- all aspects of the PA estimate of the situation
- the command's objectives

While much of the information required for a GIE presentation can be found in previously published PA Area Studies, a significant amount of situational information should pulled directly from the PA Estimate of the Situation and updated to present the latest information available.

BRIEFING FORMAT

The PA briefing format should follow an outline similar to the PA Estimate of the Situation:

- Mission - Command mission including PA objectives
- Situation and Conditions
- Information environment
- PA situation
- Media presence
- Media capabilities
- Media content
- Public opinion
- Information channel availability
- Information needs
- Current Operations/Actions
- Future Operations/Actions

Appendix G

PUBLIC AFFAIRS AREA STUDIES

FORMAT FOR PA AREA STUDIES

Public Affairs area studies provide the PA specialist and the PA planner a starting point to begin their campaign planning process. These studies are produced either to address long-term general background information or to address immediate short-term needs.

BASIC STUDY

A PA study is a document that succinctly describes the most PA-pertinent characteristics of a country, geographical area, or region. It serves as an immediate reference for the planning and conduct of PA operations.

TITLE PAGE

The title page of a Media Environment Study (MES) must show the country, area, group, or other subject of the study. It should also show the classification, the copy number, the date-time group, and command post location. (The originating unit should keep a record of the names, grades, and duty assignments of the authors.)

EXECUTIVE SUMMARY

The executive summary should address the strategic significance of the area under study. (The authors should write the executive summary last, in a clear, brief, accurate, and coherent form.)

NOTE: The commander reading the summary should us its information to decide how to employ Public Affairs in that area.

TABLE OF CONTENTS

Executive Summary

Introduction

Chapters

1. History and Ideology
2. Government and Politics
3. Foreign Relations and Policy

4. Society and Culture
5. Economy
6. Military Establishment
7. Communication Process and Effects

Endnotes

APPENDIX

A. Country Summary

B. Government Structure

C. Communications Facilities

D. Glossary

Bibliography

Distribution

INTRODUCTION

The introduction outlines the study's intent, cites the directive requiring the study, and explains the study's format.

The study's intent is to provide a summary of aspects of subject country significant to GIE. The study should identify psychological vulnerabilities, characteristics, insights, and opportunities that exist in subject country.

It is prepared as a basic source document for further development of estimates, plans, and annexes.

Although this study can help develop contingency plans, it is not tailored to any particular plan. Rather, its neutral data and insights can be used to analyze possible political and military developments in the region.

Insert here a paragraph referencing the authority directing the study and stating the research cutoff date and provisions for updating the study.

Focus the PA study on the GIE aspects of the many topics addressed. Do not view this document as a comprehensive and self-contained area study.

Instead, use it as a complement to such other standard references as the Department of the Army Area Handbook Series. In addition, the PA study should include:

Results from standard open source products.

Tries to be more analytical than descriptive in nature, making it subject to varying individual perspectives.

Works well when PA specialists compare it with studies on other countries in the region or area.

Is not a U.S. foreign policy statement or comprehensive analysis of subject country, except in areas with direct GIE relevance.

Insert here a statement of U.S. policy goals toward the country in question. This information comes from the proper USA Country Plan, Department of State Policy Memorandum, or similar document, in the priority order.

Because certain gaps exist in getting PA study material (classification level, availability of complete and timely information, or time limits on research), listing these gaps here to aid future research and guide PA study users to further inquiries.

Insert here issues or GIE-relevant material (such as an area map) not included, addressed, or completely answered elsewhere.

CHAPTER 1

HISTORY AND IDEOLOGY

In this chapter, review the evolution of the state and its people, focusing on aspects having GIE and Public Affairs significance. Do not detail chronology of the country's development. Keep in mind, however, the country's history has an important relationship to the country's historical perspective, attitudes, and current world position.

Because of its special importance to PA, in this chapter, cover historical issues thoroughly. An historical analysis of current political, economic, and military policies gives PA personnel a solid base for the study.

CHAPTER 2

GOVERNMENT AND POLITICS

In this chapter, summarize the country's political system. Give a description of its political power sources, policy making process, and the political complexities of the government.

When discussing the political system, pay close attention to the role of individuals, special interest groups, and political parties. Include the population's political attitudes, values, and view of the political system. Also discuss the government's function in society.

CHAPTER 3

FOREIGN RELATIONS AND POLICY

In this chapter, summarize the country's foreign relations. Describe its political alignment in world affairs and its relationship with the United States. Describe the foreign policy of the country. Also analyze and interpret why the country acts as it does in international affairs.

CHAPTER 4

SOCIETY AND CULTURE

In this chapter, analyze the subject country's social setting. Provide the PA personnel with the knowledge needed to understand potential audiences.

Cover the country's social organization and cultural and behavioral patterns and characteristics. Place special emphasis on the society's social values and the role of the family.

Address culture, social organization, education, customs, ethnic composition, and the interrelated effects of religion, language, and history.

CHAPTER 5

ECONOMY

In this chapter, present a brief description of the characteristics, structure, and dynamics of the national economy. Cover the subject country's economic strengths and weaknesses, current economic and labor problems, and economic potential

Describe the country's economic base and the importance of agriculture, industry, and trade. This information helps determine if the present economic structure meets the people's needs.

These economic considerations explain many of the sociological conditions that impact public opinion. Address society's perceptions of the wisdom of government economic policies.

Also describe individual or group perceptions of how members of society stand to gain or lose from those policies.

CHAPTER 6

MILITARY ESTABLISHMENT

In most countries, the military establishment involves itself in internal politics as well as external defense. Even when the military establishment does not directly compete for political power, its actions influence social and political development. Analyze the following topic areas:

- Emergency of the modern military establishment.
- Military roles in the political, economic, and social spheres, and the effects of those rules.
- Issues creating cohesion or conflict within the armed forces.

Leadership.

Conflicts within the military establishment.

Extent, quality, and influence of foreign military aid.

CHAPTER 7

COMMUNICATION PROCESS AND EFFECTS

List essential information about communication patterns for the implementation of a PA program. Include the following information:

Manner and social means of communication (not technical data on communications facilities).

Languages and language groups, nonverbal communication, and nonverbal symbols specific to the country's culture or cultures.

Distinctive styles in rhetoric or visual arts, including dramatic, poetic, and musical forms. All these forms are significant to PA.

Data on the society's formal and informal leadership positions where the incumbents are key communicators and opinion leaders.

Analysis of the reading and listening habits of the society.

Analysis of printed formats.

Analysis of media effectiveness. Also address freedom of the press issues, if any.

ENDNOTES

LIST PUBLICATIONS, DOCUMENTS, AND OTHER SOURCES USED IN PRODUCING THIS BPS. NUMBER TEXT REFERENCES SEQUENTIALLY WITHIN THE STUDY.

APPENDIX A

COUNTRY SUMMARY

Give the reader a brief overview of the subject country, its geography, and its people. These background data and statistics should include the following items:

Country. Identify the country, tell when formed, and show previous control.

Government. State briefly the type of government, method of appointing or electing leaders, and length of terms. List current leader or leaders and political power in country.

Size, location, and geographical subdivision. List the size of the area in square miles or kilometers, and give the general location. Show any geographical subdivision, such as coasts, mountains, and flatlands.

Population. State the number of people and the area density. Show the heavily populated areas.

Languages and dialects. List the official language, languages spoken by the population, percentage of population speaking each language, and areas of the concentrations.

Labor. Outline the total work force, the area of endeavor, and the percentages.

Religions and sects. List the religions of the area and the percentage of the population that practices each.

Education. List the types of systems and the primary emphasis of each.

Literacy rates. Latest statistics.

Health. List the general conditions of the populace. Describe the medical care system.

Justice. Describe the justice and court systems.

Administration. Outline the breakdown of the governmental and judicial districts, counties, or precincts.

Transportation. List the methods of transportation available and include the total capabilities. This information may include the number of airlines, airfields, kilometers and kinds of highways, and kilometers of waterways and depth.

Armed forces. List organization and strengths.

Police. List the types and areas of responsibilities.

APPENDIX B

GOVERNMENT STRUCTURE

List the formal government structure, key positions, and organizations of the country. Outline the chain of government control, including political parties, if applicable. A schematic diagram may be helpful.

Include only branches of government and their key positions, not names.

APPENDIX C

COMMUNICATION FACILITIES

Give a brief overview of the subject country's media facilities. Include the facilities' locations and levels of technical sophistication. Cover printing, publishing, and the distribution of radio and television receivers, studios, transmitters, and relay facilities. Include news service facilities. Write this appendix as if the U.S. PA units will use this equipment or contract for its services. Since PA personnel may get operating supplies or repair parts from in-country sources, provide the following information:

- Make.
- Model.
- Type.
- Series.
- Name of manufacturer.

Any other technical information on the repair or operation of this equipment.

GLOSSARY

Prepare a glossary that lists in alphabetical order all acronyms and foreign words used in the study. List also all words and terms that have special meaning and need to be defined.

BIBLIOGRAPHY

List the source material used. Include the name of the author, the title of the publication, the publisher, and the date of publication.

DISTRIBUTION

Dissemination is accomplished by the originating agency for the recipients within PA. Include in the distribution list the identification of recipient agency (by code), the number of copies furnished, and the office symbol of the recipient.

Appendix H

BRIEFINGS AND PRESS CONFERENCE FORMATS

NEWS BRIEFING AND PRESS CONFERENCE FORMAT

Before the Presentation

Know your publics
Anticipate interests, concerns and questions
Consider the latter in preparation

Prepare your presentation
Develop a strong introduction
Develop a maximum of three key messages
Assemble your supporting data
Prepare audiovisual aids
Practice

Prepare for answering questions
Anticipate what questions will arise
Prepare answers to those questions
Practice questioning and responding

The Opening Statement

A strong opening statement sets the tone for the press conference or news briefing and is crucial in attempting to establish trust and build credibility. The elements of a strong opening are:

Introduction
A statement of personal concern
A statement of organizational commitment and intent
A statement of purpose and plan for the meeting

Key messages and supporting data
A maximum of three "take-home points"
Information to support the key messages

Conclusion
A summarizing statement

Total time for all presenters should be 15 minutes or less. Do not have too many presenters. Three is usually sufficient.

Introduction

Remember that perceived empathy is a vital factor in establishing trust and building credibility and your publics assess it in the first 30 seconds.

Examples are:

Statement of personal concern: "As a resident of this community I'm interested in the safety and well-being of our families and neighborhoods."

Statement of organizational commitment and intent: " I'm here to share with you the knowledge and confidence I have in the military's ability to assist the citizens of our community. They have been trained in their occupational skill to assist with the task at hand."

Statement of purpose and plan for the presentation: Today I would like to share with you the most current information regarding the (incident.) I will also be available to answer additional questions or to continue the discussion.

Key messages and supporting data

The key messages are points you want your publics to have in their minds after the presentation. They should:

Address central issues.
Be short and concise.
Examples are:
"We have trained personnel and emergency response plans in place to aid in protecting the health, safety and welfare of the public. We are working with local and state officials to handle the incident.

"We are actively responding to the emergency...."

To develop your key messages:

Brainstorm
Think freely and jot down all pieces of information you wish to communicate.

Select key messages
Identify the most important ideas. Repeat the process until you list is down to three items.

Identify supporting data
Other information you listed probably provides support to your key messages. Organize it to reflect this.

Conclusion

Restate verbatim your key messages.

Add a future action statement --- What is your organization going to do about this problem in the short and long term?

Appendix I

WAIVER OF LIABILITY STATEMENT

Whereby, I ______**NAME**______________________ passport no:____________ am about to travel with ___________forces, and whereas I am doing so entirely upon my own initiative, risk and responsibility; now, therefore, in consideration of the permission extended to me; I do hereby for myself, my heirs, executors and administrators, remiss, release and forever discharge ________ and its member officers, agents and employees acting officially or otherwise, from any and all claims, demands actions or causes of action, on account of my death or on account of any injury to me or my property which may occur from any cause during my stay, travel as well as all ground, flight or sea operations incidents thereto.

I also agree to withhold any classified information, which may be accidentally disclosed to me, and to respect embargo restrictions, which may be imposed on information which, if disclosed, may jeopardize operational security. During my stay with _______ forces, I will not interfere with operations. I understand that failure to comply with these security restrictions will result in the loss of authorization to accompany ______and may result in cancellation of my press registration.

____________________	____________________
Signature	**Witness**
____________________	____________________
Printed Name	**Nationality**

Address:____

Please provide the following information for a person to be notified in an emergency (preferably next immediate relative):

____

____

Appendix J

SAMPLE PRE-DEPLOYMENT CHECKLIST

INTRODUCTION

A multitude of factors make each deployment uniquely different from other deployments. Each factor must be carefully examined to determine its impact on the mission, actions before deployment and what equipment is taken.

(Unit Basic Load). In all other instances the chances of drawing the UBL are remote. Certainly if the unit is deploying for war it would draw the unit's basic load of ammunition Exercise or deployment duration will affect the quantity of expendable supplies.

This checklist, like all others, is based on what has occurred before and what we have come to expect in the future. As equipment and missions change, so too must the checklist. Bottom line, expect the unexpected and react accordingly; use the checklist as a guide to help you through deployment preparation. Add to it whenever the need arises.

CHECKLIST ITEMS

1. Each HQ element and unit should have a 45-day supply of expendables identified and set aside for contingency operations. This supply should be inventoried quarterly and stockage rotated accordingly.
2. Maintain and update a master list of all areas and topics that can affect individual readiness. This includes all shots, dental status, family support plan, check to bank, etc.
3. The PAO or detachment commander (or deploying team leader) should:
    a. Review and update PA estimates annexes and plans.
    b. Obtain and comply with applicable published Public Affairs Guidance.
    c. Inventory and inspect TOE equipment for accountability and serviceability before deployment.
    d. Figure total weight and cube of equipment before moving to assembly area.
    e. Check dependency clause in TOE document or OPLAN to determine who will be supporting and who is supporting us. Ensure responsible parties know the relationship. Establish liaison with units OPCON; attached or any new parent organization.
    f. Ensure accompanying equipment not on TOE is listed on interim authorization document (such as facsimile machines or cellular phones).
    g. Prepare an internal OPORD for your element detachment or team.
    h. Review the essential elements of friendly information (EEFI) contained in the base OPLAN/OPORD and ensure each soldier understands them.
4. All leaders must consider the need for the following:
    a. Will flak vests be required?
    b. Should each soldier take a footlocker instead of a dufflebag?
    c. Will weapons' carrying/security cases be needed/available? (Will weapons and protective masks be required?)
    d. Will desert or jungle uniforms be required, and if so, what fund can be used to pay for alterations, sewing and patches?
    e. Is there an opportunity for a service contract to be initiated at the deployed location? Cameras, video, computers, etc.? if not, and cameras go down and must be swapped out, what is the plan?
    f. Will the deployment be considered TDY? If so, who prepares orders? What fund cite will be used? Are rental vehicles available? (Are we TDY under field conditions and required to carry meal card?)

g. Will a fund cite be made available after redeployment to pay for repairs? (Make this request soon after being tasked - don't wait until redeployment to find out you've got to use unit funds to repair equipment.)
h. If departing from location other than current station, what type of transportation to that location is to be used for personnel and equipment?
i. Will personnel and equipment travel together at all times? (Insist that they do whenever possible!)
j. Does each team have a credit card holder for authorized payments or purchases?
k. Has an express mail system been established to ensure timely transit of products to home station? With what frequency? (If you are the ARFOR or sub-JIB, have the division PAO's/PAD's establish a plan to get products to you for review, release or use).
l. Under what conditions will the MPAD or team(s) work? Fixed site, field conditions, etc. Will we need to deploy our own tents for sleep and work?
m. Is the heat a condition that will affect computers, cameras, batteries, etc? If so, is air conditioning available? Refrigerators for film? What about humidity problems? Air conditioning may help but can dehumidifiers do better...consider charcoal bags placed in shipping cases.
n. From what unit(s) do we draw support...rations, billets, fuel, etc.?
o. If one team is deployed for a lengthy duration, is there an opportunity for the teams to be rotated?
p. If deploying a risograph or other commercial printing machine, what quantity of copies and frequency (daily, weekly) is desired? What is the plan for distribution of copies? What is the plan for paper replenishment?
q. Has every effort been made to ensure deployed asset has commercial phones available to assist transmitting digital photographs, responding to query, accidents and incidents, fax capability, etc.
r. Has unit/team packing list been carefully checked to ensure the easy-to-forget yet must- have items are not forgotten. Glue sticks, chalk, scotch tape, blank overheads, etc.

Appendix K

PAO SOP OUTLINE

Chapter 1, Alert Notification Procedures
- Annex A, Notification Procedures
- Annex B, Section Telephone Contact Roster

Chapter 2, Individual Preparation
- Annex A, Individual Admin Checklist
- Annex B, POV Storage Plan, Procedures for Completing Post POV Storage Forms.
- Annex C, Personal Property Storage Plan
  - Section 1, Power of Attorney
  - Section 2, State of Obligations
  - Section 3, DD Form 1299, Application for Shipment/Storage of Personal Property.
  - Section 4, DD Form 1701, Household Goods Inventory
- Annex D, Recommended Personal Readiness Equipment
- Annex E, Family Member Pre-deployment Checklist

Chapter 3, Unit Preparation
- Annex A, Unit Equipment List

Chapter 4, Tactical Vehicle Preparation
- Annex A, Vehicle Preparation Standards
- Annex B, Vehicle Load Card
- Annex C, Vehicle Inspection Checklist
- Annex D, Joint Airlift Inspection Record
- Annex E, HMMWV Configuration and Load Plan

Chapter 5, Rear Detachment Operations
- Annex A, Communications
- Annex B, Logistics Coordination

Chapter 6, Public Affairs Checklists
- Annex A, Pre-Deployment Checklists
- Annex B, Guide for Media Interviews
- Annex C, METL and Tasks, Conditions, Standards
- Annex D, DoD Principles of Information
- Annex E, Policy on the News Coverage of U.S. Military in Combat
- Annex F, PA Guidance on Terrorism Counteraction
- Annex G, PA Guidance on Counter-Drug Operations
- Annex H, Standard PA Ground Rules
- Annex I, Spokesperson Guidelines
- Annex J, Command Unique Media Operations Center Guidelines

Chapter 7, Field Operations
- Annex A, Tactical Uniform
- Annex B, Installation Security
- Annex C, Tactical communication
- Annex D, Personal Hygiene
- Annex E, Morale, Welfare and Recreation

Appendix L

EXAMPLE OF MEDIA OPERATIONS CENTER

Operations of a media center will need the following support:

- Communications
- Vehicle support (day-to-day operations and media transport -- tactical or non-tactical as needed)
- Billeting and rations for media center personnel
- Admin support personnel for 24-hour operations
- Office space (hard site if possible) and power as needed

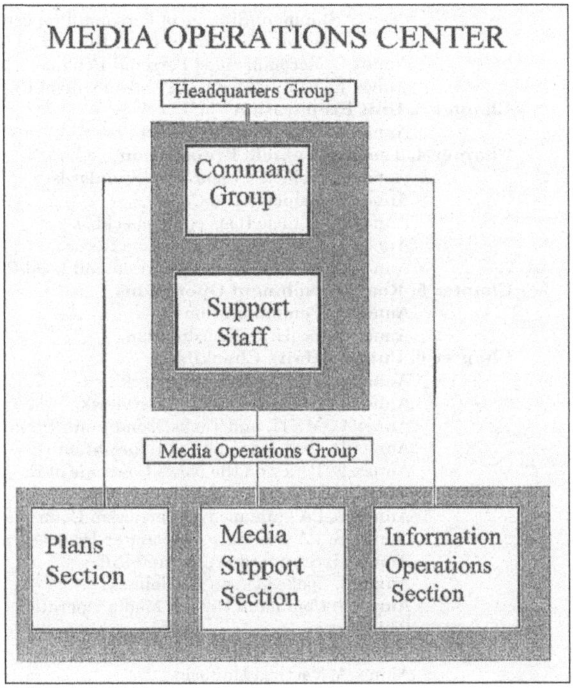

Figure L-1. Media Operations Center

Appendix M

NEW MEDIA INQUIRY FORMAT (SAMPLE)

This format is intended as an example only and should be adapted to local needs and SOPs. In addition to query forms, PAOs should maintain a separate log of all inquiries.

MEDIA QUERY SHEET

Query Number:___________

Deadline:_______________

CALLER'S NAME:____

CALLER'S NEWS ORGANIZATION:___________________________________

CALL TAKEN BY:______________TIME:___________DATE:___________

QUESTION (use reporter's precise wording):_________________________________
____
____
____

RESPONSE (if written release is made, attach a copy):________________________
____
____
____
____

SOURCE OF INFORMATION/COORDINATION (e.g., G-3, surgeon): ___________
____

RELEASED TO:____

TIME/DATE:____

RELEASE METHOD: In Person_______ Phone____________ News Release______

RELEASED BY:______________ RELEASE NUMBER:________________

. **Appendix N**

GUIDE FOR MEDIA INTERVIEWS

GUIDELINES

1. Considerations

a. When human safety or other serious concerns are involved, deal with those considerations first.

b. Communicate only information that is approved for external distribution. Always tell the truth.

c. Know to whom you are speaking. Get the person's name and telephone number, if necessary.

d. Do not be intimidated. You may tell a reporter that you need to clarify an important matter before you can answer questions.

e. Talk from the public's viewpoint. Avoid jargon. Speak within the audience's frame or reference.

f. If the questions do not lie within the framework of approved statements or within your area of expertise, find the appropriate technical advisor or spokesperson.

g. State the most important fact at the beginning. Place your own headline on the answer.

h. Attack problems in your answers, not people.

i. Do not repeat offensive or negative language. Do not let other people put words in your mouth.

j. Direct questions deserve equally direct and forthright answers.

k. Do not exaggerate the facts. Listen to how your answer "sounds" when spoken.

l. Ignore cameras and microphones. Talk to the reporter.

m. During videotaped interviews, it is all right to stop your statement and start over.

n. Do not say "no comment." Explain why you do not have an immediate answer.

o. Keep your composure, even if a news reporter gets snappy.

p. Be prepared to provide sufficient evidence for statements you make.

q. Be especially alert about photos. You have little control over photos taken off military reservation property, but you have every right to control photos taken on the military reservation.

r. Be aware of your surroundings and follow local OPSEC rules when determining interview location

2. What will be asked?

a. What happened and where? When did this occur

b. Are there injuries or deaths as a result? How many and to whom?

c. What actions is the unit taking to control the situation?

d. Have chemicals or other hazardous substances been released into the environment? What kinds? How much?

e. What types of hazards are presented to people off-site?

f. Have off-site emergency response personnel been notified? Which ones?

g. Are unit operations shut down?

h. Has the site or facility been evacuated?

i. How many people are employed at this site?

j. What do you do at this site?

k. How old is the facility? Does it meet current regulations?

l. Why did this situation occur? (DO NOT SPECULATE.)

m. Are there safety rules covering the situation? Were they violated?

n. Has a Site Emergency Response Plan been activated? What does that involve?

o. Tell me about your organization?

p. Will this situation have national ramifications, or will its effect likely be limited to a single site or region?

q. How much money is this going to cost the taxpayers?

r. Is there insurance coverage for the loss or damage? How much?

s. Are commanders handling the situation locally or is a higher headquarters taking control?

t. Has this occurred anywhere within the unit before? Why weren't you ready?

u. What do your soldiers think about this situation?

v. For accidents and incidents, don't speculate causes. Use "ongoing" investigation statements.

Appendix O

MEDIA RESEARCH AND ANALYSIS

AN INTRODUCTION

Since the early part of the 20th century, when there was almost no interest in the size of audiences or in types of people that comprised various audiences, business leaders and their organizational communicators have increasingly come to rely on research for nearly every major decision they make. This expanded demand for information has created a particular demand for media communication research capabilities, specifically -- the development of a scientific basis for media analysis and media effects.

The importance of mass communications research and media analysis is partly due to the realization that gut feelings or reactions are not entirely reliable or credible bases for decisions. Although common sense is often accurate, Army commanders and other decision-makers need additional, more objective information to evaluate problems, especially when lives are at stake. Thus, the past 50 years have witnessed continuing evolution of media analysis, combining research and intuition to create a higher probability of success.

This evolution has resulted in a "scientific" approach to media research known as media content analysis.

SCIENTIFIC RESEARCH: A DEFINITION

Scientific Research is defined as a **"systematic, controlled, empirical, and critical investigation of hypothetical propositions about the presumed relationships among observed phenomena."** This definition contains the basic terms that are necessary in defining the method of scientific research and describes a procedure that has been accepted for centuries. In the 16th century, for example, scientist Tycho Brahe conducted years of systematic and controlled observation to prove wrong many of Aristotle's theories of the universe. By gaining an understanding of the phenomena, he challenged the accepted beliefs and knowledge of the time with his own hypotheses. Thus, scientific research was begun.

Whether we realize it or not, we all conduct research as a matter of course in our day-to-day life whenever we speculate about the possibility of something -- we start with an idea or concept and test it.

All research begins with a basic question or proposition about a specific phenomenon -- for example, Why do Americans usually support the soldiers within the Army when they may not support the operation the soldiers are involved in? What factors determine why Americans will support the political justification for military involvement? What types of messages are most effective in garnering support for American forces?

The answers to these questions can be forecast to some degree with well-designed research studies. There are some difficulties, however. The Army media analyst faces the problems of determining which data collection methods can most accurately provide answers to the questions at issue, and in gaining adequate access to information prior to and during military operations. In the pages that follow, we will describe the methods and procedures PA professionals may use in overcoming these difficulties.

RESEARCH APPROACHES

There are several research approaches or "methods of knowing" which have been used to conduct studies: intuition, authority, and science.

In the **intuition** approach, one assumes that something is true because it is "self-evident" or "stands to reason." An example of this type of thinking would be if some Public Affairs leaders resist efforts to perform area studies because they believe they already "know" their AO.

The **authority** method seeks to promote a belief in something because a trusted or credible source says it is true. Here, the emphasis is on the source, not the methods the source may have used to gather his information.

The **scientific method** approaches learning as a series of small steps, with each step identifying more specific information and leading to a more clearly identifiable conclusion.

For example, one study or source provides only an *indication* of what may or may not be true; the "truth" is found only through a series of objective analyses.

This means that the scientific method is self-correcting in that changes in thought or theory must be continually reviewed, that issues and situations require constant monitoring.

The scientific method has become a valuable tool to produce accurate and useful data in mass media research. This annex focuses solely on the scientific approach and forms the fundamental basis for media content analysis in Army Public Affairs.

MEDIA RESEARCH: THE SCIENTIFIC METHOD

The goal of Public Affairs media research is to provide the methodology to support situational assessment, planning and decision-making that is fast, inexpensive, reliable and valid. The application of scientific methodologies to media research by Public Affairs personnel accomplishes this goal.

Five basic characteristics, or tenets, distinguish the scientific method from the other methods of research. A research approach that does not follow these tenets cannot be considered a scientific approach:

- **Scientific research is objective.** Science tries to rule out eccentricities of judgment by researchers. When a study is undertaken, explicit rules and procedures are constructed and the researcher is bound by them, letting the chips fall where they may. Objectivity also requires that scientific research deal with facts rather than interpretations of facts.

- **Science is empirical.** Researchers are concerned with a world that is knowable and potentially measurable. (Empiricism derives from the Greek word for "experience.") Analysts must be able to perceive, understand, and classify what they study and reject nonsensical explanations of events. For example, a newspaper editor's claim that declining readership rates are "God's will" would be rejected by scientific researchers because such statements cannot be perceived, classified or measured. Experience shows that there are usually easily identifiable reasons for declining readership.
- **Scientific research is systematic and cumulative.** No single research study stands alone, nor does it rise or fall by itself. Astute research analysts always use previous studies as building blocks for their own work. One of the first steps taken in conducting research is the review of all available literature on the topic so that the current study will draw on the heritage of past research.
- **Scientist attempt to find order and consistency in their findings**. In its basic form, scientific research begins with a single, carefully observed event and progresses ultimately to the formulation of theories and laws. A theory is a set of related propositions that present a systematic view of phenomena by specifying relationships among concepts. Researchers develop theories by searching for patterns of uniformity to explain and describe the information collected.
- **Scientific research is predictive.** Science is concerned with relating the present to the future. In fact, scientific researchers strive to develop theories because they are useful in predicting behavior. The importance of theories lies in their ability to predict an outcome or an event successfully. If a theory generates predictions that are supported by data, and the results are always the same, the theory can be used to make predictions in other similar situations.

MEDIA ANALYSIS STEPS

Evaluation of a problem must follow a standard sequence of steps to increase the chances of producing relevant data. Analysts who do not follow a prescribed set of steps increase the amount of error possible in the study.

These steps are:

- Select a topic (issue, situation, perception, or belief).
- Review existing research and other available information on the topic.
- Develop hypotheses and research questions.
- Determine an appropriate methodology, format or design.
- Collect relevant data.
- Analyze and interpret the results.
- Present the results in appropriate form (Information Paper, PA Study or PA Estimate)
- Validate and replicate the study when necessary.

The use of the scientific method of research is intended to provide an objective, unbiased evaluation of data pertaining to an issue or event. To investigate hypotheses systematically, media analysts must follow these steps. However, merely following the eight steps does guarantee that the research is good, valid, reliable or useful.

A countless number of intervening variables (influences) can destroy even the most well-planned research effort. Unanticipated events occurring during the research period may impact the results and they must be accounted for during the process. However, PA analysts must remain focused on the purpose of the research effort and not lose sight of the original objectives.

STEP 1 – SELECTING A TOPIC, DETERMINING RELEVANCE

Selecting a research topic is usually not a concern for Public Affairs analysts -- planning guidance, current situations, and most importantly, the operational issues confronting our commands, will guide the application of media content analysis. In most instances, the Public Affairs analyst will receive planning guidance well in advance of an operation, which will help determine the issues to be addressed.

Once the basic subject has been chosen, the next step is to ensure that it is relevant to the operation or situation at hand. This can be accomplished by answering six basic questions.

- What is the goal of this research effort?
- Is the subject too broad?
- Can the subject really be studied?
- Is the subject significant?
- Can the results of the research be generalized, communicated and understood?
- Does the issue lend itself to analysis?

Underlying all eight steps of the Media Analysis process is the necessity for validity. In other words, are all eight steps (from topic selection to data analysis to presentation and interpretation) the correct ones to follow in trying to answer these questions?

The answers to these questions will help focus the research you must do, make information gathering easier, and ensure the results are valid.

STEP 2 – REVIEW OTHER RESEARCH AND INFORMATION

Media analysts should always begin studies by consulting all literature, research, and other information available on the topic. The review provides information about what work has been done, how it was done, and what the results were as they apply to a given subject. It not only allows analysts to learn from (and eventually add to) previous media research, but also saves time, effort and money.

The review also helps to identify the facts pertaining to the situation being studied.

Completed media content analysis also provides a starting point for PA leaders who will follow in your position after you move on.

STEP 3 – DEVELOP HYPOTHESES AND RESEARCH QUESTIONS

After the general research are has been identified and the existing information reviewed, the analysts must state the problem or issue as a workable **hypothesis** or **research question.**

(Example: "The American public is losing confidence in the Army's ability to protect its soldiers, resulting from the media's portrayal of Army leaders as negligent and soldiers as lacking competence in avoiding fratricide.")

A hypothesis is a formal statement regarding the relationship between variables and is tested directly. In the example cited above, those variables are the news media, the messages they send, and the perception and understanding of those messages by the American public. With a hypothesis, the predicted relationship between the variables is either true or false. Identifying the degree of "trueness" or "falseness" and their implications is essential to the development of information campaign strategies.

On the other hand, a research question is a formally stated question intended to provide indications about something, and is not limited to the relationships between variables. Research questions are generally used in situations where an analyst is unsure about the nature of the problem under investigation. The intent is merely to gather preliminary information. Research questions are generally used to identify the focus and scope of a research project.

STEP 4 – DETERMINE AN APPROPRIATE METHODOLOGY/RESEARCH PROGRAM DESIGN

Given the variety of situations facing Public Affairs personnel, different approaches to media research are required. Some issues lend themselves more toward survey methodology via telephone, E-mail, or standard mail; others are best attacked through in-person interviews. Still other problems necessitate a controlled evaluation situation designed to eliminate extraneous variables by targeting analysis to specific media types. (An example of this approach would be a study of how a newspaper covered a specific story over a six-month period.)

The approach selected by the analyst depends on the goals and purpose of each particular study. Regardless of whether the problems or issue being addressed is a local one, affecting only a fraction of a community audience, or a national issue affecting us all, all research requires a design of some type. All procedures, including all variables, samples, and measurement instruments, must be pre-designed with hypothesis and research questions in mind.

There are four characteristics of research design that should be noted if a research study is to produce reliable and valid results:

- **Accurate setting**. For a study to have external validity, the study must be conducted as an historical account of the situation during the time frame studied. The analyst must have a clear understanding of the events unfolding around him and attempt to document as much related information as possible.
- **Clear cause-and-effect relationships**. The analyst must make every attempt to identify spurious dependent relationships and

weed them out. The results of a study can be interpreted with confidence if and only if all confounding effects are identified.

STEP 5 – COLLECT RELEVANT DATA

STEP 6 – ANALYZE AND INTERPRET THE RESULTS

STEP 7 – PRESENT THE RESULTS IN APPROPRIATE FORM (PA STUDY OR PA ESTIMATE)

STEP 8 – VALIDATE AND REPLICATE THE STUDY WHEN NECESSARY

Appendix P

INFORMATION PROGRAM EVALUATION

METHODS

How our audiences perceive the Army is critical to the success of all operations we are involved in. Internally, the Army's people require certain information to function effectively. The more they know and understand, the better they perform. Information about the operation, the unit's particular mission, how the commander feels about the situation, and a host of other subjects are of interest to both soldiers and civilians. Externally, the general public has specific needs for information about what their Army is doing and how they are doing it. This appendix explains methods for measuring success in the conduct of PA information programs.

Command information is communication between the commanders and those commanded. Command Information is different from the Public Affairs function of Information Provision in that it is the commander's responsibility to inform his people. Commanders must communicate their intentions and the troops, community, indeed the general public, must know his concerns and intentions. It is especially important to note that PA Information Provision techniques and procedures are just one channel that the commander may use in communicating to his audiences.

A poorly recognized fact is that the communication links between the commander and his audiences occur on various levels and assorted channels. This type of communication no longer fits the "top-down" communications model of the cold war Army. The explosion of today's digital technology has provided individual soldiers, civilians, family members, and the general public the ability to bypass rigid, controlled, vertical communication systems in favor of the common user, multidirectional, reciprocal, simultaneous, real-time transactive communications systems. Americans have the power to bypass the gatekeepers and ignore canned, shoddily produced, dated industrial age information products, in favor of accessing on-line information services or the Internet directly. Information is passed in all directions, continually. These audiences will have access to many more sources of information, which makes evaluating the effects of Army PA Information Programs all the more difficult.

This explosion of information technology has also highlighted how critical it is for PA elements to stay up to date on communications technologies, information services, and socio-economic trends of these forces at work.

Despite the rapid change in the information environment, the general steps for evaluating information program effects has remained the same:

- Determine the command's mission and the commander's method for accomplishing that mission.
- Identify all the various audiences interested in information related to the command, its members, and the mission.

- Identify a public opinion baseline -- the template against which new public opinion information will be compared.
- Gather identify all the messages communicated and identify which audiences' received such information.
- Gather information on the information program impacts. This is done reviewing unit newspapers, letters to editors, responses to information programs fact sheets, formations, surveys, and interviews. Check related bulletin boards on all on-line information services. Check related Newsgroups on world-wide-web nodes, which commonly carry related information. Monitor discussion groups on on-line services.
- Attend commander's calls, staff meetings, formations, briefings, and other gatherings where audience reaction, troop morale and like information will be discussed.
- Evaluate the knowledge of the targeted organizations. This is accomplished through in-person question/answer surveys and interviews, E-mail surveys, and electronic town hall meetings, etc.
- Coordinate with other staff elements addressing similar information issues (SJA, Chaplain, PMO, IG, etc.).
- Produce a summary of information gathered in an impact assessment.

Appendix Q

PRINCIPLES OF PA SERVICE AND STANDARDS

INTRODUCTION

1. This checklist identifies the primary tasks associated with the functions of Public Affairs, and establishes standards for successful accomplishment of those tasks. Standards of service equate to minimum exceptions of an operational commander and will be used to judge unit readiness, leader effectiveness, and individual soldier performance.

2. The following definitions apply to this checklist:

Austere: No existing PA units, assets or Army signal information infrastructure available in area of operations upon deployment. Commercial communications infrastructure is not available. PA elements must perform all missions and provide all PA support using organic personnel and equipment.

Existing: PA units, assets and Army signal information infrastructure in place before deployment. Commercial communications infrastructure is available. Deploying PA element assumes duties of or augments organic PA elements. Existing personnel and equipment augmented by additional PA elements to accomplish PA mission and provide PA support.

FUNCTIONS/TASK	CONDITION	STANDARD
Planning		
1. Perform Global Information Environment analysis	Existing	Austere
2. Develop PA Estimate (COAs)	Existing	Austere
3. Develop PA Strategy (plan)	Existing	Austere
4. Develop PA Guidance	Existing	Austere
5. Coordinate PA Annex	Existing	Austere
Media Facilitation		
1. ID media support requirements	Existing	Austere
2. Register news media	Existing	Austere
3. Coordinate media support	Existing	Austere
4. Provide media orientation	Existing	Austere
5. Coordinate news media interviews	Existing	Austere
6. Plan/coordinate news briefings	Existing	Austere
7. Establish Media OPS Center	Existing	Austere
8. Operate MOC	Existing	Austere

9. Provide media assistance/escort	Existing	Austere
Information strategy		
1. Identify target audiences	Existing	Austere
2. Identify information needs	Existing	Austere
3. Develop info themes/CMD messages	Existing	Austere
4. Gather info/develop products	Existing	Austere
5. Disseminate information to deployed forces	Existing	Austere
6. Disseminate information to families/home station audiences	Existing	Austere
7. Disseminate information to national/local news media	Existing	Austere
8. Disseminate information to general public	Existing	Austere
Public Affairs Training		
1. Identify training needs	Existing	Austere
2. Develop unit PA Training plan	Existing	Austere
3. Develop Family Spt Training Plan	Existing	Austere
4. Develop Senior LDR Training Plan	Existing	Austere
5. Conduct unit PA training	Existing	Austere
6. Conduct Family SPT Training	Existing	Austere
7. Conduct Senior LDR Training	Existing	Austere
8. Prepare SMEs for Media interviews	Existing	Astere
9. Evaluate training efforts	Existing	Austere

Appendix R

PRIVACY ACT/FREEDOM OF INFORMATION ACT

INTRODUCTION

This checklist addresses personal information about soldiers and any living persons that can or cannot be released under the provisions of the privacy act.

CHECKLIST ITEMS

- AGE (date of birth) = releasable. This information is public record.
- HOME OF RECORD/PRESENT ADDRESS = In most cases, home of record can be released if no street address is given. There is no general rule for disclosure of this information. Widely different circumstances surround each incident, and judgment is made on a case-by-case basis. In most cases, the person's present geographical location may be provided (city, state), but not the street address. In each case, the desires of the actual person or next of kin should be considered.
- MARITAL STATUS/DEPENDENTS = RELEASABLE. This information is public record, including names, ages and sex of dependents.
- AWARDS/DECORATIONS/CITATIONS = releasable.
- EDUCATION/SCHOOLING/SPECIALTY = releasable. Major area of study, school, year of graduation, degree and specialty designator is releasable.
- RACE = In most cases, NOT releasable. However, where the fact of an individual's race is relevant in providing essential facts to the Press, it may be released (such as in a racially oriented protest or altercation.)
- CHARACTER OF DISCHARGE:
- ADMINISTRATIVE = NOT releasable, unless the individual provides his written consent.
- PUNITIVE = releasable. This includes discharges resulting from courts martial.
- DUTY STATUS = releasable.
- PERSONNEL BOARDS:
- Results of promotion boards and augmentation boards are releasable.
- Results of administrative discharge boards and aviator flight boards are NOT releasable.
- PHOTOGRAPHS IN THE CUSTODY OF THE DEPARTMENT OF DEFENSE = releasable, unless they warrant an invasion of anyone's personal privacy.

Appendix S

PA GUIDANCE ON TERRORISM COUNTERACTION

In view of the continuing media interest in the subject of terrorism, the public affairs guidance contained in this message is provided to assist PAOs in responding to media queries and in developing local contingency plans.

TERRORISM IN GENERAL

a. **U.S. POLICY**. All terrorist acts are criminal. The U.S. Government will make no concessions to terrorists. Ransom will not be paid and nations fostering terrorism will be identified and isolated.

b. **RESPONSIBILITY**. Department of State is the lead agency for response to international terrorist incidents that involve U.S. military personnel and facilities outside the U.S. The administrator of the Federal Aviation Administration is responsible for terrorist incidents that affect the safety of DoD personnel or property aboard an aircraft in flight. When terrorist incidents occur at military installations within CONUS or its possessions (Puerto Rico, Guam, Virgin Islands, American Samoa, and trust territories), the FBI will be the lead agency. If the FBI declines to exercise its authority, military authorities may take appropriate action within the limits of their responsibility to resolve the incident.

c. **NOTIFICATION.** All terrorist incidents should be reported immediately through channels, to OASD/PA. No public release of information regarding a terrorist incident should be made without OASD/PA approval except for cases involving public safety.

COMBATING TERRORISM

Combating terrorism can be divided into two major areas: counter-terrorism (offensive measures) and anti-terrorism (defense measures).

- Counter-terrorism. The following statement may be used in response to queries regarding counter-terrorist forces:
- "The U.S. Government has trained forces and equipment from all four services to cope with terrorist incidents. We have also said that command and control elements for these forces exist and have been exercised. These elements report to the Joint Chiefs of Staff, as do other command and control elements for military operations. We do not comment on any details concerning the circumstances under which these forces may be deployed, their identity, or tactics."
- Requests for interviews or filming of counter-terrorism training will not, repeat will not, be approved.
- Requests for photos of counter-terrorist forces personnel or their training will not be approved.
- Because of the clear association/confusion surrounding the relationship between special operations counter-terrorism forces,

all requests for interviewing or filming special operations forces and or training will be approved by OASD/PA.

- Questions beyond the scope of the above guidance should be referred to OASD/PA..
- Anti-terrorism. The following guidance is applicable to media request for information pertaining to anti-terrorism.
- DoD officials, senior leaders, commands and knowledgeable individuals may discuss the subject of anti-terrorism as it pertains to those areas/installations for which they are responsible.(Anti-terrorism measures and procedures should be discussed in a general manner without going into a checklist of specific details.)
- Media requests to film anti-terrorist training will be approved on by OASD/PA.
- Photos of anti-terrorist training should be forwarded through channels for approval.

QUERIES

In response to queries regarding a possible or real terrorist threat at a particular base/installation/activity the PAO may acknowledge, if appropriate, that increased security measures have been/will be taken without going into specific details of all the measures taken. PAOs may, when appropriate, acknowledge the obvious.

For example, increased security measures such as increased guards at the gate or additional patrols, if they are obvious to the public may, in some cases, be acknowledged. PAOs should, however, exercise care and prudent judgment in any discussion of these or other security measures, which have been/will be implemented.

Appendix T

BROADCAST OPERATIONS

This section focuses on the radio and television services provided by Armed Forces Radio and Television Service at the unified command level and the coordination necessary by Army component commanders to ensure operational area support. It identifies and explains the AFRTS mission and its capabilities. It also discusses planning considerations and theater broadcast information requirements.

AFRTS CONTINGENCY BROADCAST OPERATIONS

The scope of the ABS mission of providing AFRTS radio and television news, information and entertainment programming to DoD personnel stationed overseas greatly expands during wartime to include support of global contingency requirements. As new contingency plans are developed based on emerging joint and Army doctrine, ABS must consider how the additional broadcasting personnel and equipment resources needed to support a rapid deployment broadcasting mission can be obtained while simultaneously meeting increased requirements in existing unified command theaters. The immediate response necessary to meet contingency requires the development of AFRTS appendices to Unified Command Operations Plans (OPLAN).

Army Broadcasting Service (ABS) is the Unified Command AFRTS Planner (UCAP) for the U.S. Southern Command and the U.S. European Command. ABS also has Geographic Area Planner (GAP) responsibilities for U.S. Forces Korea (USFK). ABS and AFRTS networks within these unified commands are responsible for updating and maintaining appendix content under provisions of the American Forces Information Service (AFIS) Concept Plan for Peacetime and Wartime Operations for the Armed Forces Radio and Television Service (AFRTS CONPLAN 98-1) and DoD Directive 5120.20-R

Contingency and/or Wartime Plans define the Mobilization/Contingency mission. Force structure to support these plans may be packaged as blocks, deployable units or detachments for ease of planning.

According to CONPLAN 98-1 and DoD Directive 5120.20-R, Unified Commanders and Subordinate Commanders, with the advice of ABS, determine the type of AFRTS Flexible Response Option (FRO) necessary. Unified and Subordinate Command support is required for the current levels of support contract and for any mission changes which affect AFRTS service in the CINC's area of operation.

AFRTS Mission

Provide live news, information and entertainment programming, free of censorship, to all DoD elements in place or deployed worldwide.

Provide U.S. military theater commanders with sufficient electronic media resources to effectively communicate DoD, Army, theater and AO command information.

Concept of Operations

Existing facilities and services will be the first AFRTS assets used to respond to AFRTS taskings.

Upon implementation of a Contingency Plan or OPLAN, AFRTS assets in the AO come under the direct operational control of the Unified Command for the period of the operation. When the operation is terminated, normal command relationships will be resumed.

The AFRTS Commander reports directly to the Unified Commander responsible for the theater of operations.

The AFRTS Commander retains direct command authority over AFRTS personnel and equipment.

The AFRTS Commander is responsible for all AFRTS matters concerning the operation, accomplishing direct coordination for the Unified Command with AFIS/AFRTS for all AFRTS issues requiring AFIS/AFRTS action, and managing all AFRTS assets involved in the operation in accordance with the AFRTS CONPLAN 98-1.

The AFRTS Commander will be a designated member of all public affairs meetings as a member of the staff.

The AFRTS Commander is authorized direct coordination with other members of the unified command's staff to work specific AFRTS support requirements.

All requests for command internal information or emergency announcements from AO organizations or personnel will be forwarded to the director of public affairs for approval.

All public information released by public affairs officials will be available for AFRTS use.

AFRTS radio and television electronic news gathering (ENG) will be dedicated for the AFRTS "on-air" mission and in direct support of the Unified Commander's internal information program.

AFRTS will provide full, factual and timely internal information and news to military audiences in the AO, consistent with national and operational security, and host country sensitivities.

AFRTS will follow Operations security (OPSEC) and communications security (COMSEC) rules.

AFRTS OPERATING PARAMETERS

AFIS is responsible for all AFRTS satellite programming services and overall policy and guidance for their use.

The AFRTS Broadcast Center (AFRTS-BC) is responsible for providing non-local and non-theater radio and television programming material to AFRTS facilities in the AO, except as outlined in Flexible Response Options (FROs).

The Television-Audio Support Activity (T-ASA) in Sacramento, California, is responsible for providing technical and logistical support to AFRTS.

The AFRTS facility in the AO will provide service based upon initial Flexible Response Options (FROs) and continue operations until directed to modify its services by AFRTS, the Unified Command or as the mission requirement dictates.

AFRTS will provide AO-wide announcements as required on both radio and television in order to facilitate unified command needs.

At unmanned repeaters and cable distribution systems, local officials may have AFRTS personnel make local announcements, if possible, in coordination with the AFRTS commander or on-site command representative, if approved by the director of public affairs for the operation.

AFRTS NETWORK COMMANDER OR DESIGNATED REPRESENTATIVE IN THE AREA OF OPERATIONS

The AFRTS network commander or designated representative will be co-located with the unified command director of public affairs.

The AFRTS network commander or designated representative will have a command function with direct operational command authority over all resources assigned to support the AFRTS mission in the AO.

The AFRTS commander or designated representative will ensure a logistics and engineering function responsible for providing advice and assistance to maintenance personnel assigned to AFRTS outlets, and maintaining unmanned equipment. This function will assist the AFRTS chief engineer in developing new equipment support requirements as changes occur in the AO.

The AFRTS network commander or designated representative will ensure internal information ENG coverage of unified command activities of interest to the members assigned in the AO. In joint service situations, Army AFRTS representatives may also be responsible for the production and duplication of radio and television internal information products for use at AFRTS outlets and television programming for DoD, or satellite cabled sites in the AO.

UNIFIED COMMAND RESPONSIBILITIES

The unified command provides logistic support for AFRTS. This includes vehicles, POL, and supply requirements in the AO as noted in the OPLAN. Also included is vehicle maintenance and POL for all AFRTS contingency vehicles. If additional forces are deployed to support AFRTS, the unified command assigns additional vehicles to support the expanded maintenance and production requirements.

If security and/or intelligence forces determine that AFRTS facilities have been identified as a potential target by hostile forces, the unified command will notify the AFRTS facility and provide security to targeted facilities.

If contract communications support is terminated during the implementation of the OPLAN, the unified command provides communications support for AFRTS use in the AO.

This includes existing long-wire or microwave systems for distributing the AFRTS signal and support for telephones, facsimile transmission and computer equipment for AFRTS.

If support cannot be obtained from existing assets, the unified command should be prepared to augment the AFRTS mission as outlined in the OPLAN.

The unified command provides personnel, administrative, vehicle and other logistic support for all AFRTS personnel assigned in the AO and those deployed to supplement that force. This includes unit line numbers and entry clearances required for all deployed personnel supporting the AFRTS mission.

The unified command obtains country clearance for construction of any temporary transmitter towers required due to expanded AFRTS service, which may occur as the operation unfolds.

The unified command provides electrical backup power for AFRTS facilities if contract services are terminated during the implementation of the OPLAN.

The unified command is responsible for obtaining necessary broadcast frequencies in consultation with the host-nation government to meet AFRTS broadcast requirements.

Flexible Response Options

Although each operation will differ, the following are general concepts of AFRTS Flexible Response Options (FROs) available for peacetime engagements, wartime operations and stability and support operations in an area where little or no AFRTS service exists or where crisis situations require a modification to existing AFRTS services. The unified command AFRTS planner (UCAP) is responsible for developing specific equipment, support and manning requirements to implement the AFRTS FROs that best support the specific operation.

The Unified Commander for the area of operations must request AFRTS radio and/or television services or for a change in present level of service before deployment. The Unified Commander requests AFRTS assistance through the unified command AFRTS planner responsible for the area of operations. The request will then be forwarded to ABS and AFIS for final approval.

FRO One: Direct to Ship (DTS) Service Support System. DTS is an U.S. Navy peacetime capability that provides news, sports, information and entertainment to audiences on ships at sea. A wartime adaptation of the service can provide immediate access to three radio and two television channels for land-based audiences including geographically separated units down to the lowest level. A deployable AFRTS kit containing an individual receiver decoder (IRD) provides service. This service provides a single-source 24-hour capability of receiving all services, but only one channel can be accessed at a time and no local or theater command information would be available.

FRO Two: Satellite Direct Radio and Television (SDRTV). SDRTV is an unmanned AFRTS satellite service that can be provided to virtually any land based audience on the globe with up to 10 stereo radio channels and six television news, sports, information and entertainment channels. In the European theater it will include a regionally generated signal. Service is provided at a single location using a deployable AFRTS kit containing a simultaneous receiver decoder (SRD) that provides a capability of receiving all services and accessing up to six radio and television programming sources at a time. As a public affairs option,

SDRTV provides an internal information data stream that can be accessed with the addition of a computer, printer and proprietary software to the SDRTV equipment package for use by public affairs activities in providing support to deployed populations. Unified command public affairs offices and the UCAP should consider coordinating the use of the additional capability whenever a manned public affairs activity is deployed.

FRO Three: Manned Radio Systems. Signal Distribution Systems. These deployable systems include audio and video transmission and cable systems that provide a capability to distribute, DTS, SDRTV or manned radio service to an expanded autonomous geographic area such as a base camp or Air Force base in an AO.

FRO Four: Manned Radio Systems. These deployable systems provide a capability for local, live internal information and radio news. Various types and sizes of local radio systems exist that can be used to establish a range of services from simple local break away "radio-in-a-box" to a full service facility with local production capability. Some of these systems will include radio transmitter that can provide limited signal distribution without deployment of FRO Three. This system can provide a limited single-source radio service to outlying populations that are not served by FRO One or Two, are not available or would not be appropriate programming sources.

FRO Five: MOOTW Management, Local TV and Network Live Radio. These deployable personnel and systems support the development of an AFRTS management function to oversee dispersed AFRTS operations and will add local television and network-wide live radio capability using organic distribution systems. The capability will establish a network to support operations in an AO comprising a large peacetime engagement of medium to long duration encompassing a large geographic area where the operations commander requires near real-time internal information capability. The system provides network administrative, computer, maintenance, engineering and operations support functions. These functions provide the unified command director of public affairs with AFRTS management expertise not normally available on the public affairs staff. The television service system is designed to produce AO information that can also be fed to the AFRTS Broadcast Center for rebroadcast to DTS/STRTV audiences worldwide.

FRO Six: Theater Satellite Radio and Television Operations (TSRTO). In a major regional conflict where large force deployments are planned the AFRTS Broadcast Center will dedicate one channel of radio and one television channel for use by the UCAP to broadcast directly to the theater of operations. Programming will include time shifting "prime time" so that each 12-hour shift receives prime time programming in the first four and-one-half hours of off time. In conjunction with FRO One or Two, this will provide a virtual network capability to the AO commander and the PAO. It will appear to the audience as if the broadcast was occurring in the AO when in fact it is originating from the Broadcast Center. Initial spots can be unsophisticated radio readers and character-generated (CG) messages on television. If there is a manned radio facility providing theater and operation-specific internal information in place, they will forward copies of all AO specific spots via computer to the Broadcast Center for use initially as television CG messages within the dedicated TV channel. As with all internal information, they will be developed in coordination with the PAO in the AO. If there is insufficient AFRTS manning in place which would be the case if there were numerous geographically-separated operating locations, the internal information will be supplied, via computer, by the AO public affairs office. Base and component command PAOs will be advised of the scheduling of theater programming and encouraged to provide service-unique spots for use in theater. Service will continue until a significant drawdown occurs, the operation stabilizes to the point where local TV spots are not needed and the presumption of pre-conflict programming will serve the majority of the deployed population or an AO based network begins serving the operation.

Appendix U

COMMUNITY SURVEY

COMMUNITY SURVEY EXAMPLE

I. The area

A. Geographical description
1. Of areas surveyed--size of cities, counties.
2. Of surrounding area, if pertinent.
3. Climate, topography, annual and seasonal temperatures, rainfall, etc. (one sentence will suffice for each.)
4. Are the industries dispersed or centralized? Attach a map of the Area indicating the location of the principal plants. The map should show the names and numbers of principal streets and highways furnishing access to these plants.

B. Population
1. Of city
2. Of area.
3. Of labor market area, if different from above.
4. Breakdown by sexes, color, native or foreign born, educational level, percentage of homeowners, etc.

C. Industrial data
1. Types of industries and number of each, labor force of each, key products of the area, and additional data as considered applicable.
2. Does one type of industry dominate the area? If so, give pertinent information regarding the industry.

II. Manpower

A. Labor market rating
1. Is department of labor market classification, a,b,c, or d?
2. Include supporting statistical data.

B. Unemployment
1. Totals and percentages of skilled, semiskilled, and unskilled.
2. What types of skills are most commonly available?

C. Source of labor supply

1. Compared to the World War II years, what are the reserves of women, handicapped, older-age groups, part-time workers, and school graduates?

2. Has there been much intermigration to total population?

D. Occupational classification of area workers. What are the most common occupations of the area? The less common?

E. Skills in shortage category. List, with numbers of each, if available. Make a comparison of this list with the national shortage list.

F. Area wage schedules
   1. List the wage schedules of major occupations and industries.
   2. How do they compare with national averages?
   3. How do they compare with neighboring areas? With competing areas?

G. Requirement of defense industry in area
   1. Is manpower available for present production schedules? Current planned production?
   2. What skills are lacking for production schedules, both present and future?
   3. Do employers ordinarily use training programs? If so, give some Examples.

H. Other pertinent information
   1. Include current work stoppages, if any; record of work stoppages During last 10 years.
   2. Are workers highly organized? Principal unions?

III.Industrial facilities

A. Facilities suited or adaptable for defense production.
B. List facilities with current and World War II products.
C. Give current and capacity employment.
D. Give types of machinery.
E. What defense contracts are held or sought?
F. Vacant factory space. Describe space and indicate production potential.

IV.Housing

A. Housing regulations.
   1. Is it a critical defense housing area under public law 96? P. L. 139?
   2. Local rent control?

B. Housing units available.
   1. Number for sale, including 1, 2, and 3 bedrooms. Price ranges. Are the prices reasonable?
   2. Number for rent, including above information. Apartments available. Number, size, price ranges. Are the rents reasonable?

3. Sleeping rooms available. Number, price ranges. Are the prices reasonable?
4. Building permits issued (in past 12 months)
5. Number for houses--1, 2, 3, or more bedrooms.
6. Number for apartments.
7. If houses, number for rent and contemplated rental prices.
8. Housing units contemplated .
9. Number and sizes. How many bedrooms?
10. Number of these for rent.
11. Estimated rental rates.

C. Builders

1. Adequate number of experienced builders?
2. Do they have trained skeleton force?
3. Can other necessary housing construction workers be secured?
4. Is land available? Under option?
5. Can materials be obtained?
6. What bottlenecks?

D. Building capital.

1. Is capital for the building of housing and rental units readily available? If so, on long- or short-term loans?
2. What are the sources of this capital?
3. Does the community object to construction of more housing units now?

V.Adequacy of housing

For present work force? For expanded production? (quote a figure or percentage, such As peak load in world war ii or 50 percent above present.) This figure should be adequate to cover planned defense expansion known to you at time of Survey.

VI.Other community facilities and services

Discuss each of the following items as to adequacy for the present work force and for an expanded work force; give specific facts for each, as pertinent (yes and no answers are not adequate.)

A. Water
B. Electric power
C. Gas
D. Sewerage

*Note: for items 1 through 4 above, describe sources of supply, capacity, reserve storage, current use, reserves on hand, plans for expansion--whether on hand or projected.

E. Transportation: types and numbers
F. Highway and road systems: are the roads serving the area adequate and in good? What is the present traffic load and the peak capacity of these roads? Describe any unsatisfactory factors. What action is contemplated or considered necessary to assure free traffic movement within the area? (survey requests will furnish, whenever possible, specified information on any industrial and

defense manpower requirement changes under consideration for the area.) Contact with local, state, and federal highway authorities should be made, if necessary, to explore fully this phase of the survey.

1. Schools: number of each type of school, crowding, shifts, new construction, etc.
2. Hospitals: number, number of beds, population per bed.
3. Doctors: number, population per doctor and per dentist.
4. Fire protection: size, ratios, and ratings.
5. Police protection: size, ratios, and ratings.
6. Shopping centers and shopping hours.
7. Recreational facilities: number of each type.
8. Churches: all denominations.
9. Sanitation service (garbage collection).
10. Laundries, dry-cleaning businesses, barber shops, beauty shops, etc.
11. Banking facilities (include arrangements for shift workers).
12. Hotels: number and number of rooms, scale of rates, etc.
13. Restaurants and other eating places.
14. Newspapers: number (morning, evening).
15. Municipal government (form, etc.).
16. Tax rate: local, county, state.
17. Cost of living index: get whatever information is available Compare local figures to national index.

*Note: include reference material, maps, booklets, etc., If possible.

Appendix V

AUDIENCE SURVEY

Audience surveys systematically gather information about the effectiveness of CI programs and products as they relate to a particular group of people. The commander and the PAO to make decisions about management and direction of an internal information program or product use the results.

PUBLIC AFFAIRS REQUIREMENT

The PAO will conduct a readership survey at least every two years (every three years for the Reserve Components). Coordination with the DOIM for possible computer and analysis support is recommended. Additionally, AR 600-46 can provide information on conducting surveys. The survey will provide data on distribution effectiveness, reader awareness and acceptance, readership and perceived usefulness of standing features and topics covered, and opinions of the value and effectiveness of the publication. Repeated surveys will provide trend data.

The survey may include any or all of the 20 questions listed in the Readership Survey (RCS: SAOSA-223) (app H) in AR 360-81. However, surveys not using these tested questions must be pretested to ensure validity before being used in a survey. Survey respondents will be selected using probability-sampling techniques.

Informal surveys, such as those included in a newspaper or conducted randomly/haphazardly with a few people, are not substitutes for readership surveys. This does not preclude an editor from periodically publishing a coupon or set of questions to solicit informal feedback that is not statistically projectable.

Before administering the survey, the survey managers must coordinate with the agency that will provide response analysis to be sure questionnaires; answer sheets, data entry program, or any other materials are appropriate and usable. Survey conduct may be included in the command's CE publication contract and may also be contracted by the command for Army Funded newspapers, providing funds are available.

When civilian employees are surveyed, PAOs should also coordinate with the civilian personnel officer for local union notification requirements. Completed questionnaires may be analyzed by the local Director of Information Management (DOIM) to provide percentages of responses to survey questions. Where computer support is available (from the local DOIM or DRM), responses will be analyzed using a program such as the Statistical Program for the Social Sciences package. Questionnaires must be constructed using the parameters of available software.

A written discussion of findings and conclusions drawn from the survey will be forwarded within 60 days after the survey is completed through the local commander, appropriate major command, to HQDA (SAPA-CI-PMN), Room 2E625, The Pentagon, WASH DC 20310-1510.

As a minimum, the report will contain the survey statistics, an analysis of the data, identification of strengths and problem areas (e.g., distribution, more sports, etc.), recommended improvements and changes to editorial policy, and an indication that the commander has reviewed the results.

Surveys may be conducted any time. However, no newspaper's survey report on file at HQDA should be older than 3 years (4 years for the Reserve Components). This allows for the time to conduct a survey.

A copy of the most recent survey will also be submitted with the annual CI Program Assessment Report (DA Form 510-R), unless the survey was previously submitted to the MACOM and OCPA-HQDA.

The PAO will conduct electronic media surveys at least every two years (every three years for the Reserve Components).

SURVEY ADMINISTRATION

Among the more common methods of conducting surveys are the mail survey, face-to-face interview, and telephone interview. The mail survey is the preferred method for purposes of this requirement, although other methods, managed properly, may be used.

Survey respondents will be selected in a totally random manner (e.g., simple random, stratified, or systematic, using probability sampling procedures). Survey managers will select samples, which achieve at least a minimum of a +/- 5 percent reliability (error margin) at the 95 percent confidence level.

Sample sizes shown for the various reliability levels (e.g., +/- 5 percent error margin) are the number of usable responses received, not the number of questionnaires to be sent out. For a population of 5,000, 357 usable responses will accurately reflect, to within +/- 5 percent, what the entire 5,000 member audience would have said, had it completed the survey.

Experience with mail audience surveys shows that they realize an approximate 30 to 35 percent response rate. Therefore, send out at least three times as many questionnaires as are needed for analysis. Remember that incentives encourage responses.

Maybe the local MWR office or similar staff agency could provide bumper stickers, discount coupons, or other incentive for completed responses.

FOCUS GROUPS

Focus groups. One of the most effective ways to learn how CI products are being received is to conduct focus group interviews. Focus group interviews are structured group discussions in which representative members of the audience are brought together to discuss one or more CI products or issues. These interviews or sessions can examine the effectiveness of products or programs, gain suggestions for improving existing products or programs, and determine the need for new products or programs.

The key to effective focus group interviews is proper planning. Focus group organizers must determine who will participate, and what are the specific objectives of the session (e.g., what topics or issues will be discussed, what specific questions will be asked, what is to be done with the results, etc.). The method of selecting participants should be determined and the location for the meeting secured. While there is no optimal size for focus groups, generally groups of six to 10 individuals are manageable. Group makeup (officer/NCO/enlisted, men/women, military/civilian, active/reserve, retired/family member) depends on the objectives of the session. Generally, homogeneous groups are preferable. Often it will be necessary to hold more than one focus group session to obtain information needed to evaluate a particular CI program or product.

The moderator or group leader should be someone skilled in interview techniques and knowledge about the product or program being evaluated. It is often best not to have a high-ranking individual as the moderator with a group of junior enlisted or young family members, as free flow of information and opinions may be inhibited. The group leader must facilitate the discussion, not serve as an interrogator.

Focus group sessions should be informal. Participants should be encouraged to speak whenever they wish; the moderator should focus the discussion on the topics without being overbearing. If participants agree, it will be useful to videotape their comments for use in evaluating the session.

It is important that all group members understand that their honest opinions are being sought, and that the session is intended as a positive method of improving CI within the organization. No punitive actions should occur as a result of these sessions.

Appendix W

PA Lessons Learned

"What this century's history teaches us is that the Army's real strength is its ability to change and adapt to the period's requirements. Our ability to change was the key to victory in two world wars and a cold war, and it will be the foundation for our future success."

--General Dennis J. Reimer

INTRODUCTION

Explosive developments in information age technology have made the prospect of sharing lessons and ideas across a wide audience a reality today. With ready and easy access to E-mail and the Internet, soldiers can distribute documents, graphics, and photographs with lightning speed.

This appendix is based on an article published in the Center for Army Lessons Learned (CALL) ***News from the Front!***

This section provides public affairs officers (PAOs) with a tool for capturing observations and an outlet for rapid analysis and dissemination of tactics, techniques and procedures (TTPs) to the force. The initial focus is on defining and narrowing the scope of TTPs. It is important for officers and soldiers in the field to clearly understand the process before proceeding to methodologies of collection. The next section provides a structure for developing a narrative product for publication by CALL. Such a product will provide a coherent article of information, which can be quickly used by the force. The final section describes the observation-gathering process. Understanding the process for collecting data will prove invaluable to the operational planner and for producing effective training tools for the future.

DEFINITION AND LIMITS

Focusing the collection effort is central to capturing meaningful observations. Although CALL regularly sends combined arms assessment teams (CAATs) to major exercises and actual operations to gather observations, units from the field, including any public affairs section or detachment, can provide great insight by planning for the collection of information. In fact, only the Army as a whole can make CALL a continuing conduit of information for use by soldiers.

By using the structure and tools described below, units can provide useful TTP by establishing a collection effort as part of the originating operation order (OPORD), with almost no interference with normal operations. Indeed, the tools will enhance planning for future (and remedial) training by incorporating the capturing of TTP into the plan.

TTP are often limited to the specific operation or exercise. The function and use of TTPs are analogous to legal precedents. In law, if given circumstances of a case are generally similar to a prior case, it is assumed that a judgmental decision for the present case should be the same. However, circumstances in law often have aspects that are unique and must be considered before rendering a new decision. When applying TTP, study prior situations in context and use the lessons prudently.

STRUCTURING THE PAPER FOR USE BY CALL

PAOs at all levels can build upon the after-action review (AAR) process in the plan by producing a publishable document. In almost all exercises, units learn and consequently implement improvement measures. By employing the structure below, units can effectively share information throughout the force -- not only from Combat Training Center (CTC) rotations but also from home-station training and exercises away from the training centers.

Do not view the structure below as a rigid construct. Rather, it should serve as a point of departure for unit writers. Although quantitative material is useful for commanders and researchers, make this document narrative in format. Use graphics to support the narrative, if possible. Bring together data into a cohesive product that other units can readily use without resorting to sifting through large amounts of charts, lists, and disjointed bullets.

- Type of unit. Describe the type of unit the PAO supported (mechanized infantry division, separate brigade).
- Context of event. Summarize the general setting for the exercise or operation. (See Exercises and Actual Operations below. More operational context information is provided in this section.)
- Commander's comments. If possible, the commander can provide a brief (one or more paragraphs) commentary on public affairs operations. Work closely with the unit's executive officer or chief of staff for such input.
- Interaction with PSYOP, Civil Affairs, Signal. As information operations continues to grow and doctrine is further developed, interaction between various agencies will also continue to expand. While ensuring coordination with PSYOP and civil affairs operations, PAOs will continue to recognize the separation in functions of the organizations required by law. Discuss the coordination measures used.
- Media Relations.
- Summary of events. Provide a summary of events. Were press conferences and interviews scheduled and executed? What was the pace of daily operations? What was the routine daily schedule?
- Command messages. In developing this section, answer the following questions in detail: What were the command messages? More importantly, did the command messages come through to print or broadcast? Were any command messages distorted or misinterpreted? How can clarity be improved for the next operation?
- Summary of higher headquarters' public affairs guidance (PAG). Write a one or two paragraph summary of the initial and follow-up PAG received from higher headquarters. (Provide a complete copy as an appendix.) Provide answers to the following questions

following the summary: How did PAG influence operations? Was the PAG clear and meaningful? Were excerpts used to create lower level command messages?

- Media contacts. Describe the types and numbers of media contacts. Did the unit encounter numerous print-journalist requests? Electronic requests? Were there patterns in the requests? What could a future media preparation package contain to answer some questions in advance? How did the PAO prioritize media access? Were major outlets afforded more opportunities?
- Summary of Media Releases. Summarize media releases in one or two paragraphs. What were the major themes? What media received the releases? Were releases used in stories? Were there any comments from members of the media about the releases?
- Media Content Analysis. During and following an event, gather press clippings and, if possible, record electronic media stories about the event. What was the nature of the coverage? What was the tenor of editorial comments? Did command messages get exposure? Was the content of articles generally accurate? What could PAOs do in the future to improve the accuracy of content?
- Command Information Products. The command information program in the field is fundamental in the minds of American soldiers. And, this phenomenon is nothing new to this culture.

> **CIVIL WAR**
>
> "When the (Civil) war entered Pennsylvania, the Philadelphia Inquirer often sold up to 25,000 copies of a single issue to the men in the field. During a lull in the Battle of Cedar Creek in October 1864, observers later remarked that the first thing the men did along the line was to sit down, boil coffee, and pull out their newspapers."

- Because of this intense internal interest in events surrounding the operation and events back home, it is imperative that PAOs adequately address methods and practices used to inform the soldiers. Answer the following questions in the narrative: Did soldiers in the field receive consistent and timely information about operations and world events? What was the distribution method? How was it evaluated for effectiveness? Were reproduction resources available and used adequately?
- Changes Incorporated in the Tactical Standing Operating Procedures (SOPs). From the TTPs gathered, what changes will the unit make to the SOP? Briefly describe why the unit is making the change and what mechanisms to put in place to test the effectiveness of the change.

EXERCISES AND ACTUAL OPERATIONS

Operational information is important for the reader to understand the context in which the PAO operated. Provide the following information for the CALL document to contribute to the reader's full understanding of the event(s):

- Mission - Summarize the unit's mission. (The focus here is not on the PAO's mission, but on the mission of combat headquarters.)
- Commander's Intent - This information is available on the OPORD. By incorporating this information, the reader will have an appreciation of the context in which the PAO operated. Emphasize the components of the intent:
  - Purpose
  - Method
  - Endstate
- New Equipment Used - Was new equipment available and used in the operation? Describe the equipment. Was it useful? What were the additional training requirements for using the equipment?
- New Techniques Used - Did the PAO incorporate techniques which are not described in doctrinal manuals? Describe the techniques employed.
- Structure - What was the structure of the PAO unit or shop? What manning -- required versus on-hand? (Note: Do not provide information that is classified under provisions of unit strength reporting (USR) regulations. Seek to provide a document that is free of classified material.)
- Operational Developments - How did the headquarters change its plan during the operation? How did the change(s) affect PAO operations? The descriptions here will bridge the gap between the original plan and its actual implementation.
- TTPs applied during the mission and for future operations - In bullet narrative, describe TTPs gathered in the operation. The bullets must contain sufficient detail for the reader to understand the situation and application possibilities for future operations. Support the bullets by providing individual observations (see TTP-Gathering Process below) as a combined appendix. The narrative in the base document must stand alone, with the appendix of individual observations providing additional detail.

TTP-GATHERING PROCESS

Units can contact CALL when developing plans for collecting TTP. CALL analysts can provide observer guidance, assist in delineating responsibilities of observers, identify documents or reference for use in developing a collection plan, and describe collection methodology.

- Observations. Individual observations assist in providing the basis for the narrative document described above. Use the form below to capture observations and develop a database for use in narrative development. Provide a copy of each observation to CALL as an appendix to the narrative. (Note: Any document published by CALL will not list units nor individuals by name. Refer to units by level ("the division" instead of "the 101st Airborne Division") and personnel by position ("a brigade chaplain" instead of "Chaplain Jones"). The purpose of CALL publications is to share ideas - not to point fingers.

- Observation Forms. The observation form (Appendix A) can be used for individual observations. A Microsoft Word version is available. Contact CALL via E-mail at call@leavenworth.army.mil or DSN 552-9571 (commercial 913-684-571) to receive a copy of the document. The document contains key components which aid the researcher in preparing analyses:
- Observer Name - The observer's name is used administratively only. No observer's name will appear in a CALL product.
- Administrative Information - Like the observer's name, unit information is used administratively only. Unit names do not appear in CALL products.
- Observation Indicators - Check all the appropriate blocks.
- Interoperability Indicators - Check all the appropriate blocks.
- Environmental Indicators - Check the appropriate block.
- File Name - Employ a system that differentiates each observation. One method is for observers to use name initials combined with sequence number and date (John Smith's first observation of May 5th would read, jsmay0105). Other systems are acceptable if plainly explained.
- Observation Title - Give the observation a brief, distinct title.
- Observation - In one sentence, summarize the observation.
- Discussion - Provide as much detail as necessary to provide a clear picture to the analyst or future reader. The length of the discussion will vary.
- Lesson Learned - In the context of your observation, provide a TTP.
- DTLOMS Implications - Describe how the observation impacts one or more areas in DTLOMS:
  - Doctrine
  - Training
  - Leadership Development
  - Organization
  - Materiel
  - Soldier Support

Include other media support, such as photos, sketches, or slide presentations in support of the narrative text.

Provide the narrative text, appendices and other material to CALL at the following locations: call@leavenworth.army.mil, or Department of the Army, Center for Army Lessons Learned, 10 Meade Avenue, Fort Leavenworth, Kansas 66027-1350.

In addition to maintaining and expanding a database of information, CALL publishes ***News From the Front!***, and a host of other publications for easy use by the force. Importantly, much of what is available has originated from the field -- from soldiers just like you. ***News From the Front!*** is published six times per year and provides a forum for a wide variety of topics of interest to the field. Other publications, including newsletters, CTC Bulletins, special editions, Handbooks, and more, focus on specific topics. Much of the published holdings of CALL can be found on the CALL website, http://call.army.mil, post libraries, or by contacting

CALL at the E-mail address listed above. Various search engines are available on the website to assist researchers.

CONCLUSION

Sharing information is possible with rapid and potentially colossal results. Leaders and soldiers who understand the TTP-gathering process can build plans for the future into every OPORD. By incorporating a plan to collect data and produce a clean narrative product for use by the force, soldiers throughout the Army gain maximum benefit from existing and future advancements in information technology. PAOs can focus on critical elements for successful media relations operations and command information programs in the field. Planners can easily adapt collection plans to exercises or actual operations -- anywhere in the world.

Learning is crucial for continued success on the battlefield. PAO planners must inculcate a practice of gaining a full understanding of the process and incorporating it into future exercises and actual operations.

Appendix X

MEDIA GROUND RULES

OPERATION DESERT STORM GROUND RULES

The following information should not be reported because its publication or broadcast could jeopardize operations and endanger lives:

(1) For U.S. or coalition units, specific numerical information on troop strength, aircraft, weapons systems, on-hand equipment or supplies (e.g. artillery, tanks, radars, missiles, trucks, water), including amounts of ammunition or fuel moved by support units or on hand in combat units. Unit size may be described in general terms such as "company-size, multi-battalion, multi-division, naval task force and carrier battle group." Number and amount of equipment and supplies may be described in general terms such as "large, small, or many."

(2) Any information that reveals details of future plans, operations or strikes, including postponed or cancelled operations.

(3) Information, photography and imagery that would reveal the specific location of military forces or show the level of security at military installations or encampments. Locations may be described as follows: all Navy embark stories can identify the ship upon which embarked as a dateline and will state that this report is coming "from the Persian Gulf, Red Sea or North Arabian Sea." Stories written in Saudi Arabia may be datelined "Eastern Saudi Arabia, near the Kuwaiti border, " etc. For specific countries outside Saudi Arabia, stories will state that the report is coming from the Persian Gulf region unless that country has acknowledged its participation.

(4) Rules of engagement details.

(5) Information on intelligence collection activities, including targets, methods and results.

(6) During an operation, specific information on friendly force troop movements, tactical deployments and dispositions that would jeopardize operational security and lives. This would include unit designations, names or operations and size of friendly forces involved until released by CENTCOM.

(7) Identification of mission aircraft points of origin, other than as land or carrier based.

(8) Information on the effectiveness or ineffectiveness of enemy camouflage, cover, deception, targeting, direct and indirect fire, intelligence collection or security measures.

(9) Specific identifying information on missing or downed aircraft or ships while search and rescue operations are planned or underway.

(10) Special operations forces methods, unique equipment or tactics.

(11) Specific operating methods and tactics, (e.g. air ops angles of attack or speeds, naval tactics and evasive maneuvers). General terms such as "low" or "fast" may be used.

(12) Information on operational or support vulnerabilities that could be used against U.S. forces, such as details of major battle damage or major personnel losses of specific U.S. or coalition units, until that information no longer provides tactical advantage to the enemy and is, therefore, released by CENTCOM.

*** Damage and casualties may be described as "light," "moderate," or "heavy."**

Glossary

AFRTS— Armed Forces Radio and Television Services

BOD—Broadcast Operations Detachment

C2 protect—command and control-protect—see **command and control warfare**

C2W—command and control warfare

command and control warfare—The integrated use of operations security (OPSEC), military deception, psychological operations (PSYOP), electronic warfare (EW), and physical destruction, mutually supported by intelligence, to deny information to, influence, degrade, or destroy adversary command and control capabilities, while protecting friendly command and control capabilities against such actions. Command and control warfare applies across the operational continuum and all levels of conflict. Also called C2W. C2W is both offensive and defensive:

C2-protection—To maintain effective command and control of own forces by turning to friendly advantage or negating adversary efforts to deny information to, influence, degrade, or destroy the friendly C2 system.

civil affairs—the activities of a commander that establish, maintain, influence, or exploit relations between military forces and civil authorities, both governmental and nongovernmental, and the civilian populace in a friendly, neutral, or hostile area of operations in order to facilitate military operations and consolidate operational objectives. Civil affairs activities (1) embrace the relationship between military forces and civil authorities and population in areas where military forces are present; and (2) involve application of civil affairs functional specialty skills, in areas normally the responsibility of civilian government, which enhance conduct of civil-military operations. These activities may occur prior to, during, or subsequent to other military actions. They may also occur, if directed, in the absence of other military operations.

combined operation—an operation conducted by forces of two or more allied nations acting together for the accomplishment of a single mission.

command information—see **internal information**

community relations—establishing and maintaining effective relationships between military and civilian communities through planning and active participation in events and processes which provide benefits to both communities.

community relations program—that command function which evaluates public attitudes, identifies the mission of a military organization with the public interest, and executes a program of action to earn public understanding and acceptance. Community relations programs are conducted at all levels of command, both in the United States and overseas, by military organizations having a community relations area of responsibility. Community relations programs include, but are not limited to, such activities as liaison and cooperation with associations and organizations and their local affiliates at all levels; armed forces participation in international, national, regional, state, and local public events; installation open houses and tours; embarkation in naval ships; orientation tours for distinguished civilians; people-to-people and humanitarian acts; cooperation with government officials and community leaders; and encouragement of armed forces personnel and their dependents to participate in activities of local schools, churches, fraternal, social, and civic organizations, sports, and recreation programs, and other aspects of community life to the extent feasible and appropriate, regardless of where they are located.

GIE—global information environment

global information environment—all individuals, organizations, or systems, most of which are outside the control of the military or National Command Authorities, that collect, process, and disseminate information to national and international audiences.

ground rules—conditions established by a military command to govern the conduct of news gathering and the release and/or use of specified information during an operation or during a specific period of time.

information age—the future time period when social, cultural, and economic patterns will reflect the decentralized, nonhierarchical flow of information.

information architecture—Description and specifications of information systems to include identification of communicators, information transmitted, equipment specifications and network designs. Includes operational, system and technical architectures.

information operations—continuous military operations within the military information environment that enable, enhance, and protect the friendly force's ability to collect, process, and act on information to achieve an advantage across the full range of military operations; information operations include interacting with the global information environment and exploiting or denying an adversary's information and decision capabilities.

information strategy—a synchronized plan for using all available and appropriate methods of communication to achieve specific goals of informing target audiences.

internal information—communication by a military organization with service members, civilian employees and family members of the organization that creates an awareness of the organization's goals, informs them of significant developments affecting them and the organization, increases their effectiveness as ambassadors of the organization, and satisfies their desire to be kept informed about what is going on in the organization and operation (also known as command information).

JIB—joint information bureau

joint information bureau—facilities established by the joint force commander to serve as the focal point for the interface between the military and the media during the conduct of joint operations. When operated in support of multinational operations, a joint information bureau is called a Combined Information Bureau or an Allied Press Information Center.

joint force—a general term applied to a force composed of significant elements, assigned or attached, of two or more Military Departments, operating under a single joint force commander.

joint operations—a general term to describe military actions conducted by joint forces, or by Service forces in relationships (e.g., support, coordinating authority), which, of themselves, do not create joint forces.

JOPES—Joint Operations Planning and Execution System

joint task force—a joint force that is constituted and so designated by the Secretary of Defense, a combatant commander, a subunified commander, or an existing joint task force commander. Also called JTF.

media facilitation—the range of activities such as providing access and interviews that assist news media representatives covering military operations.

media operations center—facility that serves as the focal point for the interface between the military and the media covering an event or operation.

media pool—a limited number of news media who represent a larger number of news media organizations for news gathering and sharing of material during a specified activity. Pooling is typically used when news media support resources cannot accommodate a large number of journalists. The DoD National Media Pool is available for coverage of the earliest stages of a contingency. Additionally, the combatant commanders may also find it necessary to form limited local pools to report on specific missions.

METT-TC—mission, enemy, terrain, troops, time available and civilians

MIE—military information environment

military information environment—the environment contained within the global information environment, consisting of the information systems and organizations—friendly and adversary, military and nonmilitary—that support, enable, or significantly influence a specific military operation.

MPAD--Mobile Public Affairs Detachment

multinational operations—a collective term to describe military actions conducted by forces of two or more nations, typically organized within the structure of a coalition or alliance.

news media representative—an individual employed by a civilian radio or television station, newspaper, newsmagazine, periodical, or news agency to gather information and report on a newsworthy event.

NGO—nongovernmental organization

nongovernmental organizations—transnational organizations of private citizens that maintain a consultative status with the Economic and Social Council of the United Nations. Nongovernmental organizations may be professional associations, foundations, multinational businesses, or simply groups with a common interest in humanitarian assistance activities (development and relief). "Nongovernmental organizations" is a term normally used by non-United States organizations. Also called NGO. See also private voluntary organizations.

PAD—Public Affairs Detachment. The smallest of the PA units.

PAG—public affairs guidance

PAOC—Public Affairs Operations Center.

private voluntary organizations—private, nonprofit humanitarian assistance organizations involved in development and relief activities. Private voluntary organizations are normally United States-based. "Private voluntary organization" is often used synonymously with the term "nongovernmental organization." Also called PVO. See also nongovernmental organizations.

psychological operations—operations to convey selected information and indicators to foreign audiences to influence their emotions, motives, objective reasoning, and ultimately the behavior of foreign governments, organizations, groups, and individuals. The purpose of psychological operations is to induce or reinforce foreign attitudes and behavior favorable to the originator's objectives. Also called PSYOP.

PSYOP—psychological operations

public affairs assessment—an analysis of the news media and public environments to evaluate the degree of understanding about strategic and operations objectives and military activities and to identify levels of public support. Includes judgments about the public affairs impact of pending decisions and recommendations about the structure of public affairs support for the assigned mission.

public affairs estimate—as assessment of a specific mission from a public affairs perspective.

public affairs guidance—normally, a package of information to support the public discussion of defense issues and operations. Such guidance can range from a telephonic response to a specific question to a more comprehensive package. Included could be an approved public affairs policy, news statements, answers to anticipated media questions, and community relations guidance. Public affairs guidance also addresses the method(s), timing, location and other details governing the release of information to the public.

public information—A general term describing processes used to provide information to external audiences through public media.

sustaining base—the home station or permanent location of active duty units and Reserve Component units (e.g., location of armory or reserve center) that provides personnel, logistic and other support required to maintain and prolong operations or combat.

Bibliography

Joint Pub 3-13. Joint Doctrine for Information Operations. 9 October 1998.

Joint Pub 3-61. Doctrine for Public Affairs in Joint Operations.

Joint Pub 1-02. DOD Dictionary of Military and Associated Terms. 24 March 1994.

DOD Directive 5122.5 Public Affairs Program. 12 February 1993.

DOD Directive 5120.20-R. Management and Operation of Armed Forces Radio and Television Service. 9 November 1998.

FM 3-61 (46-1). Public Affairs Operations. 10 March 1997.

FM 3-100.71 (71-100). Division Operations. 28 August 1996.

FM 3-07.29 (90-29). Noncombatant Evacuation Operations. 17 October 1994.

FM 3-0 (100-5). Operations. 14 June 1993.

FM 3-13 (100-6). Information Operations. 27 August 1996.

FM 3-100.1 (100-15). Corps Operations. 29 October 1996.

FM 3-07.7 (100-19). Domestic Support Operations. 1 July 1993.

FM 3-07.6 (100-23). Peace Operations. 30 December 1994.

TRADOC PAM 525-5. Force XXI Operations. 1 August 1994.

AR 200-1. Environmental Protection and Enhancement. 23 April 1990.

AR 360-5. Public Information. 31 May 1989.

AR 360-61. Community Relations. 15 January 1987.

AR 360-81. Command Information Program. 20 October 1989.

Public Affairs Guidance on National Guard Bureau Environmental Programs, National Guard Bureau Office of Public Affairs, 1994.

Commander's Guide to Environmental Management, U.S. Army Environmental Center, 1995.

Index